Experiencing New Worlds

Experiencing New Worlds

Edited by
Jürg Wassmann and Katharina Stockhaus

Berghahn Books
New York • Oxford

First published in 2007 by

Berghahn Books
www.berghahnbooks.com

Library of Congress Cataloging-in-Publication Data

Experiencing new worlds / edited by Jürg Wassmann and Katharina Stockhaus.
p. cm.
"Some of the essays brought together in this volume were originally presented at the symposium, Space Games: New Research Approaches to Spatial Orientation, Spatial Perception and Spatial Cognition', held in Heidelberg, May 2003"--Acknowledgements.
Includes bibliographical references and index.
ISBN-13: 978-1-84545-327-5 (hardback : alk. paper)
1. Human geography--Congresses. 2. Space perception--Congresses. 3. Geographical perception--Congresses. 4. Landscape assessment--Congresses. 5. Cognition and culture--Congresses. I. Wassmann, Jürg. II. Stockhaus, Katharina, 1966-
GF95.E97 2007
304.2--dc22

2007035168

British Library Cataloguing in Publication Data

A catalogue record for this book is available from the British Library

Printed in the United States on acid-free paper.

ISBN 978-1-84545-327-5 (hardback)

Contents

Figures

Tables

Acknowledgements

The realization of the publication at hand is based upon different initiatives and various kinds of support that we would like to acknowledge here.

Some of the essays brought together in this volume were originally presented at the symposium 'Space Games: New Research Approaches to Spatial Orientation, Spatial Perception and Spatial Cognition', held in Heidelberg, May 2003. This symposium was addressed especially to younger scholars coming from anthropology, psychology and geography who were planning interdisciplinary field research in Oceania. The intention was to facilitate an interdisciplinary dialogue in which the invited disciplines (social anthropology, linguistics, psychology, cognitive science and geography) present current research approaches and methods regarding the common topic of 'space'.

Financial support for this symposium was generously given by the Fritz Thyssen Stiftung für Wissenschaftsförderung. Our thanks go to all participants and especially the student discussants of this symposium who contributed with their ideas and comments to the enrichment of the papers here, as well as to the Internationales Wissenschaftsforum (IWF) in Heidelberg who provided such a pleasant place of venue.

In the process of compiling this volume, other scholars have been asked to contribute essays from their special fields of expertise, in order to provide a wider and well-balanced collection of papers from the different disciplines. We would also like to thank them for their readiness to participate, at often short notice, and for their patience during the process of publication.

We also wish to express our gratitude to many others who with their collegial and professional input helped to develop and realize this publication. We would just like to mention three events with partner institutes in Aarhus, Heidelberg and Nijmegen where – many years before this book project had been started – workshops on related themes were held with the help of Ad Boorsbom, Ton Otto, Toon van Meijl, and Eric Venbrux.

We also would like to acknowledge the generous financial support for the overall project 'Person, Space and Memory in the Contemporary Pacific' from the Volkswagen Foundation, who sponsored this research project at the Institute of Anthropology, University of Heidelberg, and therefore enabled a number of young scholars to conduct their field research in different parts of Oceania. Here, too, our warmest thanks go to many people who have contributed in one way or another with their commitment and interest for the project, and by aiding their professional and private affiliations and relations. Be it in offering language courses, be it as invited guest professors teaching topics within this general theme, be it as advisers for PhD students in Micronesia, Australia or Papua New Guinea, they all have supported this project tremendously over the past years: Pat Gesch, John Hickey, Eric Hirsch, Dan Jorgensen, Angella Meinerzag, Greg Murphy, Carmen Petrosian-Husa, Margaret Rodman, Don Rubinstein, Sesario Salalu Sigam, Mark Solon, Bob Tonkinson, James Weiner.

Finally, we are much obliged to Marion Berghahn, our publisher, who confidently took the right steps to get our series 'Person, Space and Memory in the Contemporary Pacific' on the way with this first volume, after which some promising monographs will hopefully follow.

Introduction

Jürg Wassmann and Verena Keck

This introduction seeks to kill two birds with one stone. On the one hand, we want to introduce the volume at hand, and, on the other, there is an entire book series to follow, which we also want to do justice to in this introduction. The book series 'Person, Space and Memory in the Contemporary Pacific' encompasses ethnographic publications on the South Pacific region, which are a result of interdisciplinary collaboration by psychologists, geographers and cognitive scientists. This is one reason our introduction focuses largely on this region. Nevertheless, this introductory volume contains, apart from several examples from the South Pacific (Papua New Guinea, Indonesia, Vanuatu, Philippines), contributions which deal with other regions of the world (Japan, Azores, Africa). However important it may be for our book series to establish regional similarities and differences for the areas of Melanesia, Micronesia, Polynesia and Aboriginal Australia and to accumulate ethnographically located knowledge about one area, it is an equally important concern for this volume to overcome territorial borders, together with disciplinary borders, in order to achieve a holistic and more accurate description of the research targets 'local actors', 'emplacement and landscape', and 'memory' in this project.

This achievement is mainly due to the involvement of cross-cultural psychology and social geography, as well as a range of anthropologists who already work, by the nature of their subject, in a cross-disciplinary way on topics such as agency and tradition, personal narrative, human–environmental relationships, anthropology of landscape, anthropology of the senses, and psycholinguistics. We will begin this introduction by describing the overall difficulties of anthropological research in today's Pacific and then go on to describe the three topics mentioned above in detail and summarize latest developments. This is where the authors of this current volume will appear because they successfully managed, as we will see, to come to terms with some of the major difficulties of anthropological research as well as suggesting new approaches and ways of solving problems.

Finally, we will address the topic of multidisciplinarity, which has always been a major concern of our own research and will now be addressed in this volume as well as being realized successfully in the upcoming series, so we hope.

Posing the Problem

Anthropologists now acknowledge that what they formerly treated to a large degree as distinct cultures, relatively bounded and contained units of enquiry, are dynamic sites where a multiplicity of processes intersect. We are conscious of the fact that cultures have at no time been isolated entities and many classical ethnographies have taken dynamic approaches, which investigate interaction and change, or the relations between groups, and groups and colonizers; nevertheless, interactions have reached a new level with regards to frequency and latitude. Indeed, it has become common to argue that anthropological research and ethnographic writing must take into account that – in the past and in the present – we all live in a creolizing world. Yet it remains difficult to study relationships between local contexts, phenomena and global processes without creating a false dichotomy between 'the local' and 'the global', for there are multiple globals, too. Never have notions of society, community, ethnic conflict, displacement, culture and personal identity been so intensively rethought within the social sciences (Friedman 1994; Hannerz 1996; Olwig and Hastrup 1997; Werbner and Modood 1997; Lockwood 2004; Harrison 2005); and nowhere are these trends and difficulties more apparent than in the study of the highly mobile populations of the Pacific, where people move – as work migrants, refugees, students and displaced people – through a region being transformed by the effects of the production and consumption of commodities, and a globalized mass media transmitting images, moral concepts and ideas.

In general terms, the massively increased mobility of populations and their cultures has meant that the reciprocal indexation of topology and culture, and territory and nation has become the focus of critical analysis. Critiques directed at the 'incarceration' of cultures in fixed places (Appadurai 1988), the metaphor of the 'localness' of cultures (Clifford 1988), and the latent functionalist search for cultural stability and localized equilibriums (Malkki 1995), all stress the need for more dynamic, process-oriented theories of culture, identity and space, while the anthropologist is, increasingly, confronted with a multiplicity of voices and actions in particular places (Appadurai 1996), and the multilocality and multivocality of culture (Rodman 1992, 2001). As a response some have begun to explore the notion of ethnoscapes – geographical regions that are represented less as physically specifiable entities than as constructed settings that signal particular modes of sociality (Gupta and Ferguson 1992).

In the contemporary Pacific, migration and displacement on a regional, national and transnational level have contributed significantly to mobility (Fox, Bellwood and Tryon 1995; Small 1997; Cohen 1997; Spickard et al. 2002; Morton Lee 2003; Marshall 2004; de Wet 2005), and their effects have often been construed negatively, by an anthropology focused on territorial 'anchoring', as the 'uprooting' of people involving the loss of tradition, culture and identity. These movements, however, have raised questions about the relevance of the conceptual boundaries erected between 'town' and 'country', between 'centre' and 'periphery', or between 'us' and 'others'. And as processes of urbanization and internationalization have proved to be opportunities for the movement of cultures, so they can open up new horizons for the study of space, identities and interethnic relations (King and Connell 1999; Modell 2002). Particularly significant here is the claim that displacement necessarily creates placements, or novel forms of localization and positioning. According to Epeli Hau'ofa, for example, transnational migration has contributed to the expansion of 'local' cultures – today, more Samoans are living in Los Angeles, and more Cook Islanders in Auckland, than on their islands of origin; suggesting that the so-called 'small nations' of the Pacific region may be becoming larger even as they become part of the 'global village' (McLuhan 1994), set, presumably, in a shrinking world (Harvey 1989).

Moreover, novel experiences of other places in the world can be had on the spot – if in a totally different manner. Consumer goods flow relatively freely to even the most remote villages, as do transmissions from distant radios and – often – television and the internet (World Wide Web). The effect of these flows is augmented by the efforts of the Pacific's young nations – sometimes comprising hundreds of language groups – to achieve a certain internal homogeneity through national institutions that have village-level manifestations, in the form of schools, churches, stores, civil servants or judges (Foster 1995). Here, nationally prescribed school books (postulating a common national history), national newspapers and the Bible all play a role.

Globalization and standardization, which consist in a multitude of different forces influencing the regional in various different ways, have not only resulted in a certain uniformity but – and to the surprise of many – also in new forms of cultural difference. Old and new ethnic, religious and political groups appropriate ideas, images and values from the mass media and create creolized cultural forms by synthesizing them with inherited and 'invented' traditions. Two features of this process are important: first, local actors do not experience new forms passively, but as active, discriminating and creative agents (see Ton Otto's discussion of agency and tradition in this volume); second, exposure to new experiences varies greatly, depending not only on region but also on the gender and age of individuals. If, in former times, experience and knowledge were handed down from older to younger generations, often according to gender, mass media and national institutions now tend to focus on younger, mostly male, actors. This raises the question of access – by location, power relations and sociological category – to these new forms of knowledge (Leavitt and Herdt 1998; Meusburger 2000; Trommsdorff and Nauck 2005). Moreover, migrant Pacific Islanders, individually or collectively, create, and seek to consolidate, new identities. The (re)claiming of identity symbols and property rights is central to a great many political struggles in the contemporary Pacific. As diffusion will also meet resistance, we also have to ask how traditional knowledge is defended against the influences of the outside world. This is substantiated by the ongoing copyright debates about many oceanic cultural achievements (Brown 1998; Clifford 2001; Leach 2003b; Hirsch and Strathern 2005).

Yet questions remain about what exactly happens at the local level, and how the effects are produced at the emotional and cognitive level of particular actors. In effect, little is known about this issue. Until now, research has mainly concentrated on the macro- or culturally specific aspects of globalization, while neglecting actual actors and their perspective of social change in their own local context. How do ethnopsychologies, or the local concepts of person, change under the effects of globalization? Are they shifting towards a capitalistic and Christian individuality? How do differentiated experiences of the globalized world and multilocal cultures affect ideas of personhood and belonging in a specific landscape and place? Is identity transformed by the freedom to select a certain lifestyle? How are individual decisions relevant to concepts of identity and personhood influenced by different experiences? How do cultural and personal memories work? How did the life histories of people change and who is mainly affected? Does the experience of new worlds, the experience of difference, lead to a heightened reflection about one's own culture, and to transformed temporal perceptions of changed relations between past, present and future? How does awareness of the fact that one is part of a nation, a worldregion or a transnational culture influence a sense of collective identity vis-à-vis the 'other'?

By examining the forces of 'globalization' – which itself is subject to local variation (i.e. globalism is different in Guam and Papua New Guinea) – as they manifest themselves and operate at cultural, biographical and cognitive levels of local actors, we try to find answers to the posed questions. These attempts to find answers must be based firstly on long-term fieldwork on site and secondly on the integration of cultural research with psychological methodologies, linguistics, geography and cognitive science. The analytical rubrics for this study are human agency, space and landscape as well as memory. Through this renewed perspective three research traditions are connected:

- the research on ethnopsychologies combined with practice theory and processual ethnography;
- an anthropologically informed geography of space and landscape;
- cognitive science, particularly the cognitive psychology of memory, perception and mental models.

Local Actors

The anthropological investigation of the concept of the person in non-European societies began with Durkheim and Mauss. Mauss describes a historical process featuring a developmental sequence from the 'social personage' to the psychological and individual person and, then, to the individual as a self-conscious and autonomous entity. According to these classical authors, humans should be regarded as 'total social facts' (*fait social total*) that comprise body, individual and collective consciousness fashioned principally by social processes. Western ethnopsychologies portray the person as an individual whose behaviour is explained in terms of inner motives. More generally, this tendency to ground all psychological matters in the individual person is closely connected to personhood in the West (Morris 1994).

Other ethnopsychologies, by contrast, construct personal identity, action and experience in ways that are not necessarily congruent with Western concepts. While we may assume that the phenomena of intentionality and self-awareness are constants, the sociological 'person', the psychological 'self' and the biological 'individual' are usually construed as culturally defined; and, while these three terms are imbued with a certain ambiguity, they remain the starting point for exploring concepts of the person (White and Kirkpatrick 1985; Harris 1989; cf. Hermans 2002). Consider, for example, Marilyn Strathern's analysis of exchange relations in Papua New Guinea, which is based on a distinction between the Western person as being a single and separated individual and the Melanesian person as being a related and even divisible person, a knot in a network of social relationships (Strathern 1988; cf. Leenhardt 1947; Kuehling 2005). We also note Andrew Strathern and Pamela Stewart's statement (2000) about the complexities and variations in the use of the central concepts of person, self, individual; their introduction of the term 'relational individual'; and the need to think about these in relation to newer concepts, such as those of permeability and partibility. Not a 'dividual', i.e. dividable, but a dyadic concept is discussed by Stasch (2002). Within this model the person is represented more as an element of a dyadic partnership than an entity.

In the definition of Harris (1989), a 'person' is a human being who acts within the cultural norms. Here, the person need not be a living human, but can be an ancestor or an animal that acts and therewith influences people's lives; on the other hand, a human being without social capacities can be regarded as a lesser person, non-person, or former person. Fajans (1985) describes in her Melanesian example the development of personhood as the progressive acquisition of social properties as food giver (and not only as food receiver) – and the dissolution of the same personhood with age. As a newborn, an infant is regarded as natural, one who has not yet gained social behaviour: he or she is not a person. An older person is dissocialized because he or she becomes dependent on receiving food again.

A further perspective for the exploration of concepts of person is to focus on other aspects beside the social component such as soul, body and child conception (Strathern 1996; Eves 1998). Keck (2005) describes the person as being composed by several elements: a human body, an impersonal vital energy, the two spiritual aspects 'breath-spirit' and 'shadow-soul', the quality of social relationships (cold, cool and hot), a personal name and a melody for identifying the person. A single element can act on the others and thus, for instance a disturbed social relationship can affect the body and manifest itself as illness. Leach (2003a) clarifies the direct physical relationship between humans and *habitat*, and on yet another angle the investigation of gender illuminates the person concept, because, as Barlow (1995) shows, an analysis of gendered lifecycles reveals the contrast between an abstractly conceived human existence and culturally charged concepts of persons as gendered (cf. Bonnemaison 1986). Some of these angles are taken up by the authors of this volume, and others have been added. For example, psychologist Gisela Trommsdorff (Chapter 3) explores the topic of agency from a different perspective, by discussing to what extent theories on agency depend on cultural belief systems, and what the implications of these cultural belief systems are for other agency beliefs. Here she refrains from the easy classification of 'independent' and 'interdependent' culturally constructed ideas of the self, but rather views the self in all cultures as incorporating both aspects in various degrees while the different selves become more or less accessible in different contexts.

Two other authors try to tap directly into culturally formed understandings of agency by analysing informants' narrations about themselves and others. While Andrew Strathern and Pamela Stewart (Chapter 5) take a closer look at the representations of the 'exotic' in a Papua New Guinean society, Stephen Leavitt (Chapter 4) searches for suppressed meanings and culturally compelling scenarios in personal narrative.

Processual ethnography focuses on the dynamic relationship between structures of power, social interactions, cultural dispositions, lived experiences, and individual agency based on the understandings of Bourdieu and Sahlins. Two of our authors have also worked with a processual approach and offer their thoughts on anthropology's take on tradition (Otto and Pedersen 2005; see Ton Otto's Chapter 2) and the recursive nature of human-environmental relationships (see Katja Neves-Graça in Chapter 8). It can be advantageous when approaches rely also on the analysis of diagnostic events, those 'that reveal ongoing contests and conflicts and competitions and the efforts to prevent, suppress, or repress these' (Moore 1987: 730). We assume that experiences of displacement, novel commodities, images and institutions can situate individual agents in a liminal position, a temporal and spatial borderline between familiar and as yet unfamiliar culture-specific structures, between past and present 'homes'. Those local actors who undergo such experiences alone or earlier than others of their culture often become cultural brokers, possessing an everyday knowledge (which is not always linguistically encoded) that enables them to follow everyday routines different from those to which they were once accustomed.

What is local (no matter where 'home' is) exudes continuity and familiarity – becomes part of the durable dispositions inculcated during the earliest years – exactly because it is highly repetitive, redundant, practical and often acquired in face-to-face interaction; it also has a strongly emotional aspect (Rodman 2001; see also Margaret Rodman's contribution in this volume). Here, first, formative experiences are made. In many contemporary contexts, however, the local is not bounded and autonomous, but rather a kind of arena where many different influences come together, resulting in hybrid, or creolized, cultural forms. Consequently, Zygmunt Bauman (1992) proposes connecting the actor not to a system but to a flexible sense of habitat, to habitats of meaning; Friedman (1998) uses the term 'Erfahrungsraum' (experiential space) coined by Mannheim; while Lave (1988) writes about 'settings of activity'.

Accordingly, some have argued that the concept of a complex culture should yield to the concept of a flowing cultural complexity (Barth 1993). Viewed in this way, culture is a set of specific dispositions acquired by individuals in the process of learning, and of mutual construction of identities through cultural and social encounters, distributed in complex ways across individuals of a population.

Considerations such as these lead us to endorse the challenge to the sociocentric, collectivist understanding of culture, and the related assumption that the carrier of a culture is a purely social unit. We agree with Bloch (1991), who has criticized this 'oversocialised' model of the individual, and contend that the notion of a uniform and homogeneously distributed culture within collectivities becomes problematic (Boster 1985; Wassmann 1993; Borofsky 1994) once we notice that their members have highly distinct experiences, different access to knowledge, that their biographies vary deeply (Gewertz and Errington 1991) – especially, but certainly not exclusively, among people who have spent significant periods in places beyond those normally counted as 'home'. Yet, even in common worlds, single lives differ (Keck 1998).

Emplacement and Landscape

The concept of person in Pacific societies is not only defined through local ideas about the person, their constituents and social environment, or by innate mechanisms of perception and remembering in mental schemata, but also by the topological environment. In many societies of the Pacific, people's sense of who they are as persons derives from their position within a genealogy and – this is the point we would like to make – their relation to landscape. Landscape can be, 'read' as genealogy or can be understood as historically experienced and constituted (Keesing 1982; Weiner 1991; Gow 1995; Bamford 1998).

The metaphor of landscape, taken from art history, has now become a theoretical term of art in the anthropology of space and place. The term originally referred to timeless, framed, idealized paintings; in more recent anthropological literature, however, landscape is understood as cultural process. Numerous studies of native conceptions of landscape published in the last few years indicate that knowledge and the memory of social relations are keyed to place (Bender 1995; Hirsch and O'Hanlon 1995; Feld and Basso 1996; Stürzenhofecker 1998; Telban 1999; Bender and Winer 2001; Wassmann 2001; Stewart and Strathern 2003; Kahn 2004). More broadly, the term 'landscape' refers to the dynamics of lived histories, the interpretation of time, relationship and person, kinship, history, identity, gender relations and emotions; thus, it postulates an anthropomorphic interpretation of the environment (Fox 1997). The inclusion of a parallel notion of seascape in field investigations has also proved a crucial contribution (Pomponio 1992), as have studies on the relevance of acoustic information (Feld 1982; Gell 1995) or the associative strength of smell (Engen 1991; Beer 2000b).

The use of space, patterns of settlement, and above all, the symbolism of spatial arrangements are classical issues (Durkheim and Mauss 1963; Widlok 1999; cf. Meusburger and Hennings in this volume). The traditional focus on cosmological themes and their instantiation in architectural schemes is also reflected in an extensive ethnographic literature. A related tradition focuses on the symbolism of domestic space, and continues to attract attention (Bourdieu 1977; Fox 1993), with a new emphasis on the 'built environment' (Lawrence and Low 1990) and the interactional use of domestic and public spaces.

Space becomes place through the memory of persons, and memory is constitutive of personhood. According to the ecological psychologist Graumann (2002), space has to be appropriated by a person in different modes: appropriation occurs over time, by marking, naming, defining, categorizing, and evaluating space as appropriate or inappropriate, owned or free; by locomotion, resulting in paths and roads; by cultivation; by conquest of foreign land; by building, constructing and setting. Through these and other means, space is personalized and becomes lifespace, becomes place. On the other hand, and necessarily, personhood is emplaced, and thereby spatialized. Whereas space is a term for abstract geometrical extension that makes no reference to any human activities, place, in contrast, has a strongly experiential connotation and is mediated by our memory.

Orientation in space can also be seen as part of a problem-solving process (Frensch and Funke 1995; Widlok in this volume). Where do I find what I search for? How do I find a way to move from my present to the desired location? This problem solving occurs as an interaction between a certain task and a specific person. In routine situations, the person has no problem but a routine inscribed in the habitus for doing the task. In nonroutine situations, meaning new or unforeseen, problem solving involves finding new solutions to the given task, yet this necessarily depends on the use of preexisting cognitive resources, and, therefore, on cultural schemata and other aspects of the habitus. Thus, creativity and innovation under novel cultural circumstances always involves the use and transformation of preexisting cognitive resources. By the logic of the relationship between habitus, personhood and place, the creative reconstitution of the place leads to a reconfiguration of the habitus.

While we are interested in all these issues, our main aim is to explore how space is perceived in non-European cultures. Spatial conceptualization is constrained by the nature of the physical world, as well as by the nature of human psychobiology with its visual system and upright posture. These environmental and cognitive constraints form the basis of current models of our conceptualization of space. Mainstream cognitive science seems to hold that, universally, humans take an anthropomorphic perspective on space, and use the kind of egocentric constructions familiar from European languages. These suggest that a person's body stands in the centre of the universe and that spatial coordinates radiate out from it. The Piagetian child struggles to extend this framework across the environment, learning slowly about the constants of the world beyond his perception. This idea of one's own body as the centre of the universe, and the conviction that what is called the egocentric (relativistic) conception of space is 'more natural and primitive' (Miller and Johnson-Laird 1976: 381), are deeply rooted in Western traditions of scientific and philosophical thought (Kant 1768; Piaget and Inhelder 1967; Levelt 1989).

However, cross-cultural psychological and linguistic evidence raises questions about these perspectives and one may suspect that cultural factors have been underestimated. Thus, many languages lack egocentric words like left/right, in front/behind or on top/below (of the body) (Gumperz and Levinson 1996; Senft 1997; Wassmann and Dasen 1998, 2006; Bennardo 2002; Levinson 2003). Non-Indo-European speakers often conceive humans as not located in the centre but, rather, as part of the environment. Therefore, orientation is referred to in geocentric, not egocentric, terms: on the basis of fixed external points which correspond approximately with our cardinal system (e.g. uphill/downhill; sunrise/sundown; to the mountain/to the sea, etc.). It is known that these systems, used in opposition to our cardinal points, also have an effect on the micro level: in Pacific everyday life, one asks for 'the rice bowl over there at sunset', or asks a child before sleeping to turn his head 'uphill'.

Therefore, we need to learn how space is encoded linguistically and cognitively by local actors and investigate how this affects their understandings of their lived environment. This task has been taken up by the last three authors of this volume and they describe and critique in detail research methods and results of the experimental 'Space Games' (see Gunter Senft's, Joachim Funke's and Thomas Widlok's contributions). One resulting question was, for example: Are one's own concepts of space 'exported' or are they adapted to new localities? The Yupno in Papua New Guinea 'take along' their geocentric system of orientation with 'upriver/downriver' to the town of Madang and there equate the main road leading to the sea (downriver) with the Yupno river – but turned by approximately 180 degrees (Wassmann 1994). And there exist many more examples where other Pacific people either duplicate or adapt their system of orientation when abroad.

Memory

Within the context of cognitive theory, human experiences and thought are understood as information processing. In addition to perception, mental representations, and memory, it is important to render evident those processes through which people are enabled to acquire a mother tongue and to internalize the sociocultural specificities of the particular society in which they are enculturated. The latter themes are elementary to anthropological knowledge as a central concept within the discipline. According to Bloch, 'anthropologists' concerns place them right in the middle of the cognitive sciences, whether they like it or not, since it is cognitive scientists who have something to say about learning, memory and retrieval' (Bloch 1991: 184).

The definition of cognition used by anthropologists is often extremely sociological. Jean Lave (1988) sees cognition as originating in social practices of individuals. As Clifford Geertz put it: 'thought is, in a great part anyway, a public activity' (1973: 45). Perhaps all this may seem commonplace but it is striking how often anthropologists' theories of thinking, meaning, memory and retrieval have not been compatible with those of cognitive scientists.

In everyday interaction with our environment we experience a constant flow of new impressions that increase our efficiency in processing further information. This position assumes an information-processing mind by means of which information that we require to be part of our environment is encoded, stored and retrieved.

> Without the possibility to process information there cannot be life nor evolution, no understandable perception of objects or situations, no conceptual understanding, no language, no culture, and also no identity .... Without memory we could not even think, since without internalized schemata, concepts and categories, there is no basis for mental representation (Reimann 1998: 146 [our translation]).

Memory is constitutive of personhood and is a central aspect of cognition (Baddeley 1997). Our memory is not uniformly organized. Different tasks are performed by different systems: the modal, the short-term memory and the long-term memory. Atkinson and Shiffrin (1968), who were the first to formulate a cognitive psychological model of memory, proposed that these three elements correspond to three distinct locations in the brain (localized representations). Insofar as culture is a phenomenon of memory, through which we organize our knowledge of the world, we are interested in exploring the concept of long-term memory. In turn, we understand long-term memory as containing several subsystems. Memory testing is an increasingly complex research area, thus it is becoming more important for anthropologists to differentiate between different levels or subsystems for their research. As Edgar Erdfelder and Martin Brandt suggest in their contribution in this volume, memory psychologists need to take greater care in choosing appropriate methods of memory measurement, as well as in the interpretation of the data from memory tests, because in cross-cultural research a seemingly obvious thesis often influences the interpretation of results strongly and may lead to the wrong answers.

The procedural memory contains knowledge in the form of implicit rules of behaviour which cannot be expressed through language but rather are enacted through practice (for example, try to explain to someone how to ride a bicycle). Implicit memory is constituted by abstract principles which organize an array of perceptions into seemingly coherent patterns. The origin of these patterns emerges from repeated associations of closely related everyday experiences (Schacter 1992; cf. Whitehouse 2001). These experiences are habitual, embodied, and they match Bourdieu's notion of *habitus* of the body (habitual set of dispositions). Squire, who has studied the underlying structures of the brain, contends that:

> Non-declarative memory includes information that is acquired during skill learning (motor skills, perceptual skills, cognitive skills) habit formation, simple classical conditioning including some kind of emotional learning, the phenomenon of priming and other knowledge that is expressed through performance rather than recollection. (1992a: 233)

And Schacter formulates about the same topic: 'Implicit memory is an unintentional, nonconscious form of retention that can be contrasted with explicit memory, which involves conscious recollection of previous experiences.' (1992: 559).

The anthropologist Borofsky (1994) speaks of 'knowing' (how to do) that is contrasted with 'knowledge', the so-called declarative knowledge (knowing that). According to Tulving (1985), the declarative memory consists of episodic and semantic memory, where information stemming from personal experience (always linked to place and time) and universal knowledge are stored.

Information from recurrent, appropriately structured daily episodes may develop and solidify through time into so-called mental cultural models or schemata (Strauss and Quinn 1997; cf. Bartlett 1932; Piaget and Inhelder 1967; Minsky 1975; Schank and Abelson 1977). Schemata are mental models that organize the acquisition and storage of knowledge in stereotypical and prototypical sequences of thoughts and actions. They are presupposed, taken-for-granted models of the world that are partially shared (although not necessarily to the exclusion of other alternative models), and they are organized in ways fundamental to the way human beings think, remember and forget (Squire 1992b; Squire and Kandel 2000; Schacter 2001). These schemata may be of two different, equally basic forms: proposition schemata and image schemata, to use Lakoff's terminology (1987). In both forms we may think of them as simplified worlds, from which the complex, comprehensive knowledge is reconstructed spontaneously. Much of our knowledge in the schemata is fundamentally nonlinguistic (Eysenck and Keane 1997), having the form of neither sentences nor words. There are neurological constraints on the storage of schemata that influence what and how we think. Moreover, since basic dispositions that inhere in these schemata have considerable inertia – *hysteresis* is Bourdieu's term – their modification and adaptation tends to be slow – but they adapt continuously to new experiences made by the actor and because of that they interpret subsequent experiences in a different way.

A promising new field of work, albeit under debate, for both anthropology and cognitive psychology could be the new PDP models (parallel distributed processing), that is, the connectionist or neuronal networks, which were developed as computer models by Rumelhart and McClelland (1986; McCloskey 1991; Bechtel and Abrahamsen 2001). The human brain and its composition of billions (10^{10}) of nerve cells, so-called neurons, serve as a model for neuronal networks. Connectionism has no serial process of analysis carried on along a single line by a single processor. Instead, connectionism suggests that we access knowledge, either from memory or as it is conceptualized from perception of the external world, through a number of processing units which work in parallel and feed in information simultaneously (distributed representations). Information received from these multiple parallel processors is analysed simultaneously through already existing networks connecting the processors. Knowledge is not located in specified parts of the brain, but stored in these pathways of connections (synapses) between the neurons or processors. Again, there must be a significant number of social interactions and positive assessments in order that some connections get a higher 'weight' and so become stabilized for a period of time. In this case, they resemble schemata or cultural models. This packing works very well for quick and efficient operations in familiar domains, but occasionally the network can also be unpacked into linear sentential sequences which can then be put into words. This uncertainty and dependency on know-how from our everyday life, that is, on cultural phenomena, makes these models attractive for anthropologists (Shore 1996).

The ways in which cultural knowledge is organized in memory has fundamental implications for anthropology.

> If an anthropologist is attempting to give an account of chunked and non-sentential knowledge in a linguistic medium (writing) she [sic] must be aware that in so doing she is not representing the organisation of the knowledge of the people she studies but is transmuting it into an entirely different logical form [i.e. the verbally articulated form]. (Bloch 1998: 15)

The notion of 'cultural memory' goes beyond the notion of tradition, belonging to an external dimension of human memory. The cultural memory can be stored externally and intermediately, such as in specialists, experts, shamans, griots, priests, 'old crocodiles' or in systems of notations such as knotted chords, *churingas*, or in writings. Here, memory works as data storage. Experts need to be instructed and to control the diffusion of knowledge. Certain people can be excluded from this knowledge or receive only parts of it. Thereby, emplacement is the most basic medium of mnemonics. There are memory places and even landscapes can serve as a medium for cultural memory. How best to access these memory places is the topic of Susanne Kuehling's contribution in this volume. She proposes the use of different qualitative anthropological methods in combination with more quantitative geographical experiments (see Chapter 10).

Thus the question is how a place is perceived and experienced, how an impersonal geography (at home or abroad) gains significance and can become 'home' (Myers 1986; cf. Augé 1995). In this struggle to find one's roots, a sense of place is not so much about residence in a locale as about belonging. 'The poetic power of place' (Fox 1997) is often based on the knowledge of names of places, or the power to name them. A landscape-based cosmology is one way in which Aboriginal identity has been maintained, especially since the European colonists and the Aboriginals created such divergent landscapes out of the same pieces of geography. Landscape is a mnemonic for past generation (Morphy 1995).

A different field of memory storage is explored by the anthropology of senses. Here, as Bettina Beer contends in this volume, a whole new access can be created to culturally relevant and emotionally connected memories. Her access point are smells, which trigger very different memories and feelings in different cultures, as the examples from the Philippines and Papua New Guinea illustrate (see Chapter 11).

Whether in Europe or elsewhere cultures seem to be increasingly hybrid and dynamic. People move and make new experiences, and new images and values arrive even at the most remote village. But as people reject or adapt the varying 'globalisms', from Guam to Goroka, their experiences differ. It is the particular interaction and experiences with these 'new worlds' that this book series is concerned with. Local ethnic psychologies may change in just the same way as individual life histories. Concepts of person and agency, space and landscape, as well as memory therefore have to be reconsidered. We plead here for precise local studies whereby the local is understood as the realm of experience where different influences meet. The contributions of this volume make excellent use of the three aspects listed above – of person, space and memory – as tools for analysis and categories in which to think about change, at a level of detail not known before. Therefore, we trust that not only readers with an anthropological background, but also geographers, linguists, psychologists and representatives of cognitive science will find this volume and the following series appealing and useful for their own research.

A Personal Last Word[1]

There is no doubt: multidisciplinarity is the trend today.

We suggest research on the concepts of person in changing Pacific societies, with the units of our analysis being local actors, emplacement and memory – all this in combination with a cooperation between anthropology, human geography and cognitive psychology. This implies that – whenever possible – researchers from different disciplines should work together on the studies proposed.

They extend from the side-by-side working of different disciplines uniting in poly-disciplinary research to bringing together disciplines, establishing contact, and cooperating, as, for instance, in the cross-disciplinary communication at scientific conferences, in joint research projects or joint institutes. Alternatively, a common paradigm is developed that bridges several disciplines – transdisciplinarity as a research requirement dictated by the increasingly more complex developments of problems for whose solution the knowledge of any individual discipline is no longer sufficient. Coordinated cooperation can, finally, also mean thinking that spans several academic fields, and leads to a standardization of the comprehension of phenomena by trying to combine the partial explanations of various disciplines. Research is organized in a multidisciplinary manner whenever a research problem exceeds the framework of an individual discipline – as in our case – and whenever the scientists from the individual disciplines complement each other and cooperate.

'Classical fieldwork' was crucially defined by Bronislaw Malinowski, a youngish European who, between 1915 and 1918, lived for a total of two years as a 'single' in the village of Omarakana on the Trobriand Islands. As a one-man enterprise, he compiled extensive ethnographic material and at the same time confided his frustrations in his diary. This 'lonely wolf' image influenced generations of male and female anthropologists with regard to the organization of their fieldwork. 'Ethnographers have been trained to think of themselves as the single instrument of data collection, analysis and presentation' (Schensul et al. 1999: 86–87).

But anthropology has changed, not least because its 'object' has also changed: it has become a 'shifting anthropological object' (Olwig and Hastrup 1997). With increasing frequency, research today is being initiated by the indigenous communities themselves (Tuhiwai Smith 1999) – in part as a reaction to European colonialism and the associated question of the authorship of knowledge or as an expression of turning back to one's own traditions in the sense of cultural resistance. There are Trobrianders today who themselves work as anthropologists. Even though the single fieldworker is still the rule, we observe that anthropologists increasingly work in teams, either as married couples, as teams of anthropologists, interdisciplinary groups, or together with indigenous researchers (Mishra and Dasen is one example found in this volume).

As to teams of researchers, the most frequent (though not always the least complicated and best-functioning composition) appears to be the husband-and-wife team, either consisting of two trained scientists or of a male or female anthropologist and the accompanying partner, with or without children (Flinn, Marshall and Armstrong 1998).

> Fieldwork as a husband/wife team worked extremely well. Some of this may have to do with our particular circumstances. Whereas I was doing anthropological fieldwork for a Ph.D., Lind was taking an extended break from her career as a senior medical administrator. The competition that might have

> arisen between a husband/wife team both of whom were intent on carving out anthropological careers from the same fieldwork, never arose....The advantages of doing fieldwork together in this way was not only that we could share ideas (and arguments), but the various periods of fieldwork themselves could be reflected on retrospectively with someone who understood its complexities, poignancy, mistakes and humor. (Personal communication, anthropologist Michael O'Hanlon)

The linguist Gunter Senft notes:

> This situation again improved immensely when in 1989 we returned to Tauwema with our children, two and four years old at the time. Without Barbara and without the fact that only with children were we respected by the Trobrianders as proper adults, I would not have had access to multitude of information. (Personal communication, linguist Gunter Senft)

Apart from these private considerations one could name several convincing scientific reasons in favour of a man-and-woman-team during fieldwork, particularly if both are trained anthropologists. The anthropologist Pierre Lemonnier has said:

> Fieldwork experience as team: good and helpful. ... Anga are such people that you do need to have two (M and F) anthropologists if you want to have an idea of what's going on. To take the most obvious domain, it would be totally impossible for a female anthropologist to take part in and witness the secret part of the male initiations. A female anthropologist cannot enter the forest at that time: hence the absolute necessity to have a male colleague working at the same time to be able to talk about these sort of rituals. Symmetrically, as far as I know, no male anthropologist ever paid attention at what the women were doing during the male rituals. And that's why Pascale's finding about the key-role of women-mothers-sisters during the Ankave initiations are so important. If Pierre had been alone, he would probably have carried on with the entirely wrong idea that Anga male initiations are a men's story. (Personal communication, anthropologist Pierre Lemonnier)

But let us turn to the cooperation of researchers from different disciplines. Multidisciplinarity, with its multiple social and institutional complexities, is not really something new. One of the first major multidisciplinary field studies in anthropology was the well-known 'Cambridge Anthropological Expedition' (1898–99) to the Melanesian Torres Straits islands. Today the increasingly strong demand for cooperation is a result of the particularization of the disciplines themselves, of the growing specialization in one's own discipline and the steadily accelerating expansion of knowledge that can no longer be managed.

We would like to conclude with reflections on our own interdisciplinarity experiences, which could highlight the problems with potentially incompatible approaches and disciplines.

The Guam '*lytico-bodig* Project' was created in cooperation with the National Institute of Aging–Mayo Clinical Study of Neurodegenerative Disorders in Micronesia. In these large-scale projects, anthropologists are employed as 'staff' together with scientists from other fields.

> Opportunity waits on Guam to reveal the cause of an obscure malady. On Guam, Chamorro natives suffer *lytico-bodig*, an endemic, paralyzing disease that resembles Lou Gehrig's, Parkinson's and Alzheimer's disease, sporadic neurodegenerative diseases affecting millions of the world's aging population. For forty years foremost physicians have sought the silent cause of this common disease on this small tropical island in the far western Pacific. ... This is your opportunity to be part of the discovery, in the same village Magellan discovered four centuries ago.

With this advertisement in the *Anthropology Newsletter* of the American Anthropological Association published in May 1992, the group of researchers stationed in Guam was looking for a medical anthropologist. On the island, neuroscientists had been seeking intensely – and to this day in vain – with the help of a variety of medical and quantitative epidemiological methods after the cause of this neurodegenerative disease.

In this project dominated by biomedicine, anthropology was only a small and – compared with 'hard-facts'-producing medicine (at least that was the conviction of the doctors) – somewhat 'exotic' and rather 'vague' science represented by a young researcher (Verena Keck) from Europe with far from excellent knowledge of English who, moreover, obstinately insisted on living in one of the affected villages instead of visiting the villages from a more comfortable university apartment, i.e. from an Americanized world – as the numerous doctors flown into Guam for a short time usually did.

> I worked particularly closely with a neurologist who also lived in Umatac, I was able to accompany him on his rounds of the patients and had access to all the medical histories and the results of the previous research. Our working together didn't end with the time we had spent in Guam either but was continued with meetings in Europe, the U.S.A. and Indonesia as well as with joint lectures on the occasion of conferences of neurologists. It was – and still is – difficult writing publications together; so far the outcome is a series of unpublished manuscripts. At the point where it came to scientific writing, the differences between disciplines and customs, for instance how an article is to be structured, sometimes manifested themselves with quite some tensions. (Verena Keck)

But smaller multidisciplinary teamwork can also make sense, and it is certainly a lot easier to organize. In numerous cases of working together it is the anthropologists who invite representatives of neighbouring disciplines: they have already been in the field longer, they know the language and they have familiarized themselves with the culture.

> Being in the position of a professor who cannot get away more than two or three months, outside of teaching semesters, I could not have started new field sites by myself, because it takes too much time. Being able to move into the field where an anthropologist has already established the groundwork, has all the connections, etc. is of course a great solution to this problem. Also to determine which topics are interesting for a common study. (Personal communication, psychologist Pierre R. Dasen)

We would like to add a fictive conversation: a psychologist and an anthropologist converse in the field. This is to illustrate the collision of different research agendas and methods and to lead us into the volume's first chapter, which presents the differences between the two disciplines systematically. The conversation is based on minutes recorded from memory during the fieldwork of Pierre R. Dasen (PD, psychologist) and Jürg Wassmann (JW, anthropologist) in the Yupno Valley, Papua New Guinea, some years ago; similar talks also occurred during subsequent joint fieldwork in Bali. The discussion has been going on for some time and the tone of the conversation is slightly tense but still friendly.

PD: You are continuously trying to find an ideal-typical presentation of the system [of the Yupno culture]; what's missing is the occupation with actual everyday life. For instance, in West Africa [where I did fieldwork before] there are real artisan villages. I could observe the people there while they were working and ask them direct questions.

JW: Here, there just don't happen to be any specialists, and what is produced, anyway? String bags, arrowheads, perhaps. And during the day the people are scattered all over the place. Some are gardening, others hunting in the bush. Good. In the evening they're all in their houses. But these are pitch dark. What am I supposed to be able to watch there?

PD: It's true that the Yupno are very individualistic. But because of this, we also have the chance of obtaining individual answers. The disadvantage is, of course, the general 'speechlessness' when faced with us since they are not used to talking about their own cultural behaviour.

JW: One simply mustn't ask direct questions, but must carefully delineate specific subjects for the people, we have to listen and e.g. write down the relevant vernacular key words so we can later ask more directly.

PD: In addition, the answers are much too uncertain and we can't react to them quickly because your knowledge of the language is still too limited. Also, many questions (e.g. about 'multiplication' or 'subtraction') can't, apparently, be posed adequately because the respective words don't exist.

JW: We have noticed that the men all count a bit differently, another example of individualism. Good. Nevertheless, I'd rather sit for hours with a single informant – he trusts me and he talks.

PD: No! You have to question a lot of people, collect samples, for example, we still need ten old women for counting.

JW: Impossible! How am I going to find ten old women? They won't come, because they're afraid to give answers. They have the feeling that they're being 'tested'. And you know very well that women don't do any counting in public because of social rules. Look, in the attempt to classify 'hot' and 'cold' objects they ran away. They found the whole thing absurd.

PD: It went quite well [with male informants] with the numbers.

JW: Yes, but they're not particularly important to the Yupno.

PD: I find it simpler if one first knows which cognitive problem one wants to examine and, accordingly, then looks for a 'suitable' culture: for conceptions of space, for example, I go to Australia, this is a suitable place to study geocentric space; for problem-solving regarding handicraft, I go to Mexico – that makes sense.

JW: Not at all. First of all, I want to describe a culture; as far as I'm concerned, the individual variations in everyday life too. In any case, that with which the people are concerned, otherwise they just don't come [into our house] at all. The methods [of psychology, for example] are therefore only an aid. The Yupno tell me themselves what presents a problem.

PD: I am more interested in investigating fundamental topics, such as orientation in space or the relationship between language and cognition. But in order to do this I have to be able to do tests or other tasks and these may well be strange for the Yupno, or even considered to be childish or even insulting, I am aware of that and that is why we want to proceed in a playful manner [referring to the 'space games'].

JW: These tests, my impression is that these tasks are disturbing my confidence relationship with my informants; we have to find other solutions!

(And so it went on for quite some time.)

Note

1. Some materials in this section have been adapted from the article, 'The Final Requiem for the Omniscient Informant', published in *Culture and Psychology* 1995 1(2): 167–201, and the chapter 'Interdisziplinäre Projekte und Teamarbeit', published in B. Beer (ed.), *Methoden und Techniken der Feldforschung* (Berlin: Reimer 2003), pp. 203–23, with the permission of the editors.

Part I
Local Actors

Chapter 1

The Methodological Interface of Psychology and Anthropology

Ramesh C. Mishra and Pierre R. Dasen

Introduction

How do we experience new worlds? This question can be taken as the basis of much of social science enquiry, especially in (cross-)cultural psychology and in anthropology. Indeed, either our informants experience novelty in a situation of acculturation (and the very presence of researchers is part of it!), or the researchers can be said to experience new worlds when they are confronted with cultural diversity, or even with disciplinary diversity if and when they try to interact with colleagues of other social sciences. This chapter, written by two cross-cultural psychologists, attempts to show, in an interdisciplinary context such as this volume, some methodological questions and answers that can be derived from the experience of cross-cultural psychologists, and that could possibly be of use or interest to other social scientists, particularly anthropologists. To provide such advice in a few pages is quite a challenge, and we are well aware of the pitfalls, in particular that of over-generalization. There are, of course, many different anthropologies with their respective epistemologies and methods, just as there are many psychologies, and this has to be kept in mind, lest we be accused of naïve stereotypes.

Various disciplines of the social sciences are relevant to the study of experiencing new worlds, among them sociology, linguistics, cognitive science, and of course particularly psychology and anthropology. All of them are primarily concerned with the study of human beings, although they do it from different vantage points. While one can find considerable evidence for the explanation of social or cultural phenomena in psychological terms early in the history of anthropology (e.g. Rivers 1914; cf. also Jahoda 1982), the explanation of behavioural phenomena in cultural terms has generally been less attempted in psychology (Jahoda and Krewer 1997). The newly developed field of (cross-)cultural psychology is an exception, where culture 'is taken seriously' (Dasen and Jahoda 1986) by psychologists in understanding human behaviour.

Cross-cultural psychologists accept the principle of 'the psychic unity of mankind'. They generally believe that basic psychological processes underlying behaviour are universal, but the way in which they develop and find expression across human populations varies considerably according to cultures. In order to demonstrate this, they obviously need to compare across several cultures. They think that culture (as an independent variable) influences behaviour (as the dependent variable). Cultural psychologists refute this, and claim that culture and the psyche mutually constitute each other. Also, they tend to concentrate on one single culture, and eschew comparison; they are hence often closer to anthropologists.

Despite facing several criticisms, psychological perspectives have gained substantial recognition in anthropology. The emergence of several subdisciplines within anthropology has particularly been instrumental in this process. Much of the work carried out in cognitive anthropology, psychological anthropology and symbolic anthropology seems to have a strongly psychological or quasi-psychological flavour. While the sharing of perspectives has been partly reinforced through research collaborations between some anthropologists and psychologists, some of them feel uncomfortable with the use of the methods and perspectives of the other discipline. The problems that bother anthropologists quite often relate to the nature of psychological inquiry, the (mainly quantitative) methods used by psychologists, including the level of analysis and the ecological validity of behaviour observed in unnatural circumstances. There are also concerns about the level of understanding of behaviour reached by anthropologists and psychologists even if they engage in the study of the same group of people at the same locations.

While the mutual influence of these disciplines seems to be imbalanced, in this chapter we will try to argue for an active collaboration between anthropologists and psychologists. We strongly feel that both of them can learn a lot from each other. Psychologists can learn the ways of analysing the larger eco-cultural context in which human behaviour occurs in order to add greater meaning to their conclusions, and anthropologists can learn to employ methodological rigour in their studies and analyses of human behaviour in order to enhance the objectivity, reliability, validity and generality of their conclusions.

In the following sections we would like to discuss some of the issues that are still alive as matters of debate between anthropologists and psychologists. None of these issues is new; researchers in the respective disciplines faced them time and again. The purpose is to bring them together in the hope that researchers will find some advantage in an interdisciplinary approach when designing, executing and reporting their research studies. In this paper, we commit the very crime we are quick to denounce, that of overgeneralization: We speak of anthropologists and psychologists as if these were homogeneous categories! Of course, we know they are not, yet it is quite impossible to take all the distinctions and complexities into account. Each reader will have to tune the statements according to his or her own particular situation.

Levels of Understanding

There is no doubt that the kind of presuppositions anthropologists and psychologists bring to the enterprise and the kind of observations and inferences they are led to make in their studies of human behaviour often lead to difficulties. These difficulties can be well illustrated by narrating a brief but extremely interesting dialogue (cited in Cole and Scribner 1975) that took place between Margaret Mead and Jerome Kagan at the conventions of the American Association for the Advancement of Science held in 1972. Professor Kagan presented a paper summarizing the research that he had carried out with children in Guatemala for almost a decade. Assessing the responses of young infants to stimulus novelty he had noted that rural Guatemalan children, as compared to Americans, were several months retarded in one particular cognitive function, the 'activation of hypotheses' (an increase in attention and active mental work) at around 1 year of age. On the other hand, using recall of familiar objects, recognition memory, embedded figures test, and other standard measures of cognitive development, he reported only minimal differences between Guatemalan and American children by the age of 8 or 9 years. Such findings led Kagan to conclude that 'infant retardation seems to be partially reversible and cognitive development during the early years more resilient than had been supposed' (Kagan and Klein 1973: 957). It appeared that there was a period of 'catch-up growth' during which Guetmalan children grew up at a faster rate, so much so that by late childhood their cognitive performance was fairly comparable with that of American children.

Following Kagan's presentation, Professor Mead took the floor as the discussant and asked: 'Why has it taken psychologists so long to discover what anthropologists have known all along?' Professor Kagan replied by saying that perhaps the psychologists were a little slow. This unusual response was followed by great laughter, and obviously a great variety of discussions. This little dialogue does raise a number of basic questions that are alive even today with respect to the level of understanding of a phenomenon by anthropologists and psychologists.

Was the reply Professor Kagan gave in response to Professor Mead's question really appropriate? The fact that Kagan took ten years to discover what Mead already knew is not really surprising. If psychologists are really slower than anthropologists in understanding a behavioural phenomenon, it could be by virtue of the respective methodological demands of their disciplines. Does one consider informants' reports or casual observations of behaviour on a few cases to be enough? Or does one require a lot more time in setting up the study in a manner that fulfils the conditions of sampling and measurement?

A related question is whether anthropologists and psychologists reach the same level of understanding of a phenomenon by using the 'methodological armoury' of their respective disciplines. In other words, did Mead and Kagan really know the same thing about Guatemalan children? The basic issue is whether the differentials of perspectives and methods lead to the same kind of understanding of a behavioural phenomenon. It may be noted that Mead and Kagan shared their opinion about cognitive competence of Guatemalan children without examining the suitability of each other's data before reaching that conclusion. They had also not examined other research evidence in which the performance of Guatemalan children was found not to be up to the 'American standards' on tasks similar to those that Kagan had used. Does it mean that what others have reported about these children's competence is absolutely wrong, or is there something else involved in these studies that makes such apparent contradictions in understanding inevitable? As cross-cultural psychologists we keep our fingers crossed in answering any one of these questions.

We believe that both anthropology and psychology can be characterized by certain strengths and weaknesses in the study of human beings. We also believe that a joint venture of researchers belonging to these disciplines can reinforce the strengths, tone down the weaknesses of each other's research and promote a better understanding of behaviour.

Levels of Analysis

When attempting to understand human behaviour, anthropology deals with cultural-level processes, whereas psychology is confined exclusively to individual-level processes. The former focuses its analysis at the phenomena that operate at the group level, whereas the focus of the latter is primarily at the phenomena that operate at the individual level. Adherence to these practices has often remained so strong that some anthropologists (e.g., Tiger and Fox 1974) have even managed to interpret behaviour in terms of biology and culture without any reference to psychology. If psychology claims to be the science of behaviour, then how was this possible? The same is true for psychologists who, in their attempts to be 'scientific', have altogether ignored culture in their studies. Shotter (1975) argues that in the relationship between man and nature, there is a third term, '*culture*', which is not genetically inherited but communicated to man after birth as a '*second nature*'. This implies that any analysis of human behaviour will be incomplete and its description lopsided if culture is not brought seriously into psychological research.

Considering culture as a group-level phenomenon, anthropologists have largely remained occupied with the search for and description of 'uniformity' in behaviour of individuals of a given culture. They believe that cultural-level processes must be available to and expressed in all individuals of that culture more or less in the same manner. Hence, it is possible to discover a particular pattern of behaviour of individuals in a given culture. Psychologists, on the other hand, are largely engaged in studying behaviour by controlling for any possible influences of cultural variables. This concern can be easily discovered even in their studies carried out away from the laboratory. They believe in individual differences in behaviour, and try to demonstrate these variations within the same cultural setting. They show that individuals participate in cultural processes in variable degrees, and argue that different patterns of individuals' behaviour are likely to be observed within the same culture. These differences cannot be easily glossed over, and a major task of psychologists is to find out appropriate explanations for them.

The preoccupation of anthropologists and psychologists with the study of group-level and individual-level processes respectively has led to different strategies for organizing research studies. Anthropologists have preferred to focus on the analysis of customary achievements of people (e.g., in hunting, navigation) in their 'natural-cultural habitat' (the broader ecological and cultural setting in which people negotiate their day-to-day life) and interpret them in terms of people's recurrent experiences, which provide the basis for all learning. In doing so they do engage in some observations, but their reliance on informants is often the key issue in the data collection process. Thus, the analyses of behaviour are carried out at a global level, and the conclusions remain specific to the group of people studied. They lack the component of generality, which is a major goal of scientific inquiries.

On the other hand, psychologists, by virtue of their training in laboratories, generally feel uncomfortable in dealing with the 'natural-cultural habitat'. They prefer to design situations according to the needs of the study, and carry out observations in contrived situations that often do not match with real-life situations. The focus is more on the effect of variables that are proximal and immediate in space and time than on those which are embedded in the life of people but not so easily observable. Although this approach fails to meet 'ecological validity' (Berry 1980), generalizations are advanced for all situations and for all kinds of people. Thus, while anthropologists do not intend to come down to the level of designing observational or experimental situations, psychologists do not intend to go beyond contrived situations to observe people's behaviour as it occurs in natural settings. Hence, the concerns expressed in studies and the levels at which analyses of behaviour are carried out differ sharply between anthropologists and psychologists. Some rapprochement between the two is essential for producing a unified science of behaviour.

Use of Instruments

The scientific study of any process or phenomenon requires the use of instruments. These may vary along a dimension of sophistication with crude or less sophisticated at one pole of the dimension and highly sophisticated at the other. Instruments add accuracy and precision to any observation or assessment. While psychologists cannot think of a research study without using instruments and quantitative measures, anthropologists often find most of them to be too superficial, so as to deny their usefulness in research. Psychologists are in possession of innumerable instruments that range from simple paper and pencil variety to the most sophisticated electronic equipment. These have been found to be extremely helpful in assessing the processes underlying behaviour. The advantage is that they can be designed and used in ways that permit the elicitation of standard responses in a variety of situations. These responses can be recorded and preserved for thorough analyses at later points. Instruments can also serve other purposes. In particular, they can relieve us from dependence on memory, and render valuable help in identifying the processes involved in a particular behavioural phenomenon or in checking and ensuring the reliability of our observations.

As an example, we may consider the classical study of Gladwin (1971), who documented the remarkable navigation skills of Puluwat navigators in Micronesia. The study showed that they used a very complex system of calculation based on a star compass, winds, currents and several other cues to keep track of their distance and direction while navigating at sea. In terms of data, the study relied on past memories of the navigators as well as participant observation. However, Gladwin's descriptions of navigation processes are not precise enough to satisfy cognitive psychologists. And inconsistencies between what people report doing and what they actually do have been noted in both psychological and anthropological studies (e.g. Wassmann 1993).

Gladwin did try to use an instrument to tap the navigation process of Puluwat navigators in an experimentally designed situation, in particular the Piagetian task of working out the number of possible permutations among a set number of coloured tokens. This situation was inspired by the fact that navigators know the alignments for departure from one island to another in a large group of islands (and rote learning is used to instruct young navigators). The navigators were unable to perform this task, while young non-navigators who had been to school had no problem with it. This tells us that the navigators' skills do not easily transfer to novel situations, and also that it is difficult to adapt tasks to make them meaningful to unschooled informants, but it does not help in specifying the psychological processes actually used. Had Gladwin succeeded in devising some appropriate psychological tasks, one would have been able to place greater confidence in Gladwin's conclusions about the cognitive processes of the navigators. The availability of instruments not only encourages testing of the reliability of verbal reports, but also provides us with an opportunity for cross validation of data obtained from different sources. Here again, the psychological approach can be meaningfully integrated with anthropological studies to enhance their reliability and, more importantly, to describe the processes underlying the behavioural phenomenon under consideration.

Methodological Problems

As a discipline, mainstream psychology is still characterized by the use of rigour of scientific methodology in its research – more than any other discipline within the category of social sciences. The concern for variables, their manipulation and control, precise observation of behaviour using some kind of instrument and sophisticated statistical treatment of data still dominates the design of psychological studies. Since these concerns are not shared by anthropology, most psychologists consider its approach less sound. Methodological inadequacies of anthropology have often been criticized. Bronfenbrenner's (1979) comment in this respect is quite pertinent. In his book *The Ecology of Human Development* he observed:

> There exists a body of scholarly work in which external environmental contexts are described in considerable detail and their impact on the course of development graphically traced. Such investigations are carried out primarily in the field of anthropology … . But the descriptive material in these studies is heavily anecdotal and the interpretation of causal influences is highly subjective and inferential (1979:18).

The problem seems to be inherent in anthropology because of its focus on the study of historical changes. Here, experimentation has to be excluded from research, and the approach is bound to be inferential. On the other hand, it is not surprising if anthropologists easily dismiss the vast mass of carefully conducted research by psychologists, charging that it is narrow, artificial, parochial and unconcerned with behaviour in ordinary life.

A critical examination of the manner in which anthropologists and psychologists go about their research would indicate a number of methodological problems with both of them. In the following section, we will discuss them briefly.

Sources of Data

Because both anthropologists and psychologists collect data to substantiate their research inquiries, the sources of data surface as a major point of concern for both of them. While anthropologists prefer to observe behavioural episodes in natural ordinary life situations, psychologists mainly prefer to design experimental or semi-experimental situations to study behaviour. Besides observation, anthropologists also depend a lot on their informants for data that forms the basis of ethnographic descriptions, whereas verbal introspection by informants does not constitute a valid source of data in mainstream psychological studies. Which one(s) of these various sources may provide us with good quality data is a matter of serious debate.

Data obtained from different sources are often fraught with the problem of inconsistency, e.g. between data obtained from interviews and observations, and also between those obtained from observations and experiments. Given this state of affairs, the acceptability of data emerges as a major problem for researchers in both disciplines. Mishra, Sinha and Berry (1996) noted these difficulties when they interviewed parents (and also children) about child socialization practices with a standardized interview schedule, and also observed parent-child interactions in a specially designed situation in which parents and children had an opportunity to work together. Data obtained from these two sources did not harmoniously match, suggesting that what people reported and what they did in real-life like situations were not the same thing. The question is, 'which one of these two data sets is more reliable and worth reporting?' One cannot find a definite answer unless the question is directed at the 'predictive value' of data in terms of theoretical assumptions that initially might have set the stage for the particular research study.

In our studies of spatial encoding (Wassmann and Dasen 1998; Mishra, Dasen and Niraula 2003; Mishra and Dasen 2005) we have encountered similar difficulties. In tasks designed by Levinson (2003) and his team at the Max Planck Institute for Psycholinguistics (see the chapters by Senft and by Widlok in this volume), informants are asked to reproduce a display (such as three animals in a row) on a second table after a rotation of 180°; these nonverbal tasks assess whether they use a geocentric or an egocentric frame of spatial reference. On some of the items, the informants are also asked to explain how they remembered the display. Some informants encoded spatial arrays of stimuli in a relative frame, but explained them in absolute terms, or vice versa, in a significant number of cases.

Thus, the value assigned to data obtained from different sources (including primary and secondary sources) comes up as a major issue that requires some agreement between anthropologists and psychologists. We must remember Campbell and Fiske's (1959) advocacy of the use of a 'multi-method and multi-measure' approach to search for a deeper understanding of phenomena even if it turns out somewhat problematic in terms of generating inconsistent data sets. This practice would at least save researchers from producing method-bound conclusions, which may be acceptable to the scholars of one discipline, but not to those of others.

Task Specificity

We indicated earlier that instead of observing behavioural phenomena in natural contexts psychologists prefer to design experimental situations in order to study behaviour and infer the underlying psychological processes. A large variety of tests and tasks have been developed for this purpose, but one can also design new tests/tasks in case the standard ones appear less appropriate. While the main purpose of these tests/tasks is to make valid inferences about the psychological life of individuals as it unfolds in their day-to-day settings, psychologists hardly ever show any concern about the search for correspondence between the behaviour observed in laboratory settings and that displayed by people in real-life settings.

Psychologists also find difficulty in drawing inferences about the daily life of people based on their performance on a single task. The usual practice is to use a number of tasks supposedly aimed at measuring the same process. Confidence in the assessment procedure is enhanced if high interrelationships in performance on various tasks is demonstrated. Although the hope is that similar tasks would demonstrate similar patterns of performance, this is not always confirmed. Minor changes in task settings can change the pattern of performance. This has been demonstrated, for example, in studies of pictorial perception (Serpell 1979), and on the spatial encoding tasks by Widlok (this volume) and by Troadec, Martinot and Cottereau-Reiss (2002).

In our studies of spatial encoding mentioned above, we systematically used (among several other instruments) two spatial encoding tasks. While both tasks had the same structure, one of them (called 'The Animals') was easy to encode verbally ('All three animals look north"), while the other (called 'Steve's Maze') elicited a more iconic encoding (the memory of a shape, or movement of the hand). With the first task, we found systematic geocentric encoding, and relatively more egocentric encoding with the second one. This 'task specificity' was found in all our studies, in Indonesia, India and Nepal, which speaks for the reliability of the finding. In this case, making changes to the task settings (such as changing the wording of the instructions, or the orientation of the display along main geocentric axes or not) made little difference in terms of results.

Evidence for 'task specificity' in performance suggests that the nature of the task is an important factor in determining the type and the level of performance of subjects. Thus, any general conclusion about the behaviour of individuals based on their performance on psychological tasks has to be drawn with great caution. It is important to use several tasks, several types of instructions and settings, in order to make sure that the tasks really 'measure' what is intended by the experimenters. The research orientation developed by Cole (Cole et al. 1971; Cole and Scribner 1974) under the name of 'experimental anthropology' has been very telling in this respect (for a review, see Segall et al. 1999).

Domain Specificity

Besides task specificity, research provides considerable evidence for 'domain specificity' of behaviour. Studies of cognitive behaviours of people carried out in diverse cultural settings support the notion of 'domain specific competence' (Berry 1984), which means that people in some societies can competently deal with stimuli related to behavioural domains that are reinforced in their settings. For example, hunters and gatherers develop high levels of differentiation in the perceptual domain because of the day-to-day demands placed on their life for disembedding (both visually and tactually) fruits, nuts and animals in the camouflaged forest environment (Berry 1976; Mishra, Sinha and Berry 1996). Contrary to this, the agriculturalists develop high levels of differentiation in the social domain because of the high premium placed on social relationships in agricultural societies.

Similarly, Dasen (1975; see Segall et al. 1999 for a review), using tasks derived from Piaget's theory, found that the children in hunting and gathering societies developed spatial concepts much earlier than quantitative concepts (quantity, weight, volume), while the opposite was true of children in agricultural societies. This is no doubt due to what is adaptive in each type of ecological setting, and hence what is valued in each type of culture: spatial concepts that help in way finding for a nomadic type of existence, and quantification in a setting where agricultural produce has to be stored and exchanged.

Thus, in order to study behaviour in different eco-cultural settings, one has to be sensitive to the kind of pressures that operate on people who negotiate life in those different settings. This essentially requires a prior knowledge of such features of groups in order to design or select appropriate tasks to make valid assessments of their behavioural characteristics. Failing that, the assessments are likely to end up estimating behavioural features that are less strongly valued among them. Informants, who are commonly used by anthropologists as important and valid sources of data, may be meaningfully utilized by psychologists as 'information providers' on such important aspects of people's life and as 'assistance providers' in selecting/designing appropriate tasks for behavioural assessments. In this way, the 'grasp of natives' point of view' (at which anthropologists are highly competent) can be meaningfully integrated with psychological experimentation and testing.

Wassmann and Dasen (1994) suggested a research procedure in three steps, combining anthropological and psychological methods. First comes what people say, obtained through interviews with informants, but taking care to include a variety of informants from different segments of the society, i.e. not only experts but also 'jpfs' (just plain folks). But, as mentioned above, people do not always do what they say. Hence observation is the second necessary step. And then, if the process of interest cannot be easily observed, it may be necessary to induce it, by using 'artificial' situations – which is what psychologists usually do. These induced situations, however, have to be ecologically and culturally valid, which is only possible if the researcher has a sound culture knowledge. Step three is therefore dependent on steps one and two, i.e. in cross-cultural psychology, psychological methods can only be used in combination, and after ethnographic ones.

Reliability of Data

The accuracy of behavioural data obtained through observation, interview or experimentation has always been a matter of serious concern in psychological research. While there are chances for misrecordings, omissions and even distortions (e.g., where there was great reliance on memory), it is also likely that the researcher does not record all behavioural episodes to the same extent. A related question is whether two researchers engaged in observing and recording a given behavioural phenomenon would do it in the same way. What confidence do we have that they will record the same events in the same manner? In other words, what is the reliability of the observations and recordings made by a particular researcher?

While anthropologists may feel quite easy with, and confident about, the observations and recordings (made by them or their assistants), psychologists will always have doubts about them. In fact, they would hesitate in accepting any data that have not been tested for 'inter-observer reliability'. The problem appears more serious if the sources of data (e.g., observation, interview) are largely unstructured and primarily focused on qualitative features of behaviour that have either to be dealt with in their original form (i.e., as they are recorded), or require some kind of transformation in an attempt towards quantification. Psychologists use several techniques by which they try to ensure the reliability of data irrespective of whether they are of a qualitative or quantitative nature.

The question of reliability is also linked to the number of observations made of a particular behavioural phenomenon. Some anthropologists feel satisfied by asking just a few key informants about their behaviour, or by observing them on a few occasions. They may advance generalizations based on those few collections of interviews and observations. This practice has, of course, been criticised from within anthropology (e.g. Wassmann, 1995), but has not completely disappeared.

A basic tenet of scientific enquiry is to seek confirmation (replication) by a second researcher. While this is unfortunately not a systematic practice in psychology, it is almost never even attempted in anthropology. There are, of course, good reasons to claim that 'field work' implies a very personal relationship between the researcher and his or her informants that cannot easily be handed over to someone else. Another argument is that societies change over time, so that a strict replication is impossible unless carried out within a very short time frame. The world of American cultural anthropology has been shaken somewhat by the Mead versus Freeman controversy over adolescence in Samoa (Mead 1928; Freeman 1983; Côté1994), a fantastic example of nonreplication, with all kinds of political and epistemological ramifications.

Some anthropologists tend to assume that all behavioural phenomena are cultural in origin and are expressed almost in the same way in all or most of the individuals who share that culture; there is no scope for questioning its variability across individuals within the same culture. Hence, grand generalizations were sometimes made by anthropologists about the behaviour of people of certain cultures without enough empirical evidence for making that high level of claim. The 'culture and personality' school, exemplified by Kardiner and Linton's (1945) work, provides a good example of this problem. They analysed material from several cultures provided by anthropologists, without undertaking any fieldwork themselves (hoping that others had undertaken it seriously). Thus, they described some 'modal personalities' as characterized by almost entirely negative and almost pathological traits, which were attributed to child-rearing practices such as maternal neglect. It is hard to believe a finding like that, but Kardiner and Linton did not question it because it was not at odds with the Freudian theoretical notions through which they were trying to view cultural systems and behaviours. The question is whether a culture whose members live in such miserable psychological states could survive for long. The answer would certainly be 'no'. Such conclusions would not have been drawn had any attempt been made at seeking reliability of data through either direct or indirect methods.

Similar fallacies exist in the field of psychology as well. For a long time, psychology was characterized by its pursuit of general principles of behaviour. A vast majority still believes that the laws, principles and theories advanced in the discipline of psychology are 'universal' even if they have been developed on the basis of data provided largely by undergraduate students of psychology in the United States. They plead with forceful arguments that psychology must continue with its search for generality or universality. In doing so they perhaps believe that the world is homogeneous. They never care for or pay attention to cultural variability across populations around the globe. Hence, it is no surprise that we have several grand theories of behaviour that are claimed to apply to everybody everywhere. These theories started faltering when they were taken to test in cultural settings other than those in which they had been developed. Their inadequacies were realized, limitations pointed out, and modifications were suggested for their reformulation. Cross-cultural research carried out in different parts of the world indicates that nothing like a 'universal psychology' exists so far. However, the development of a universal psychology can certainly be set as a goal of psychological research, and attempts should be made to achieve that goal through systematic research in different cultures and across different psychological (behavioural) domains. Berry (1993) has suggested a scheme for integrating research in order to arrive at a psychology that may be called universal in its real sense.

The question of reliability is not confined to data collection alone; it is also crucial with respect to the inferences drawn from data. The development of a unified science makes this imperative for anthropology as well as for psychology. Psychologists may play a crucial role in convincing their anthropologist friends to address the issue of reliability by engaging them in repeated observations of the same phenomenon, using more than one observer/researcher for data collection, carrying out their study with a larger sample size, and searching for some correspondence in their data sets to enhance the credibility of their conclusions.

Systematization in Research Data

The search for systematic relationships in data is another important goal that anthropologists may like to welcome in their research studies. Of course, some schools in anthropology do this, for example those that use wide ethnographic data sets such as the Human Relations Area Files (Barry III 1980; Naroll, Michik, and Naroll 1980; Levinson and Malone 1981; Segall 1989). However, the use of such a comparative stance, including inferential (correlational) statistics, is far from achieving consensus in anthropology, and has been more of interest to cross-cultural psychologists; it seems to be a vanishing enterprise. Similarly, some anthropologists influenced by John and Beatrice Whiting (Harvard Graduate School of Education) have not hesitated to appropriate the toolkit of quantitative methods. Excellent examples are Munroe and Munroe (1997) or Super and Harkness (1997). Note that these colleagues partly identify with cross-cultural psychology, and that they are quite often husband-and-wife teams, one of each discipline, which is possibly the closest way to establish an interdisciplinary collaboration!

On the other hand, ethnography based on 'radical cultural relativism' is still quite popular. Adherents believe in and try to deal with each culture in a unique way without attempting any kind of systematization of elements at either the cultural or behavioural level in an empirical manner. In fact, as Leach (1967: 87) observes, 'he [the anthropologist] purposely chooses a small field within which all the observable phenomena are closely interrelated and interdependent' (quoted in Jahoda 1982: 56). They also believe that each experience in a particular cultural setting is unique to a group of people. They adopt a similar approach in studying the behavioural domain of people, and believe that each cultural experience is linked to the development of a particular kind of behavioural repertoire in a unique way. This approach to the study of cultures and behaviours can be described as 'butterfly collection.'

The metaphor indicates that assuming that each behaviour is specific to a particular culture would lead to a great collection of behaviours without making any attempt towards examining their interrelationships. Hence, they can tell us about very specific aspects of different butterflies (cultures), but cannot suggest any general principle that could be used in making these descriptions more meaningful. Thus, while the collection of specific descriptions can document cultural diversity, it is inherently incapable of telling anything about commonalities. What is required is the extension of research towards the search for generalities across cultures. Since the very beginning, cross-cultural psychologists have committed themselves to the pursuit of these twin goals (Berry and Dasen 1974).

This is not in any way undermining the strengths of anthropology as a discipline. It simply reflects a kind of preference on the part of many anthropologists due to their adherence to traditional approaches of research in their discipline. But the boundaries of various disciplines are not so rigid today, and there is a considerable amount of dialogue and sharing of perspectives and methods among them.

Active collaboration between anthropologists and psychologists will be extremely rewarding for both. While psychologists may learn the use of wider perspectives (e.g., ecological, cultural) and contexts (e.g., learning environments) to examine the behavioural characteristics of individuals or groups, anthropologists may learn the use of certain statistical techniques through which they may go about searching for patterns in data and achieve greater precision and generality in their findings.

Anthropologists interested in integrating psychological methods could possibly start with the more qualitative approach of cultural psychology. Greenfield's (1997) chapter on 'empirical methods for cultural psychology' would be an excellent start; she demonstrates that there can be different takes on the concept of 'validity' (see also Kirk and Miller 1986), and that qualitative and quantitative methods are complementary. This has become the most generally accepted methodological stance in social sciences such as education (Miles and Huberman 1994; Pourtois and Desmet 1997), although it is still not fully accepted in mainstream psychology. A further step would then be to consult some writings on methodology within cross-cultural psychology (for example Brislin, Lonner and Thorndike 1973; Lonner and Berry 1986) or social psychology (Camilleri and Vinsonneau 1996).

Considerable progress has been made in recent years, especially under the impulse of Ype Poortinga and his team, in rethinking psychometric methods for cross-cultural use (Poortinga 1983; Poortinga 1997; van de Vijver and Leung 1997; see also Berry et al., 2002 for an introduction). These authors are concerned with the appropriate use of psychological tests and scales across cultures: how can we make sure that these instruments really measure the same underlying construct; i.e., can we really attain comparability in test scores? This is obviously a very complicated issue. In short, the idea is to assess the construct validity of the instruments separately for each culture through factory analysis; items can be rephrased or removed until comparability is achieved. It is then only on the basis of comparable structures that cultural differences can also be documented.

Conclusions

With a primary focus on the study of humans, both anthropology and psychology have made remarkable achievements in their respective fields of knowledge. Commitment to disciplinary practices has resulted in a lot of advancement in respect of the understanding of cultural and behavioural features of groups in the discipline of anthropology. On the other hand, psychologists have made considerable advancement in the understanding of behaviour both in micro and macro settings. In anthropology, researchers have made sweeping generalizations concerning the role of culture in the behaviour of individuals. In doing so they have altogether ignored the fact that human behaviour is highly variable in response to various stimuli, and that within the same culture, a number of behavioural patterns can be found. On the other hand, psychologists have largely turned their backs on the analysis of the role of cultural factors in behaviour. Except for a small number of cross-cultural psychologists, they have mainly focused on the study of the variables that are present in the immediate environment of individuals. The domination of an experimental approach has led to an adherence to refined methodology in terms of research design, specification of variables including their linkages, selection of samples, techniques of data collection and their analyses. Such refinements in methodology accord them credibility within the social sciences. However, the concern for methodological rigour on the part of psychologists has also led to the criticism of producing decontextualized and asocial knowledge about human behaviour, which does not make much sense if one wants to address larger issues of the society (e.g., population, health, environment).

We believe that anthropologists and psychologists need to work together to examine the strengths and weaknesses of each other's disciplinary practices more closely. While psychologists need to read more widely in history, ethnology, and the centuries of writings on man and society to develop wider perspectives on human behaviour, anthropologists may borrow from them some methods in order to grasp the notion of controls and correlations, including systematization and generalization of knowledge, to produce research that receives a wider audience and greater interdisciplinary acceptance.

Ramesh C. Mishra and Pierre R. Dasen can be contacted at, respectively: rcmishra_2000@yahoo.com and pierre.dasen@pse.unige.ch.

Chapter 2

Rethinking Tradition: Invention, Cultural Continuity and Agency

Ton Otto

Introduction: From Tradition to Agency

Reflecting on the concept of tradition, many people will spontaneously link it with cultural continuity and repetition; the heavy weight of tradition or the 'dead hand of history' is seen as compelling human beings to act in the same way as their parents and ancestors. There is little space for human agency in such a view. Anthropological and historical research into traditions reveals, though, that the situation is much more complex. Traditions appear highly changeable and their maintenance or adaptation involves the active involvement of human actors. The modern understanding of tradition is greatly influenced by the academic debate about the *invention of tradition*, which was initiated by the celebrated anthology edited by Hobsbawm and Ranger (1983) and the equally innovative but more regionally focused special issue of *Mankind*, edited by Keesing and Tonkinson (1982).

In this chapter I want to contribute to an anthropological understanding of human agency by analysing the concept of tradition. I begin by reviewing the 'invention of tradition' debate and then go on to analyse the concept of tradition in relation to the question of cultural continuity and invention. I show that culture is a multiplex concept that refers to various modalities of patterned meaning and action, which can be more or less explicit for the human actors involved. I discuss 'habit' and the process of habitualization, 'custom' and the process of institutionalisation, and 'tradition' and the process of legitimation. Finally, I identify invented traditions as a special kind of tradition. All modalities of culture imply human agency and can be the object of explicit knowledge and influence, but the routine involvement of conscious action varies enormously. Traditions are the most explicit modalities of culture and require, therefore, the most conscious engagement of human actors. Since traditions are subject to human choice and strategic action, I will suggest that we can profitably study them as cultural, political and economic resources. I end this chapter by reflecting on the political implications of researching tradition: many anthropologists and other researchers have found that their descriptions and evaluations of traditions have had some very real and often unintended consequences for the people involved. This raises questions of an epistemological and moral nature.

In relation to the theme of this book – person, space and memory – the theoretical move in this chapter is thus in a way a reverse one. Instead of starting from the individual actor, I begin my analytical work by focusing on collective phenomena – traditions – to find out to what extent these collective phenomena presuppose the active agency of human actors, to what extent they determine the specific character of human agency, and to what extent they are open to human creativity and strategic manipulation.

The Invention of Tradition

Two decades ago, Hobsbawm and Ranger published their celebrated anthology *The Invention of Tradition* (1983), which included articles on the emergence of new traditions in Europe, colonial India and Africa. A year before, Keesing and Tonkinson edited a special issue of the journal *Mankind* (1982) under the title 'Reinventing Traditional Culture: The Politics of Kastom in Island Melanesia', which deals with similar issues in a different region. One can only speculate about the causes of this synchronous discovery of a new field of research, but one possible explanation is that this was prompted by cultural developments in the recently decolonized regions in the Pacific, Africa and elsewhere as well as a revived interest in regional traditions in Europe, as the continent moved towards greater economic and political integration. Whether this is a sufficient explanation for the sudden academic interest in the phenomenon or not, the idea that traditions can be newly constructed and serve political and social functions released an enormous amount of scholarly energy among historians, anthropologists, ethnologists and political scientists, and led to the production of a substantive corpus of literature on similar processes all over the world.[1]

The word 'tradition' has a long history in Western thought and usage (Williams 1983, Wiedenhofer 1990) and evokes a range of connotations that have provided opportunities but also caused problems for theoretical understanding. Since the primary popular meanings all refer to continuity with the past,[2] the provocative strength of the title 'invention of tradition' (or 'reinventing traditional culture') hinges on the contradiction between innovation and continuity.

Eric Hobsbawm's introduction to the *Invention* volume summarizes and clarifies the theoretical implications of the included studies in an exemplary way. He defines the term 'invented tradition' as

> a set of practices, normally governed by overtly or tacitly accepted rules and of a ritual or symbolic nature, which seek to inculcate certain values and norms of behaviour by repetition, which automatically implies continuity with the past. In fact, where possible, they normally attempt to establish continuity with a suitable historic past. (Hobsbawm 1983)

Note that in Hobsbawm's definition invented traditions establish continuity with the past in two ways. Firstly, Hobsbawm observes how the repetitive nature of the practices itself creates a sense of continuity. Secondly, he notes that invented traditions 'normally' evoke a more explicit link with a specific past. The word 'invention', on the other hand, stipulates that the practices are in fact new and therefore do not constitute a direct continuity with past practices, and, following directly from this, that the claimed continuity with a certain historic past is 'largely factitious' (ibid.: 2).

Hobsbawm's notion of 'tradition' is developed from an analytical distinction of tradition from the apparently similar concepts of 'custom', 'convention', and 'routine'. Whereas traditions characteristically claim or represent invariance, customs allow for flexibility and change even though by nature they engender continuous repetition. And in contrast to traditions, conventions and routines do not have a significant ritual or symbolic function. I do not quite agree with Hobsbawm's observation that '"custom" ... dominates so-called "traditional societies"' whereas modern societies are more likely to invent traditions (p. 2). I will argue that customs – as well as traditions – are central to all kinds of societies, even though differences can be found in the type of traditions that prevail.

In my reading, the major theoretical contribution of Hobsbawm and Ranger's volume lies in three central assumptions. In the first place, the invention of tradition is seen as a process of formalization and ritualization that may probably be found in all times and places. But it is likely to occur with particular frequency in times of rapid social change, when the strength and legitimacy of old patterns are weakened or destroyed. The editors claim that such changes have been especially momentous in the past 200 years (pp. 4–5). Secondly, one of the great transformations and innovations of this period is the emergence of the 'nation' as a political project (associated with nationalism and the rise of the nation-state). It appears that the invention of tradition has been particularly relevant and widespread in relation to this transformation and has often involved an element of 'social engineering' (pp. 13–14). Thirdly, the editors assume that invented traditions perform certain social functions. As already mentioned in the above definition, traditions inculcate certain values and norms of behaviour by repetition; thus, they are devices for socialization and the production of consensus. In addition, traditions can serve to legitimize institutions and relations of authority. Finally, traditions have the capacity to symbolize and establish social cohesion and membership of a group or community. This third function implies the other two as a means for creating social identities (including those of 'imagined communities' such as the nation). It is particularly this function that gives such an impetus to the invention of new traditions in times of change (pp. 8–9).

Apart from being co-editor of the 1983 *Invention* volume, Terence Ranger contributed to the same volume with a chapter on colonial Africa. Because his contribution spurred an extensive discussion within African studies,[3] and Ranger himself returned to the debate several times in writing,[4] it is useful to see how he modified his original positions. In a chapter published in 1993, 'The Invention of Tradition Revisited: the Case of Colonial Africa', Ranger defends his earlier work by highlighting that it dealt with a certain historical period characterized by the rise of romantic nationalism, in which traditions were particularly frequently invented. In Africa, this historical divide corresponded with the cleavage between precolonial and colonial societies. Ranger observes, however, that newer work – including his own – points out that his original focus on European inventions of colonial state ceremonies and so-called African traditions (ethnicity, customary law, traditional religion) was too narrow. For one thing, colonial hegemony was far from absolute and involved the contestation of invented traditions by various agents. Africans also participated quite actively in formulating and creating the new traditions and this involvement was part of the development of the colonial society. Finding that the word 'invention' has gotten in the way of a fully historical investigation of these two processes, he suggests that we use the expression 'imagination of tradition' instead. He also suggests that there should be more research into the relationship between postcolonial states and African tradition.

Reinvented Traditions in the Pacific

The debate on the invention of tradition in the Pacific began with a special issue of the journal *Mankind*: 'Reinventing Traditional Culture: The Politics of *Kastom* in Island Melanesia' (1982). 'Kastom' is the Melanesian Pidgin equivalent of tradition. In their respective introductions, the editors, Keesing and Tonkinson, set the tone for many of the ensuing discussions concerning the Pacific area. Keesing emphasizes the ideological dimension of *kastom*, which serves as a 'symbol' for anticolonial struggle. Melanesian societies gained their political independence rather late (in the 1970s and later[5]), and as they advanced towards autonomous statehood the discourse of tradition was an important factor in uniting local groups and evoking a sense of national identity. Keesing (1982) argues that the discourse has older roots in earlier anticolonial movements. The power of the discourse lies in its potential to disguise and mediate contradictions. The key symbol, kastom, is vague and vacuous and therefore it facilitates the establishment of a superficial consensus, which disguises underlying differences:

> The diversity of meanings Melanesians attribute to kastom underlines the way such symbols do not carry meanings: they evoke them. Their very abstractness and lack of precise content allow a consensus that would otherwise be impossible, among peoples whose material circumstances, class interest, and ethnic affiliations are different and often deeply divided. (1982: 299)

Interestingly, Keesing ends his introduction by cautioning his readers against assuming that political ideologies, like kastom, come into existence only as a result of the colonial impact: 'We err, I think, in imagining that spurious kastom is radically different from genuine culture, that the ideologues and ideologies of the post-colonial present had no counterpart in the precolonial past' (ibid.: 301). In a later, seminal article on the issue he develops this point further (Keesing 1989). Culture in all its forms has an ideological dimension in the same way as reified kastom. So-called 'authentic' cultures are the cumulative product, at least partly, of the political myths of the ancestors (1989: 24–25): 'The symbolic material of cultures – rules imputed to ancestors, rituals, myths – serves ideological ends, reinforcing the power of some, the subordination of others' (ibid.: 36). Keesing called upon Pacific islanders to develop a critical scepticism about representations of 'traditions' by Western scholars and indigenous elites alike. This politicization of the tradition debate was not equally well received by all indigenous scholars, and Keesing's article provoked an attack by the Hawaiian scholar Trask (1991), who accused him of perpetuating colonial arrogance and racism by questioning indigenous representations of their own tradition (see Keesing's reply 1991). Similar indigenous reactions against anthropological studies of the politics of tradition have characterised the Pacific debate ever since (Hanson 1989; Hanson 1991; Linnekin 1992; Friedman 1993; van Meijl 2000). In particular, the word 'invention' has caused a stir because it undermines exactly what it describes: the political use of 'age-old' traditions to defend or establish rights, status and privilege. These examples demonstrate that researchers into the political use of culture always risk being entangled in the phenomena studied.[6] I will try to unravel some of the ethical and epistemological issues at the end of this chapter.

In his introduction to the special issue of *Mankind*, Tonkinson (1982) makes the crucial observation that one has to distinguish between the political use of tradition at the local and at the national level. At the national level in regions such as Melanesia where cultural diversity is marked, the concept of *kastom* has to be largely empty of specific content, as argued by Keesing, the rhetoric of kastom being invoked to support ideas of unity, homogeneity and national identity. At the local level, though, kastom normally has detailed content and specificity and is used for defining differences and marking boundaries between competing groups (1982: 302; see also Tonkinson 1993). A central theme in Melanesia is the tension between kastom and Christianity, which was deeply felt in the village context, but mediated or resolved in the national identity project. In later articles, Tonkinson (1993, 2000) argues for the conceptualization of tradition as 'a resource, strategically employed (or not employed) by certain (but not all) of a community's members in pursuit of individual or collective goals' (Tonkinson 2000). This perspective emphasizes the agency of human actors in relation to cultural processes and calls for detailed contextual analysis of the role, distribution and strategic uses of 'traditional' knowledge.

Tradition as Culture

Sahlins's (1999) spirited attack on what he sees as unfortunate tendencies in the invention of tradition debate is to a large extent based on his reading of the Pacific material, his own region of specialization (ibid.: passim and 416, note 5). His main complaint appears to be that this debate is an example of a larger 'rhetorical shift to morality and politics that has overtaken all the human sciences' (p. 403). Culture is thereby reduced to being merely instrumental and ideological, a disguise for 'more fundamental interests, principally power and greed'. Sahlins wants to rehabilitate the concept of culture and reestablish its central position within anthropological studies. Referring to Leslie White's credo that the symbol is the origin and basis of human behaviour, he writes: 'In all its dimensions, including the social and the material, human existence is symbolically constituted, which is to say, culturally ordered. That is a position I was not then, nor have ever been since, prepared to give away' (p. 400). The key to understanding culture is that it is structured, and it is exactly this characteristic that is disappearing in the contemporary debates: 'invented' traditions are political fabrications and thus functional. But, Sahlins argues, Hobsbawm's invented traditions are in reality the same as Malinowski's mythical charter, which is also functional (p. 403). Functionality, in this sense of instrumentality, cannot stand on its own; it has to be based on structure (p. 407). The new 'inventions' are permutations of older forms of relationships, adapted to novel situations. Instead of the invention of tradition we see the 'inventiveness of tradition' (p. 408).

> From what I know about culture, then, traditions are invented in the specific terms of the people who construct them. Fundamentally, they are atemporal, being for the people conditions of their form of life as constituted, and considered coeval with it. It follows that if such traditions are authoritatively narrativized, or when they contingently rise to consciousness, they will be aetiologized: that is as charter myths. (Sahlins 1999: 409)

Sahlins's plea for a structural approach is without doubt an important reminder of some fundamental characteristics of cultural process, even though I feel that he is exaggerating the situation with regard to the invention of tradition debate. Most of the authors quoted above are well aware of the continuities involved in the creation of traditions. More seriously, Sahlins's focus on the structure of culture appears to make him imperceptive of the real theoretical contribution of the invention debate. This is to be found in the relationship between cultural process, human agency and power. Even though Sahlins claims, 'Of course I do not mean to deny historical agency', the agency implied by his approach is that of the competent language user, who makes grammatically correct sentences out of the available idiomatic repertoire when negotiating new situations (1999:

possibility for habits may thus be biologically given, they function in a different way from DNA, which controls reproduction in biological organisms. Habits are usually less precise in their prescription of action and are, generally speaking, open to conscious monitoring. Even though habits refer to processes of automation, situations of complete, machine-like reproduction rarely occur. In most cases the conscious mind has a vital reflective and monitoring role (p. 90). The conscious mind (or intelligence) forms intentions that can subordinate habits to other considerations in specific situations and choose between alternative actions. This combination of reason and habit produces a system for reproduction or cyclical time that is much more adaptable and fast-changing than the genetic system. Having reached a societal scale in my discussion of habit, I shall now introduce a new term. Young sees habits as characteristic for individuals, whereas he defines custom as 'the habit of a group, a social habit'. The force that is responsible for cycles in society is custom (p. 95). Young states that social custom interacts with individual habit to make up and perpetuate the structure of society (p. 99). The binding force of custom is continually recreated by cyclical social behaviour (pp. 100, 239). Young presents us with an essential insight into the interplay between individual habit and social custom. As he does not really develop this further, I shall move on to other authors who have also inspired the perspective presented here.

The historian and social theorist Paul Connerton is especially interested in the social dimension of memory. In *How Societies Remember* (1989), Connerton argues that there are three types of memory: personal, cognitive and habit-memory, the last of which is largely ignored by researchers. He finds this type of memory in collective commemorative ceremonies and locates it in the bodily movements and actions of the performance. On that basis he defines habit as 'a knowledge and a remembering in the hands and in the body' (ibid.: 95). Rather than the memory inscribed in textual productions and laws, it is the bodily habit-memory produced in rites and ceremonies that ensures stability and resistance to change in societies (p. 102). The force of habits can be observed in the case of so-called 'bad habits', which are hard to fight. In this context, the ethnologists Jonas Frykman and Orvar Löfgren find it enlightening to reflect on how contemporary Swedes (or Western people generally) constantly appear at war with their bad habits, such as sitting still, smoking, drinking and overeating. This attempt to control habits is seen as an aspect of modern rationality. But the best way to fight a bad habit is to replace it with a good one. In their introduction to an anthology appropriately called *Force of Habit: Exploring Everyday Culture* (1996), they provide an interesting discussion of the concept of 'habit' that in many ways resembles Young's points of view. Habits, according to Frykman and Löfgren, shape dispositions to act in a certain way. They operate on 'less tangible levels of our consciousness' and are a way to save energy. They become ingrained from early childhood and are difficult to examine, precisely because they are constantly applied (ibid.: 10). It is in encounters with people from other cultural backgrounds that habits become apparent, and this happens more in modern, multicultural societies (p. 14). Like Young, Frykman and Löfgren define habits as private and customs as public. The latter are practised by many people and maintained by a collective (p. 15). I choose to maintain this useful distinction between habits and customs and will now investigate how these terms relate to a concept of tradition.

Custom and Tradition

In order to identify the special character of tradition I find it useful to adopt a perspective developed by the sociologists Peter Berger and Thomas Luckmann in their well-known book *The Social Construction of Reality* (1967). They combine a sociology of knowledge approach, based in particular on the phenomenology of Alfred Schutz, with a dialectical understanding of human society. The key terms in their conceptual framework are habitualization, institutionalization and legitimation, which I will discuss successively. Even though Berger and Luckmann do not use the term 'habit', they have coined a term that denotes the process of forming habits, *habitualization*, claiming that all human activity is subject to this process. Repetition of actions engenders a pattern that limits the need to make choices all the time and provides for a greater economy of effort in the performance of actions (ibid.: 70–71), so habitualization provides predictability and a specialization of activity that is lacking in man's biological equipment. Because a lot of activity thus moves to the background of human consciousness through the process of habitualization, a foreground opens up for deliberation and innovation. The term *institutionalization* is used for the reciprocal typification of actions by types of actors (p. 72); hence, the mutual acknowledgement and assertion of typical actions by different actors, who thereby confirm their roles vis-à-vis each other, is the basis for the creation of social institutions. These institutions, once created, achieve a degree of objectivity in relation to the actors. They are the product of a shared history of interaction and exercise by their very existence a certain control over human conduct, channelling it in one direction as opposed to other directions that would theoretically be possible (p. 72). According to Berger and Luckmann, institutionalization is incipient in every social situation continuing in time (p. 73). It creates social worlds that are experienced as an objective reality by individual actors because most of the content of these social worlds antedates the individual's birth and conscious recollection.

The dialectical process of society that Berger and Luckmann propose contains three elements: externalization, which involves the human production of a social world of institutions through the routinization of interactions; objectification, which has to do with the externalized institutional world assuming the character of objectivity for human actors; and, finally, internalization, which refers to the process whereby this objectified product of human activity acts back on human actors by directing and forming their actions through habitualization. The following three statements are therefore all true: '*society is a human product; society is an objective reality; man is a social product*' (p. 79). This social dialectic is fully realized only in the transmission of the social world from one generation to the next. Internalization thus acquires the character of socialization and this is where the third key term enters the picture: *legitimation*. It is not enough for the new generation simply to grow up in the institutional context created by the parent generation. The latter wants to ensure the continued validity of their social world by means of legitimations that 'explain' and 'justify' the institutional order and therefore help the new generation internalize these patterns.

How do these processes relate to the terms we want to define: habit, custom and tradition? Connecting 'habit' with the process of habitualization appears unproblematic since this primarily refers to the individual actor. Institutionalization creates a social reality that exists in between individuals. According to Berger and Luckmann, this sedimentation of intersubjective experiences only becomes truly social through a sign system of one kind or another that facilitates the reiterated objectification of the shared experiences. It is at this level that I place the concept of custom, which refers to social activities that are repeated and have a certain recognizable and fixed form. Berger and Luckmann see language as the decisive sign system and one could certainly argue that (concepts of) most repetitive, recognizable social activities will be expressed in language: being in mourning, going to work, being married and so forth. However, following Connerton, I also believe that a class of social customs consisting of bodily practices is objectified in rituals rather than in language. The knowledge of these customs lies in the body rather than in the mind, if I may use this somewhat worn metaphor.[11] Tradition, finally, can be found in the social process that Berger and Luckmann call legitimation, which they describe as a 'second-order' objectification of meaning. Legitimation serves 'to make objectively available and subjectively plausible the "first-order" objectivations that have been institutionalised' (p. 110). It is thus a kind of commentary on the institutionalized social order that serves to justify it by giving it cognitive validity as well as normative authority (p. 111). Traditions can have many forms such as explicit moral instruction, religious allegory and inspirational literature (p. 86).[12] Again I would like to add that they do not necessarily have to use the verbal mode, as other visual and auditory symbols can comment on and capture insights into social relations.[13]

Traditions are repetitive practices just like customs, but their reoccurrence is more subject to conscious orchestration. Traditions are organized and executed by the proper category of actors, often experts; they are taught and performed on the proper occasions, whereas customs belong to the repetitive actions that are the stuff of everyday life. As argued above, customs are to a large extent executed automatically – when writing this, for instance, I am no longer consciously applying the rules of the English language I learned at school – but they do not fully escape from control by the human will. Of course, in order for human reason to have an impact on customs, they have to be made conscious. Since customs are social objectifications expressed in signs, they are in principle more accessible to human consciousness than individual habits. The continuous concern with socialization through various forms of legitimations also indicates that a lot of our customs are close to our consciousness and can be easily challenged. Traditions serve to legitimate customs, providing a reflexive space that confirms the customary patterns. This reflection is both cognitive and normative because it explains, justifies and provides moral authority to the activities concerned. Returning to Hobsbawm's conceptual framework, it is clear that I do not agree that custom has a special, dominant function in so-called 'traditional' societies (Hobsbawm 1983: 2). Customs and traditions are the fabric of social life in all societies, without which these societies would not exist as recognizable units. If there is a difference, this may possibly be found in the character and frequency of certain types of traditions. To explain this I have to develop the concept of tradition further.

Tradition, Historical Continuity and Invention

Traditions imply continuity with the past. As second-order objectifications they are passed on from one generation to the next. This is true to the central meaning of tradition, which refers both to the act of handing down and to that which is handed down: a belief, a ritual practice, expert knowledge. The continuity lies therefore in the practice of maintaining traditions and in their function of enforcing the institutional world to which they are connected. There is, however, a class of traditions which explicitly refers to the past. This could be the commemoration of the origin of a nation-state by connecting it to a certain historical event (for example, the oath of the original Helvetian 'Eidgenossen' in 1291) or the celebration of an ethnic, regional or national identity by evoking a more unspecified 'time immemorial'– real or fictitious (cf. Roepstorff and Simoniukstyte 2005; Scarduelli 2005). As mentioned above, traditions generally confer cognitive validity and moral authority to customary practices by commenting on them in various ways. The principles involved in this legitimation may refer to particular sources of truth and right – for example, God, nature, the king, reason, or ... the past. If the assumed continuity with the past or a specified historic period is part of their validating principle, I suggest that we are dealing with a special class of traditions whose 'traditionality' is an explicit part of their legitimizing function. As observed by Hobsbawm and Ranger, referring to the past as a criterion of validation appears to have particular attraction in times of great social change. So-called invented traditions are mostly but an especially pregnant case of the class of 'traditionalist' traditions, since their origin is known to the researcher and their claimed continuity with the past can be debunked as fictitious.[14] The past is thus used to legitimate a new practice, not an existing one.

I will now return to the question of whether we can distinguish between societies on the basis of the occurrence and/or frequency of certain types of tradition. Or, to put it differently, can we defend the fact that (a high frequency of) invented traditions are a feature of modernity, as argued by Hobsbawm (1983)? I believe that in addition to the cases presented in the *Invention of Tradition* book, and their ensuing elaborations (for example Ranger 1993, 1999) there is also evidence from other regions pointing in this direction – for example, in the ethnohistorical research I carried out on the Papua New Guinean island of Baluan. On this island the awareness of cultural difference appears to have old historic roots, as is the case in other Melanesian societies (Otto 1991). Baluan people have concepts to indicate that a certain way of acting is typical for a group or a person. This may be called a tradition but not a traditionalist tradition, since there is not any explicit reference to the past, which is a central point that is often forgotten in discussions about tradition. A number of authors tend to conflate the local awareness of cultural difference with an indigenous concept of tradition (Keesing 1982; Linnekin 1992; Babadzan 2000) but these conceptions have to be differentiated. A concept of tradition allowing for 'traditionalist' traditions also exists on Baluan, but it is of more recent origin. Its genesis can be traced as an element of indigenous people's reactions to Western penetration and colonization. They objectified their own cultural ways in opposition to Western ways and related them to the recent past: the indigenous ways belonged to the ancestors and had to be abolished in order to come on equal footing with the dominant colonizers. Hence, tradition first appeared in the 1940s as a negative value, as something that stood in the way of social change (cf. Thomas 1992). But from the 1970s onward, the valuation of tradition has gradually changed, and nowadays the concept mostly has a positive meaning (Otto 1991, 1992a, 2002). Harrison (2000) makes a similar argument using material from several Melanesian societies (including Baluan). In this region it was quite common to objectify cultural practices such as rituals and other traditions as exchangeable prestige goods. One could obtain certain ritual knowledge and performance rights through established exchange relationships. As a more recent development these cultural practices have often become objectified as a legacy from the past, that is, as traditionalist traditions. Once they gain the status of heritage, they are no longer exchanged with other groups but preferably kept as markers of identity, status and legitimacy.

Hobsbawm (1983) has argued that invented traditions have emerged particularly frequently in modern times in connection with the rise of nation-states, and Ranger (1993, 1999) has defended a clear transition in the forms of identification between precolonial and colonial Africa. Observations from Melanesia suggest that a particular focus on the past as a source of validation and authority first originated in colonial and especially postcolonial times. This may warrant the conclusion that there is a special relation between invented 'traditionalist' traditions and modernity. Whether this is a question of frequency only or whether the invention of traditionality is irrelevant in premodern and precolonial contexts remains an open question. The only thing we can be certain about is that the occurrence of traditionalist traditions presupposes that the past is imbued with a certain value.

Tradition and Agency

In order to develop a definition of tradition and to understand its relationship to cultural continuity, I have described 'cultural process' as complex and multilayered. Let me briefly review the different levels of culture while paying special attention to the aspect of agency. The first level I have identified is that of *habitualization*, which produces habits in individuals. Even though habitualization is a strong tendency in human beings, possibly with a genetic underpinning, and even though habits predominantly operate at an unconscious level, this process of behavioural reproduction is not independent of human consciousness and will. Individuals can consciously inculcate habits through training and repetitive action and they can also become aware of existing habits and try to control them. The second level of cultural process, that of *institutionalization*, produces customs shared by social actors and is a precondition for social behaviour since it secures mutual predictability and communication between actors. Like habitualization, institutionalization is a continuing process, as there are new situations and unsolved problems of interaction all the time. Discrepancies between established customs and actual social situations lead to miscommunications and to new processes of finding acceptable patterns of interaction. Customs are generally connected with sign systems such as images, gestures, and, of course, languages. This gives them a certain solidity and objectivity in relation to social actors, who are the actual bearers and reproducers of customs.

When social actors meet other social actors living by different customs, their own customary patterns become conspicuous and the object of reflection. The same may happen in the process of socialization, when the parent generation tries to transfer their customary patterns to their offspring. Both situations, transfer of customs and intercultural contact, may prompt the third level of cultural process: *legitimation*. This process produces second-order objectifications, or traditions, which are developed and maintained by reflective subjects. The process of tradition making thus requires a high level of conscious involvement. Metaphorically one could imagine the space where traditions are constructed and performed as a cultural foreground that operates upon and in contrast to a cultural background of customs, which maintain cultural patterns at a much more unconscious level. Traditions can be more or less elaborate, from simple proverbs to comprehensive systems of knowledge that require specialists and special organizations for their maintenance and reproduction.

Traditions relate to general social conduct by referring to norms and values. They may explicitly formulate rules of conduct or more implicitly indicate what is desirable in social interaction. As Holy and Stuchlik (1983) have very clearly stated, norms do not automatically generate behaviour, but they present rules to which social actors have to relate. This involves conscious choice and strategy:

> First of all, we have to realise that to invoke or disregard a norm within the course of an interaction is an action in itself and must be treated as such. In order words, to invoke a norm has to be seen as having a specific purpose. … It is irrelevant whether an individual invokes a specific norm because of his personal commitment to it or because it is convenient to him. What is important is that norms do not bring about behaviour by themselves but are brought to bear on actions by the actors in the course of their attainment of specific goals. (Holy and Stuchlik 1983: 85)

Whereas the force of customs lies in the fact that their regulation of conduct is mostly hidden from consciousness, the explication of norms that happens in traditions involves conscious action. However, because traditions are normally repeated more or less frequently, their legitimations tend to become more custom-like and less exposed to conscious monitoring and challenge. Through both explicit legitimations of customs and repetitive performance, traditions may therefore have a conservative function, enforcing customary action. Invented traditions, though, are by definition intended as innovative forces; they invoke the authority of the past and of repetitive action to legitimate something that, in fact, is new. Therefore, they involve a high level of conscious action or 'social engineering', as Hobsbawm calls it (1983: 13).

The relationship between customs and traditions can be complicated. A good example is the case of a group of Aboriginal people living at the top end of the Northern Territory, as described by Britta Duelke (2005). These people moved from their original homelands to a mission station at the beginning of the twentieth century and have lived there ever since. According to the Northern Territory Land Rights Act of 1976, which in many ways can be considered as an invented tradition encoded by law, a group of Aborigines can claim land rights if they belong to a local descent group representing the 'traditional owners' and if they have traditional knowledge and ceremonies that demonstrate primary spiritual responsibility for the land. The so-called Mission people obviously have a problem, since their history of migration is known both to themselves and to others. They nevertheless see themselves in many ways as traditional owners because they have lived on the land so long and have established cultural relations to it through birth and naming ceremonies. Their strongest claim, though, appears to lie in the fact that they present themselves as 'traditional' Aborigines because they observe traditional kin relations and because they are 'traditional owners' of land that lies elsewhere. With reference to

Berger and Luckmann, Duelke argues that kinship is the key cultural category for this group of Aborigines, a 'first-order' objectification that represents cultural continuity for them. We can thus observe an interesting shift of level or cultural register. A set of customs that operates in daily life to sustain social interactions without much reflection becomes as a whole objectified as a cultural domain that demonstrates a conscious value or norm, namely, 'traditionality', which is an important aspect of claims to land in the political and legal context of the Northern Territory. In my terminology, the kinship system therefore becomes a tradition, not so much by legitimizing the actual kinship obligations but by legitimizing something else, the right to live on land that did not originally belong to this group.

Tradition as a Resource

The enactment of traditions, as 'second-order objectifications', involves a high level of conscious involvement by human actors. This subjective engagement is the foundation of the moral authority of traditions to enforce customary patterns or to legitimize new inventions, but it also opens them for strategic use by interested parties. Therefore, I fully agree with Tonkinson (1993, 2000) that traditions can and should be analysed as resources for the attainment of particular, individual or collective goals. These goals can have a clear political character, as in the case of nation making, or they may be of an economic nature, for example securing access to land as the basis for livelihood. Before I give examples of traditions as resources, it is important to make some general observations. The strategic use of traditions implies the presence of a choice, which may involve the interpretation of the content of a particular tradition or opting for or against that tradition. In the above example of the Mission people, there was a discussion about the meaning and expression of 'traditionality' within the larger framework of the law in the Northern Territory, which provided the undisputed context for all parties. Very often there is a choice or a dispute between traditions. If we stay with the issue of access to land, one can argue that many land disputes in Papua New Guinea involve negotiations about which tradition is most pertinent to the situation: local 'traditional' law, English common law, or Papua New Guinean national law. Or if we take up the issue of the desired political system in a number of Pacific countries, one can observe a tension between traditions that include hierarchy and the right of primogeniture and those that comprise Western democratic values (Lawson 1997).

Most, if not all, human societies have different traditions that provide alternative explanations and legitimations,[15] so there is ample scope to invoke traditions in the pursuit of competing goals and interests. In this context, it is useful to remind ourselves of the theoretical work of Fredrik Barth (1993, 2002), who has developed a model of so-called 'traditions of knowledge' as part of his project to design an anthropology focusing on the study of knowledge. In Barth's understanding, a tradition always has three aspects or 'faces': a corpus of substantive assertions and ideas about the world, a range of media of representation, and a social organization. The particular norms or 'criteria of validity' of a particular tradition are shaped in the interaction between the three aspects. Thus, the ways in which the reproduction and distribution of a particular tradition are organized and the media through which it is enacted (for example, ritual, images or text) have an impact on the ideas promoted by the tradition and vice versa. In Barth's model, some people, the experts or office bearers who are centrally involved in the maintenance of a tradition, will have an obvious interest in its reproduction and development, whereas others will be more likely to choose between various options according to their situation. Following Weber, one can argue that the strength of a tradition depends on the number and the social position of those who 'bear' it (cf. Otto 2002). Proselytizing may therefore be a viable strategy to promote a certain tradition, as can be commonly observed among churches and nationalist movements. As argued by Keesing (see above) and others (Wolf 1982; Guha 1989), a particular tradition may serve ideological functions by 'disguising' power inequalities or by 'persuading' those who are subordinated that the inequalities are in their very best interests. The better its persuasive qualities, the more people will accept its content and the greater political power a tradition can muster.

Traditions can function as resources for political power in different ways. They can be used to legitimate or naturalize existing relations of power, but they can also be employed to mobilize a group of people for political change. Scarduelli (2005) describes how Lega Nord, a northern Italian political party, fabricated the idea of a separate Padanian identity with ancient Celtic roots. This radically right-wing party, founded in 1991, had as its main aim the secession of northern Italy. To rally supporters the party invented collective rituals which were broadcast on television, adopted special symbols such as the six-rayed 'Celtic' sun, and produced a magazine which published partly forged evidence for their historical identity claims. Although no real evidence existed for the historical continuity of Padania, Roepstorff and Simoniukstyte (2005) show that in the case of Lithuania a concept of the nation was based on some real continuities of language and history. Interestingly, these continuities were discovered by foreign academics before the formation of a nation became a political project of the local population in the last quarter of the nineteenth century. Once the nation was established as an idea, its key symbols remained remarkably constant in the country's troubled history of division, independence, Soviet occupation and recent independent statehood. Since the circumstances and actual realization of the nation varied substantially in this period, the authors emphasize the importance of tradition as the conscious act of handing down as a central element in creating continuity with the past and the future.

Both examples show that traditions as political resources are often invoked in the context of nationalist claims. This resonates well with

Hobsbawm and Ranger's observation that traditions were particularly frequently invented in the period of emerging nationalism. Babadzan (2000: 132), writing about Pacific societies, nevertheless stretches this connection too far with this claim: 'To become fully comprehensible, traditionalistic discourse and the politics of tradition in Oceania need to be analysed as phenomena directly or indirectly related to some ethno-cultural variant of nationalism.' While disagreeing with his equation of tradition with culture (ibid.: 134; see above), I agree that nationalisms characteristically invoke particular traditions to define the nation as a community sharing a common cultural background. However, several works dedicated to the issue of nationhood in the Pacific emphasize that this form of explicit nationalism with connected traditions is weak in many Pacific countries and that concepts of nation are often implied in other discourses and practices such as those of religious conversion, tourist promotion and consumption (Foster 1991, 1995b; Otto and Thomas 1997). In addition, it is not difficult to find instances of 'traditionalist' traditions – the ones with which Babadzan is concerned – that do not have an obvious connection to nationalism. A good example of this is a newly created tradition in a Papua New Guinean village, the storyboards, as described by Colombo Dougoud (2005).

The villagers of Kambot in East Sepik Province developed a new art form, so-called storyboards, under the influence of missionaries and expatriates. Since their invention in the 1970s, these carved boards have mainly been intended for sale to tourists and art dealers. Whereas in the late 1980s the carvers still freely mentioned the recent origin of this type of artefact, in the early 1990s they emphasized that the storyboards were traditional and presented continuity with the past. Colombo Dougoud shows that this change of perspective was induced by the expectations of tourists, who wanted to buy 'authentic' objects of primitive art. The carvers have accommodated these expectations through their choice of designs (for example, their exclusion of modern scenes) and colours and by explicitly presenting their artefacts as traditional. This emphasis on presenting the storyboards as a continuous tradition has only become important for them in their relation to the tourists. One can therefore conclude that this 'traditionalist' tradition was developed primarily as a resource with an economic purpose and not in relation to nationalism.

Researching Tradition

Anthropologists and other academics can become involved in the study of traditions in other ways than as detached researchers. In particular, they can be employed as experts in cultural research with the task of evaluating the genuineness of certain traditions in cases where this assessment is relevant to bureaucratic or judicial decisions (see for example Chevallier 2005 on France). Kolig (2005) discusses in detail the problems this involvement can entail with special reference to Australia and New

Zealand, where there is a widespread practice of using anthropologists as consultants in cases involving Australian Aboriginal or Maori claims. He sets out by reminding us of a general premise of the sociology of knowledge, namely, that knowledge is always interested: its formation is shaped by the practical interests a society or group may have at a certain time. The same applies to the formation of traditions, which he sees as a special kind of knowledge with a clear normative dimension. In Australia and New Zealand there is a vigorous revival of traditions among segments of the indigenous populations, which has to be understood in the context of the wider, white-dominated society. The revival of traditions is an expression of the increasing self-awareness among indigenous groups looking for ways to emphasize their separate identity. But reviving traditions also serves other interests. In the arenas of legislation and politics in the two countries, special reference is made to indigenous traditions, particularly in relation to land rights and privileged access to state resources. The possession of certain traditions can therefore be an important economic and political resource for indigenous groups.

Kolig provides a series of ethnographic examples that highlight the tension between the expectations of the dominant white society concerning traditions as unchanging and the ways in which indigenous people understand their cultural practices. Moreover, in the particular situations of conflict, traditions have to be articulated in relation to completely new phenomena such as large-scale mining or tourist development. This necessarily leads to a reformulation or development of the tradition. This may well be in line with the original ideas and practices, but there is also the potential for manipulation and fabrication. Kolig argues that it is often impossible to find final and conclusive evidence about the authenticity of traditions. Anthropologists, hired to give an assessment, cannot come up with an unambiguous judgement about the dividing line between sincerity on the one hand and opportunism on the other. He concludes, therefore, that there has to be a clear division of labour. Anthropologists hired as experts have to limit themselves to providing a truthful description of a situation in all its complexity and ambiguity. The task of making decisions in individual cases should be shouldered by the elected (and appointed) representatives of a society on the basis of the political and moral criteria that are forged and shared by the society as a whole.

I agree with Kolig's careful conclusions, but I should emphasize that in practice it is often difficult to implement these guidelines. The Hindmarsh case, which Kolig also discusses (Brunton 1996; Tonkinson 1997; Weiner 1995, 1997) is a good example of the complexity involved. Very briefly, it concerns a conflict between a private company that wanted to develop Hindmarsh Island near Adelaide, among other things by building a bridge, and a group of Aboriginal women who claimed to have a secret tradition involving the island that would be violated if the development project went ahead. On the basis of a report by a consultant anthropologist the project was first put on hold, but later a Royal Commission found that the evidence concerning the tradition was fabricated. The project went ahead but failed economically and the consultant anthropologist was sued by the company for the financial loss suffered. The judge dealing with this case ruled, however, that the anthropologist appropriately and justifiably had acted as an advocate for the Aboriginal women. The case is dramatic and extreme because it involved large monetary stakes, state and federal governments, and extensive media coverage; however, it does highlight some central dilemmas. First, the judge ruled that the anthropologist acted as an advocate, but is this a desirable role? One could argue that such a role will affect the status of anthropology as an academic discipline producing objective and reliable knowledge (Merlan 2001). On the other hand, one cannot prevent anthropologists from engaging in issues towards which they feel a moral responsibility. It is important to be absolutely clear about the role and context in which one acts – for example, is one's primary responsibility towards a client (as an advocate) or to the court (as an expert witness)? – and, secondly, to maintain professional standards. This brings us to the second dilemma, as the Hindmarsh case involved a number of anthropologists on both sides of the dispute attacking each other. A central issue was the validity and reliability of the consultant anthropologist's conclusions. Being exposed to critical examination in court has demonstrated that anthropologists need to become more explicit about their own methods and criteria of validity. The entanglement in issues of legal evidence has brought about a wave of criticism and self-reflection in anthropological circles in Australia and New Zealand, which ought to contribute to increased methodological and epistemological awareness.

I shall end this discussion by noting that it is not necessary to be a consultant anthropologist in order to become caught up in moral dilemmas and political conflicts when writing about traditions. Above I mentioned Keesing's dispute with the Hawaiian scholar Trask, and there are many other examples. Anthropologists get entangled in politics and morality, not, as Sahlins believes, because a rhetorical shift to morality and politics has overtaken the discipline (1999: 403), but because, as I have argued in this chapter, the subject matter of anthropology – culture and traditions – is inherently political, and because researchers cannot avoid being actors in the world they investigate. Their intellectual products can and often will be used, also in ways that were not intended. The best way to guard oneself against potential misuse is, on the one hand, to work according to the highest professional standards for knowledge production and, on the other, to be explicit about moral choices whenever necessary.

This chapter is based on the introduction to the book *Tradition and Agency: Tracing Cultural Continuity and Invention*, which I coedited with Poul Pedersen (Otto and Pedersen 2005). I cordially thank Poul Pedersen for his permission to use this text in a solo-authored contribution and gratefully acknowledge his intellectual input. I would also like to thank Nils Bubandt, Helle Vandkilde, and Robert Tonkinson for their constructive criticism of the earlier text.

Notes

1. See note 3 and the following, regionally biased list, as well as other references in this chapter: Handler and Linnekin (1984); Babadzan (1988, 1999); Errington (1989); Jolly and Thomas (1992); Thomas (1992); White and Lindstrom (1993); Briggs (1996b); Errington & Gewertz (1996); Rutherford (1996); Otto and Borsboom (1997); Apter (1999); Otto and Pedersen (2000); Hamelin and Wittersheim (2002).
2. The dictionaries I consulted all minimally referred to the following two meanings of the word tradition: (a) the act of passing down from the past to the present (especially by word of mouth or practice); and (b) that which is passed down in this way (beliefs, practices). See for example *Oxford English Dictionary Online* (Oxford University Press, 2004) under the heading 'tradition': particularly gloss 4.a. 'The action of transmitting or "handing down", or in fact of being handed down, from one to another, of from generation to generation; transmission of statements, beliefs, rules, customs, or the like, esp. by word of mouth or by practice without writing. Chiefly in phrase *by tradition*, and gloss 5.a. 'That which is thus handed down; a statement, belief, or practice transmitted (esp. orally) from generation to generation.'
3. See Vail (1989), Gorgendiere (1996), Guyer (1996) and Harneit-Sievers (2002).
4. See Ranger (1993), Ranger (1999) and Ranger (2005). He returned more often to the debate in presentations and discussions.
5. New Caledonia is still a French dependency.
6. In his monograph *Custom and Confrontation: The Kwaio Struggle for Cultural Autonomy* (1992), Keesing develops his understanding of kastom as an aspect of resistance to colonial domination. The Kwaio of the interior of Malaita staunchly resisted Western impact, in contrast to the coastal tribes. They engaged in a process of codifying their tradition (*kastomu*), in which the anthropologist Keesing played a welcome role. Thus, long before Keesing became aware of the ideological dimension of culture, he was helping Kwaio tribesman in their cultural identity project. Apart from resistance theory, Keesing draws inspiration from Gramsci's ideas on counterhegemonic discourse as well as from Guha's subaltern studies.
7. In *Islands of History* (1985), Sahlins is particularly concerned with understanding historical change in relation to human action. This book contains brilliant analyses and insights, but I find that Sahlins still reduces human agency to the skilful execution of cultural patterns. In the so-called 'structure of the conjuncture' the cultural patterns may conflict with real situations and become reevaluated and reordered as a result. But apart from the capacity to create 'grammatical' adaptations, human agency also includes the potential to become aware of conflicting patterns and strategically use the possibilities this provides.
8. Think of the theoretical efforts of evolutionism, (neo)Marxism, transactionalism, the Manchester School and historical anthropology.
9. One notable exception is Boyer (1990), who takes a cognitive perspective on tradition as a certain form of interaction.
10. If Young is right in assuming this complementariness of habit and memory, this must have important consequences for our understanding of social knowledge. Our inquisitive mind has a positive bias towards perceiving 'persisting changes in

society' and is bad at perceiving 'persisting continuities'. This is why it is so difficult to interview people about their culture!

11. Following insights from cognitive anthropology, one could perhaps more accurately say that both types of customs are stored in the brain: the one type, bodily actions, in the form of action schemata; the other in the form of conceptual schemata.
12. Berger and Luckmann use the term 'tradition', albeit not systematically (1967: 86).
13. See, for example, the way in which Baktaman rituals contain and communicate knowledge about society and cosmos (Barth 1975, 1987).
14. Hobsbawm's (1983: 1) definition allows for 'invented traditions' that do not explicitly refer to the past as part of their validation. He writes: 'In fact, where possible, they normally attempt to establish continuity with a suitable historic past.' The words 'where possible' and 'normally' clearly indicate that, according to Hobsbawm, there are also invented traditions that do not establish this continuity explicitly but only by virtue of their own repetition. This latter kind of 'invented traditions' are therefore not part of our class of 'traditionalist' traditions.
15. It is very difficult to imagine a society that does not contain at least a minimum of this kind of cultural variation, in that life experiences and situations are always different, such as between young and old people, men and women.

Chapter 3

Intentionality of Action in Cultural Context

Gisela Trommsdorff

Introduction

Presently, much discussion is going on about whether human beings possess a free will; whether human behaviour is intentional and goal-directed or rather automatic and primarily a product of neuronal processes; whether human beings can be held responsible for their behaviour; and to what degree behaviour is based on rationality and self-reflection. These discussions do more or less explicitly deal with the concept of agency and its psychological basis.

Human agency – a topic of research in philosophy, social sciences and anthropology – is discussed here from a psychological perspective taking culture into account. Presumably the development of psychology as a discipline and many of its theoretical approaches would never have been possible without the fundamental assumption of human agency. However, widely differing views on agency exist, going back to debates about dual processes of mind and body. Proponents of behaviourism studied reactions of participants to certain stimuli in experimental studies, rather than agency. Cognitive theories brought about a shift from the mechanistic paradigm to an organismic and interactionistic view, followed by emotion theories dealing with emotion and motivation as part of human agency. From the point of view of these latter theories, human behaviour cannot be regarded as simple deterministic reaction but as a process of goal-directed action. Intentionality became an important aspect of human agency while agency was regarded as the dominant characteristic of the human being. The question is, how can the person's agency be characterized? One approach is to ask for reasons why the person acts in this and not in another way. Does the person act in the observed way because he/she has decided and planned to do so after having chosen among alternatives in order to pursue certain goals? Or does the person believe that she cannot act differently due to internal or external constraints? Both alternatives are related to agency even though certain constraints (which may have resulted from the person's own previous behaviour, from external conditions, or other factors) may limit the scope and direction of agency. Agency is based on intentionality (including choice among several alternatives). However, agency does not necessarily mean 'free' choice since internal or external, stable or variable factors can constrain behaviour to a certain extent. The person may be more or less aware of these constraints, and may intentionally take into account these constraints by processes of assimilation or accommodation. The question arising here is whether the person practices agency even if she believes that factors beyond her control guide her behaviour? Thus, agency is a theoretical concept of psychologists and other scientists who aim to describe and explain human behaviour; furthermore, agency beliefs can be seen as part of a person's naïve theory on the self and other persons.

Human agency has been studied in psychology by taking into account intentions, goal setting and planning, future orientation, self-regulation, self-efficacy and coping (e.g., Bandura 2001). In the same line, Bronfenbrenner and Ceci (1994) conceive of agency as particular kinds of developmental outcomes: 'those that represent the actualization of potential for (a) differentiated perception and response; (b) directing and controlling one's own behavior; (c) coping successfully under stress; (d) acquiring knowledge and skill; (e) establishing and maintaining mutually rewarding relationships; and (f) modifying and constructing one's own physical, social, and symbolic environment' (1994: 569). A related view suggests a close relation between agency and intentionality. Tomasello et al. (2005) conceive of intentional action as a larger adaptive system in which goals and action are components that serve to regulate the organism's behavioural interactions with the environment. According to Bratman (1999), intentions include a plan of action the person chooses and commits herself to in pursuit of a goal; thus, intentions include the goal (environmental effects or self actions, or both) and the means (action plans). Another psychological approach to agency is the study of the subjective agency beliefs. One of our questions here is whether and why people differ with respect to agency beliefs and whether this affects individual and social behaviour.

The history of psychology indicates that the underlying theoretical framework of the researcher (and the dominant theoretical '*Zeitgeist*') defines whether intentional or deterministic (internal or external) causes of behaviour are used for the respective psychological conceptualization of agency. Agency beliefs are influenced by personal characteristics and by situational factors. Presumably, both scientific and lay theorizing on agency are influenced by general cultural belief systems.

Most scientific psychological approaches on agency are based on Western theorizing, assuming an individualistic orientation of the person and the basic motivation for autonomy (e.g., Ryan and Deci 2000). Only very recently have culture-informed approaches questioned the Western concept of agency based on individuality, independence and autonomy, and asked whether relatedness and social orientation guide people's behaviour in certain cultural contexts (e.g., Markus and Kitayama 1991,

2002). Therefore, our question is whether and which kind of cultural differences exist with respect to people's naïve theories on reasons for human behaviour, and whether such cultural differences affect human behaviour and social interaction.

Culture is regarded here as an important factor affecting agency while agency and intentionality influence culture. This process is mediated by social interactions which require the ability and motivation to understand the intentions of others and cultural learning (Tomasello et al. 2005). Cultural learning develops on the basis of joint attention and shared intentionality and involves the development of agency. In line with Bruner (1996), we conceive of culture as a belief system providing meaning to what we perceive. Cultural rules and norms can be understood as ways to solve fundamental problems of social existence such as understanding oneself, one's relationship to other people and to the community, and one's orientation towards people and authorities in power (Shweder 1982). Accordingly, we assume here that culture evolves as a result of individual and shared agency in processes of social interactions; culture provides solutions for problems and structures the way we understand and make sense of the world, the self, and the relation between them. Thereby, culture organizes intentional behaviour and agency.

In the following pages we deal with theoretical approaches to agency from a developmental psychological and a cultural point of view. In the first part, we examine selected psychological theories which stem from assumptions on agency and the subjective belief in agency. In the second part, we discuss to what extent theories on agency depend on cultural belief systems and what the implications of these cultural belief systems are for other agency beliefs.

Basic Elements of Agency and Its Development

Human Need for Understanding and Explanation

First, some preliminary remarks on human beings' need to understand the self and the world are necessary. This need underlies the belief in human agency and the tendency of human beings to engage in causal attributions. According to Heider (1958), subjective causal attributions indicate a belief in agency; their function is to help the person understand the world and the self.

A subjective theory on causes of observed behaviour of others usually refers to internal or external causes (or to their interaction). This holds for self-perception and for person-perception. For example, people tend to believe that the causes of their success are rather based on internal (stable or variable) attributes (ability and/or effort) and causes for other people's success are based on external attributes (and vice versa for failure). These attributions are based on specific assumptions regarding person-environment relationships; causal attributions structure one's experiences and can affect motivation and related behaviour (e.g., achievement, aggression).

Often, such attributions are not based on validated personal experience but are adopted from certain cultural beliefs. For example, a pregnant mother may interpret the felt foetus signals as the unborn infant's intention to receive care and attention, or she may interpret the same behaviour as caused by biological factors determining reactions of the unborn baby. Preschool children employ magical thinking to interpret physical processes until they have achieved the level of concrete operational thinking. In certain cultures, sickness and physical problems are interpreted as being caused by bad spirits or by bad behaviour of the person. The Shaman is asked to serve as mediator and make the bad spirits leave the sick person. In the Batak culture a practice was used where in the case of a person's transgression the victim or Shaman could give power to a magic artefact to hurt or kill the aggressor without physical contact (see Kornadt 2007). Here, the artefact is believed to possess agency.

Attributions of behaviour and of psychological or physical conditions of a person can take the form of rather complicated theoretical assumptions of agency, based in part on subjective experiences, shared cultural beliefs, and wishful thinking or fears. In this chapter, agency is dealt with as a specific characteristic of human beings. In the following section, the structure, function, and sociocultural conditions of personal agency are discussed.

Development of Agency

From a psychological view, the question is how far the development of agency is influenced by biological and environmental processes and their interactions, and how individual differences in agency develop and affect further development. Development of agency requires an understanding of intentional actions of others. Infants begin to understand intentional action by 14 months of age, including their selective attention to goal-relevant aspects of situations. This can be seen as the basis for cultural learning (Tomasello et al. 2005). The infant observer now understands the means-ends relations of the observed act which is a precondition for imitation – an act of learning about intentionality which contributes to the development of understanding others' agency in the sociocultural context.

The development of agency is a life-long process which can be seen as an ongoing active construction of the person's own development depending on the respective developmental tasks and the individual and contextual conditions. Successful agency in later life can be described as gains in some areas which compensate for losses in others (Baltes 1997). Thus, agency can be viewed as a phenomenon of considerable individual variability and plasticity.

One branch of theorizing has regarded development as being primarily influenced by (inborn) biological factors, while another branch conceived of development as primarily affected by environmental conditions. This rather simple nature-nurture controversy dominated the field for quite a long time. Since advocating one-sided positions did not improve the theoretical discussions, *interactions* between both environmental and biological conditions are now taken into account (e.g., Rutter and Silberg 2002).

Furthermore, in their bio-ecological paradigm of human development, Bronfenbrenner and Ceci (1994) have suggested an extension and redefinition of the behavioural genetics paradigm beyond the assumed interactions between genes and environment, which may be helpful to better understand the development of agency. The authors assume proximal processes through which genetic potentials for effective psychological functioning are activated. They are interested in how far persons differ in their 'innate capacity for realizing individual talents and buffering against dysfunction' and 'under what circumstances such constructive potentials find expression' (1994: 570). This view on interactions between innate and external conditions conceives of the person's agency as particular kinds of successful developmental outcomes while the authors also assume that agency of the person is part of the process of development.

In early development, the infant's behaviour can be seen as a precursor of agency. Though the infant can very early influence his or her own environment and transform aspects of the immediate setting, this capacity is limited not only in scope but also in quality. When infants influence the reactions of parents and caregivers by, for example, expressing certain needs, they are only reacting to their needs instead of acting on intentions. Acting on own intentions develops somewhat later (about after 8 months of age). A basic developmental achievement is the development of shared intentionality based on the rudimentary ability to understand the intentions of the other person – an important precondition for imitation and cultural learning (Tomasello et al. 2005). The later development of a 'theory of mind' plays an important role in the understanding of agency and intentionality and varies cross-culturally (Lillard 1998). As for the development of agency, this is influenced by innate processes during the sensorimotor development and the social and material environment and can be analysed according to the above-mentioned developmental outcomes which represent the actualization of various potentials.

To summarize, agency includes internal mental processes and the intentional expression of behaviour. Furthermore, agency can be directed to the inner self and/or to the environment. It develops in early childhood as a necessary part of human development.

Development of Intentionality and Goal-directed Behaviour

A central characteristic of agency has been seen in intentionality and the related processes of goal-directed and self-directed behaviour. Early indicators of intentionality can be observed in sensorimotor development.

Based on results from their innovative experiments on toddlers, Meltzoff and Brooks (2001) concluded that infants tend to attribute intentionality to living rather than to physical mechanistic objects, thus giving priority to the social environment and to human behaviour as the basis for the development of agency. The early understanding of physical and social agency in infancy gives rise to more mature understanding of the self as a mental agent. This is a product of complex developmental processes.

In their developmental theory of early understanding of the self as agent, Fonagy et al. (2002) differentiate five levels of agency of the self: physical, social, teleological, intentional and representational. With ongoing development, actions are viewed as related to causality. Around 8 to 9 months of age, infants differentiate actions from their outcomes; infants now represent actions as means to achieve certain goal states. During the second year, infants view the self as an intentional agent whose actions are caused by previous intentional states such as desires (Wellman and Phillips 2001) and whose actions can change physical and also mental aspects of the environment. Understanding agency as mental causation gives rise to including the representation of epistemic mind states such as beliefs (development of theory of mind). This understanding of the self as a 'representational agent' is a precondition for the development of the 'autobiographical self'. According to the authors, early in development, agency is represented as part of causal relations that connect actions to their agents and also to the world. Therefore, research on the development of agency and the self also needs to focus on the effects of agency on further development.

Several factors including context and the present developmental situation can influence the direction and onset of agency during development over the life span. From a life span perspective, agency and the related pursuit of goals is influenced by the person's past experience, expectations, hopes and fears regarding the future, belief in self-efficacy, the experience of developmental tasks, constraints and chances. Also, the belief in one's agency and self-efficacy influences goal-directed behaviour such as the choice of goals (and related plans to achieve this goal), the termination of a goal-directed behaviour after successful goal attainment, the choice of an alternative goal in case of anticipated difficulties with goal attainment, or giving up a goal after unsuccessful pursuit. From a developmental perspective, continuities and discontinuities in goal-directed behaviour indicate ways of flexible coping and optimization of development. Selection of goals and optimization of a goal pursuit thus constitute preconditions for successful development over the life span especially in times of stress or (normative or nonnormative) transition. The ability to maintain long-term goals even when facing immediate temptations is one of the major aspects of human agency. Therefore, self-control is an important and necessary ability for all children to learn.

In our own studies on the intentional duration of goal pursuit, we have induced conflicting motivations such as prosocial and achievement motivation, thereby affecting the intensity and duration of goal pursuit in preschoolers. When children who are distressed recognize the chance to shift their attention to another goal such as achievement, they do so and cope with distress effectively. This indicates a certain 'rationality' in using distracting stimuli and switching goals in order to cope with distress effectively (Trommsdorff and Friedlmeier 1999). In line with other studies which have shown that attachment security fosters the ability to delay gratifications (Sethi et al., 2000), attachment security predicts the ability to regulate emotions in an achievement situation (Trommsdorff and Rothbaum in press). Thus, the question arises whether the quality of goal pursuit (degree of persistence and flexibility) as a part of intentionality and agency is affected by the person's agency beliefs. This question will be discussed below with reference to attachment theory.

Attachment and the Development of Autonomy

Autonomy has been regarded as a main characteristic of agency in mainstream psychology. Self-Determination Theory (SDT) (Deci and Ryan 1995; Ryan and Deci 2000) assumes that striving for autonomy is one of the basic needs of human beings. From a developmental perspective, autonomy is indicated by exploration early in development, and it develops on the basis of secure attachment relationships. Attachment theory (Bowlby 1969) assumes a biologically based, universal need for attachment, and universality of preconditions for and consequences of attachment. The central assumption is that sensitive, responsive caretaking fosters a secure attachment relationship between the caretaker and child, which, in turn, allows for exploration and related social and cognitive development. Children who have a secure attachment relationship with a caregiver, as compared to children who have ambivalent or avoidant attachment relationships, are described as differing in the way they construct person-environment relationships which are represented in their internal working model (evaluating oneself as more or less competent and other persons as more or less trustworthy). Assumptions of attachment theory thus include viewing agency as based on autonomy and relatedness.

The assumption of attachment theory of a close relation between attachment security and autonomy (exploration) has recently been challenged by cross-cultural studies (Rothbaum et al. 2000; Rothbaum and Trommsdorff 2007). So far only a few cross-cultural studies on attachment have been carried out (see Van IJzendoorn and Sagi 1999). Most of these studies report a higher number of children with secure attachment relationships as compared to the other groups. It should be noted that, in several non-Western cultures, exploration and independence/autonomy are not as highly valued as in Western cultures. Furthermore, most attachment theorists have not taken into account cultural conditions of development. The caretaker's sensitivity is thought to be a basic precondition for the development of secure attachment but culture-specific variations have been ignored. Similarly, differences in the culture-specific value placed on exploration, the assumed consequence of secure attachment, have not been taken into account. Exploration and autonomy are not necessarily desirable aspects of agency in non-Western cultures; other aspects of agency may be more desirable, such as dependency and relatedness. The attachment theory is a fruitful approach to explain the development of agency when cultural factors are taken into account.

Bidirectionality in the Co-construction of Agency

Since goal pursuit and internal control seem to be influenced by the subjective interpretation of the self and the world, one may assume that the cultural context affects the way intentionality and agency develop by providing interpretive perspectives and schemata (such as the inner working model as suggested by attachment theory), which arise by co-constructive processes in the interaction between the child, the caretakers, and other relevant persons. From a unidirectional approach, socialization conditions and/or biological dispositions were focused upon as central factors for development. These approaches have largely ignored the active participation of the child in his or her development, and the role of agency in the process of active development (Lerner 1982).

When the unidirectional paradigm was opened for a bidirectional perspective, the activity of the child became a target of research, e.g., with respect to affecting the parent's behaviour (Bell 1979; Kuczynski 2003). However, bidirectionality in parent-child relations is not symmetrical, since infants have only limited resources to change their environment. Furthermore, cross-cultural studies have shown that bidirectionality is asymmetric in parent-child relations in other cultures (Trommsdorff and Kornadt 2003), and it remains asymmetric during the life span in many other cultures, especially in East Asia (e.g., Trommsdorff 2006; Trommsdorff in press). Therefore the question arises whether this implies a different extent or even a different quality of agency as part of human development in parent-child relations in other cultures?

Agency can be regarded as the ability to *change one's environment*. However, this is only one aspect of agency. Another aspect of agency can be seen in the ability to *change the self*, e.g., by learning processes which are basic for development. The child's active involvement is needed for learning and development. In their theory of internalization, Grusec and Goodnow (1994) have extended the idea of interactive processes between parents and children and the active role played by children. They assert that children have to both understand and accept their parents' message as a prerequisite for internalization. These cognitive, emotional and motivational processes are part of both children's and parents' agency; these processes affect the children's ability to change the self (e.g., by the development of self-regulation), to change their environment (e.g., by

'independent' and 'interdependent' self. The authors describe the notion of the self as an independent, separate entity in Western, individualistic cultures, in contrast to the notion of the self as interdependent with other persons and inseparable from the social context in non-Western collectivistic cultures. More specifically, the authors maintain that the *independent* view of the self includes the assumption of autonomy and distinctiveness (in attributes and processes), internal attributes and processes as determinants for behaviour, individual differences in behaviour on account of different internal attributes and processes, the value of distinctiveness (as expressed in behaviour). The *interdependent* view of the self includes the assumption of interdependence and relatedness – the self is part of social relationships, relationships with other persons and responsiveness to the others with whom one is interdependent as determinants of behaviour, variability of behaviour due to the social context and the sensitivity to social expectations in different contexts, the value of relatedness (as expressed in behaviour). There is ample evidence from several ethnographic studies of Asian and Pacific Island cultures which describe a more interdependent conception of the self (Geertz 1975; Hsu 1985; Lebra 1994). Hsu (1985) describes the meaning of the Chinese word for 'man' (*yen*) as the place of the individual in the network of interpersonal relationships; the person's wishes and behaviour are evaluated according to the way they inhibit or foster these relationships.

The assumption that cultures can be characterized according to differences in self-construals has been criticized, however, since cultures are not homogenous systems and their complexity cannot be accurately described according to these broad dichotomies (Kagitcibasi 1996; Matsumoto 1999). It is rather assumed that the self in all cultures can be viewed as incorporating both independent and interdependent aspects in various degrees while the salience of the different selves becomes more or less accessible in different contexts (Matsumoto 1999; Church 2000; for a critical meta-analysis see Oyserman et al. 2002).

Cultural Differences in Social Explanation and Self-Construal: Attribution Bias

Conceptualizations of the self are connected to thinking, feeling and behaviour, including beliefs about agency. In the following we focus only on self and person-perception (social explanation) in order to discuss the role of agency. According to the assumptions by Markus and Kitayama (1991), self-construal and person-perception vary in line with cultural values. Therefore, it can be assumed that in individualistic as compared to collectivistic cultures, and for persons with independent as compared to interdependent self-construal, explanations of one's own and of other people's behaviour will focus more on internal, global and context-free causes as compared to external situation-specific causes.

Attribution bias: individual versus group dispositions. In mainstream social psychology, it is traditionally assumed that people attribute actions to properties of the actor. There is ample empirical evidence that people tend to attribute observed behaviour to personal dispositions of the actor. This bias has been called the 'fundamental attribution error' (Ross 1977). It is discussed in the literature as 'correspondence bias', which has been described as a universal human tendency (Gilbert and Malone 1995).

As a matter of fact, the empirical studies in Western settings and with Western persons are in line with the assumption that people tend to explain observed behaviour by internal and stable properties of the actor (e.g., Jones and Harris 1967). However, it is unclear whether this attribution bias also holds for persons from other cultural settings. In India, this result could not be replicated (Miller 1984). Hindus (children and adults) explained various kinds of observed social behaviour (transgressions and prosocial behaviour) more often by external, situational conditions as compared to persons from the United States. In several further experimental studies on judgements of social stimuli, participants from the United States showed a bias towards personal dispositions (traits) and Chinese favoured social situations (environment) to explain behaviour (Morris and Peng 1994).

Several cross-cultural studies on person-perception and social explanation in East Asia produced somewhat inconsistent results with respect to the effect of discounting dispositions (Morris and Peng 1994). Most studies confirmed Heider's (1958) description of a bias towards attribution to personal dispositions as a bias held by Western (American and European) but not by Asian (Chinese) people. Possibly, this is a result of the generalized assumption in Western cultures that individual autonomy is most important. This exaggeration of individuality is not shared in Asian countries. Especially when people have to explain ambiguous behaviour, Chinese as compared to American people are less likely to attribute these to internal traits as opposed to contextual factors (Morris and Peng 1994; Morris et al. 2001). When taking into account culture variance, some interesting results appear. In the case of a cognitive style associated with conventionality, a preference for the culturally preferred attribution style can be observed (attribution to personality traits in the case of Americans; attributions to social conditions in the case of Chinese) (Chiu et al. 2000; Nisbett et al. 2001; Norenzayan, Choi and Nisbett 2002).

According to the above-cited studies, the general result is that people all over the world tend to attribute behaviour to internal dispositions at least to some degree (in Western cultures more than in Asian cultures). As for the attribution of behaviour to external causes (situations), a significant effect of culture can be observed: people from Western cultures prefer situational inferences significantly less compared to people from Asian cultures.

Agency in Cross-cultural Perspective

The main questions addressed in the next section are: how does agency affect development and behaviour? Have individual and cultural differences to be taken into account? According to our conceptualization of agency, we will deal with autonomy, achievement and control as main aspects of agency.

Autonomy

Most recent developmental theories, at least implicitly, assume that agency underlies the process of development. In Western theorizing, agency is regarded as an expression of autonomy, self-efficacy and control (see the above-cited summary by Bronfenbrenner and Ceci 1994). A prominent example of this view is the Self-Determination Theory (SDT) (Deci and Ryan 1995; Ryan and Deci 2000). According to SDT, autonomy, competence and relatedness are defined as the basic needs of the person; fulfilment of these needs increases well-being, psychological growth and personal integrity. Cross-cultural studies on SDT focus on the ways various cultural forms support and fulfil these basic needs (Ryan and Deci 2000).

The assumption that autonomy constitutes a universal basic need has stirred controversial discussions based on culture-relevant studies. Several empirical studies indicate that differences in the value of autonomy exist among Western and Asian countries. These differences are in line with cultural variations in the preference of independence. Thus, socialization strategies differ significantly between East Asian and Western families (for a summary see Choi and Choi 2002; Trommsdorff and Kornadt 2003; Rothbaum and Trommsdorff 2007). In Western (European and American) countries, independence and autonomy of children's behaviour are expected at an earlier age as compared to East Asian countries (Japan, China). The age of autonomy expectations in the development of children and adolescents is influenced by cultural values (Feldman and Quatman 1988; Feldman and Rosenthal 1990). Autonomy expectations of Chinese, as compared to American mothers and teenagers, are significantly delayed and at the same time more correlated between the generations (Stewart et al. 1999). The same pattern of results was found by Schwarz et al. (2005) for German as compared to non-Western mothers.

A problem with these empirical studies is that the authors used different conceptualizations for defining autonomy. Some authors claim that autonomy is closely related to individualism and independence. Since individualism is related to giving priority to the individual's needs and goals rather than to the group's needs and goals, autonomy is usually regarded as part of individualistic orientations and independence while independence is usually understood as being separate from or not relying on others. However, such a conceptualization of autonomy cannot be used in cross-cultural studies, especially not in cultures favouring interdependence, tradition and duty-based relationships (Miller 2003). From a culture-informed perspective, autonomy is not the same as independence and is also different from individualism and separateness (Rothbaum and Trommsdorff 2007; Trommsdorff and Rothbaum in press). Autonomy can be expressed as behaviour of a person serving the needs of the group. Accordingly, a person may be autonomous either when pursuing goals of an embedded self or of an individualistic self. The person may be autonomously independent and individualistic or the person may be autonomously interdependent and collectivistic. According to SDT, autonomy is based on the degree of internalization of values (no matter whether individualistic or collectivistic) when pursuing a goal. SDT theory regards culture as providing more or less support for fulfilling these basic needs ('cultural fit'). A relatively poorer fit with basic needs may explain why some cultural values are less easily internalized than others.

Achievement

Attribution, self-efficacy and control beliefs influence decision making and behaviour. One example is the influence on achievement. In his influential studies on achievement motivation Weiner (1992) demonstrated clear differences in achievement depending on the attribution of success and failure (to internal, external, stable and variable factors). For example, persons who prefer to attribute their success to internal and stable factors show higher achievement; the reverse is true in the case of attribution of failure to internal and stable factors. This 'fundamental attribution error' refers to a bias towards explaining one's own failure by external factors. This self-serving bias is related to a tendency to reinforce a positive self image. However, this result holds only for Western cultures.

Recent cross-cultural research has demonstrated that these results cannot be replicated in non-Western countries (Campbell et al. 2004). This may be explained by a culture-specific meaning of achievement, success and failure (see Kornadt et al. 1980; Maehr and Nicholls 1980; Yan and Gaier 1994). Achievement for personal benefit is valued in the United States but not in Asian countries where in-group benefits are more important. Also, the concept of failure has a different meaning in Western and Asian cultures. It is considered a failure in the West when personal standards (which result from social and individual comparisons) are not met; in Asian cultures, it is considered a failure when expectations of the in-group (e.g., own family, colleagues) are not met. Furthermore, for Asians it has often been demonstrated that the self-serving bias based on the need for positive self-regard is not common; rather, the reverse tendency can be observed (Heine et al. 1999). This is a result of the socio-centred thinking and belief that one should not present oneself as superior to others.

Another approach to the effects of attribution on achievement has been suggested by Dweck (1999). Her approach to self-theories and their role in motivation and agency is based on the assumption and substantive empirical evidence that 'entity theories' (belief in stable factors of self) as compared to 'incremental theories' (belief in malleable factors of self) set up conflicts between learning and performance goals; e.g., effort is avoided since it gives a negative message about one's ability. In contrast, in incremental theories effort is in the service of learning. Therefore, the emphasis on the malleable factors (intelligence, effort) allows for more learning and achievement, as can be seen in East Asian cultures (Trommsdorff and Rothbaum in press).

Control

Studies on control beliefs have their origins in social learning theory. Locus of control theory differentiates between internal (personal) and external control over outcomes and assumes individual differences in this belief and effects on learning and achievement (Rotter 1966). Cross-cultural studies, however, have demonstrated that empirical results on the relation between internal control and achievement could not be replicated in non-Western societies (e.g., Lefcourt 1973; for a summary see Weisz, Rothbaum and Blackburn 1984). Several authors have therefore suggested that control beliefs are influenced by the cultural context, and that the individual belief in control can be understood as a sociocultural construction of reality (e.g., Mirowsky and Ross 1984; Fung, Abeles and Carstensen 1999; McKean Skaff and Gardiner 2003;). In their 43–country study Smith, Trompenaars and Dugan. (1995) showed that three dimensions emerge on the basis of the Rotter items, which represent internal/external control, individual/social orientation, and a luck/chance dimension.

Other cross-cultural studies which can be related to cultural differences in agency beliefs have demonstrated the usefulness of differentiating among cultures in terms of individualistic and collectivistic orientation (e.g., Triandis 1995; Hofstede 2001). Furthermore, agency as conceptualized from a culture-informed view should be differentiated with respect to *primary and secondary control* (Rothbaum, Weisz and Snyder 1982; Azuma 1984; Kojima 1984; Weisz, Rothbaum and Blackburn 1984; Weisz 1990; Essau and Trommsdorff 1996; Trommsdorff and Essau 1998). *Primary control* is defined as control based on the intention of inducing changes in the environment. Here, the target is the environment, and assimilation is the dominant behaviour. *Secondary control* is understood as based on the intention of controlling the reactions of the self, and of altering the self rather than the environment. Here, the target is the actor, and accommodation is the dominant behaviour. Both, primary and secondary control, have been further differentiated into subcategories such as vicarious, illusory, predictive and interpretative control (Azuma 1984; Weisz, Rothbaum and Blackburn 1984).

Other authors apply the distinction between primary and secondary control to the development over the life span and the study of aging (Schulz and Heckhausen 1999). This approach is confined to studies in Western cultures and clearly favours primary control as a superior form of control as compared to secondary control. This view overlooks differences in the preference of control in different cultural groups and for different life domains. Accordingly, Gould (1999) questions the ecological validity of giving primacy to primary control. This criticism is in line with the view by

Weisz et al. (1984) and Rothbaum et al. (1982), who distinguish between primary and secondary control as two culture-specific pathways to control (their studies deal with the Japanese and the European American cultural contexts). In further culture-informed studies, socio-instrumental control based on interpersonal relations was shown to describe the specific control orientation of Chinese but not Anglo-American workers (Spector et al. 2002). Morling and Fiske (1999) prefer to use the concept of 'harmony-control' (separate from primary and secondary control) to describe the specific tendency of Hispanic as compared to Anglo-American persons to maintain harmonious interpersonal relationships – a primary goal which is pursued by harmony-control behaviour. Empirical studies are needed to test whether harmony control is more closely related to secondary control.

In further studies, three dimensions of control have been elaborated: agents of control, means and mechanisms of control, and the ends or targets of control (Skinner 1996). The agent of control may reside within an individual or a group with which the person identifies (Markus and Kitayama 1991). Here, the notion of internal locus of control is extended by taking into account the fact that the controlling self can be an individual or a collective actor (group as collective actor) (Miller 2003). Also, one may exert control through the help of other persons (powerful others). The target of control varies depending on whether the individual directs control efforts towards the environment or towards the self (including inner attributes such as needs, wishes and beliefs). This again is related to primary and secondary control.

Proxy control and self-efficacy. Self-efficacy theory suggests that the persons' beliefs about their abilities to act in a certain way influence their goals, their outcome expectancies, and thereby their actions, such as the effort they put forth to achieve their respective goal. According to Bandura (2001; Bandura and Locke 2003), the concept of self-efficacy allows for assessment of an individual's control belief. Negative self-efficacy is related to the belief that the person does not have a certain (domain-specific) competence. According to Bandura's work, self-efficacy has turned out to be a powerful concept for predicting well-being in many parts of the world.

Control and coping. Recently, the concept of primary and secondary control has been successfully applied to different coping styles of persons preferring an independent versus interdependent self-construal in different ethnic groups (Lam and Zane 2004). Here, the self-construal mediates cultural differences in primary and secondary control when coping with interpersonal stressors. This underscores the importance of studying agency with respect to the relation between cultural variables and coping.

Control beliefs and social behaviour. The preference of control beliefs as part of the self-construal and agency influences social behaviour. So far, consistent cultural differences have been documented with respect to the preference of conflict and cooperation (Markus and Kitayama 1991). Group harmony plays an important role in social interactions especially within the in-group in East Asian cultures, while the reverse is the case for social interactions in

North America. Less conflict and competition is observed in social interactions in East Asian groups as compared to the United States. However, there are cultural differences in social behaviour which even though they might appear contradictory can be explained by differences in control beliefs. Depending on culture and related control beliefs, prosocial tendencies may be inhibited. In cultures valuing interdependence and harmony, prosocial behaviour should be more dominant especially within one's in-group, as compared to cultures where independence and negotiation of individual interests are highly valued. Interdependence should foster empathy, which has been widely recognized as activating prosocial motivation and reducing aggressiveness (e.g., Trommsdorff 2006). However, prosocial motivation does not necessarily induce prosocial *behaviour*. The behavioural outcome depends on situational constraints which may induce cognitive processes such as attributions, self-concept and control beliefs blocking prosocial behaviour, even in the case of an interdependent self-construal and high empathy for the other person. Here the belief in secondary control and accommodation is higher than in the case of an independent self-construal (Trommsdorff, Friedlmeier and Mayer in press).

Thus, agency comprises more than motivation; it includes cognitive processes of evaluations of the self and the environment (including the relation between actual and intended behaviour). This cognitive evaluation can give rise to different cultural meanings of behaviour. Prosocial behaviour usually means to act on behalf of the other person (by exerting primary control); it can also mean to make the other person lose face. Empathy and related prosocial motivation can therefore be in conflict with the belief in secondary control and thus intentionally constrain prosocial behaviour. Or empathy could stimulate prosocial behaviour, which means to intentionally choose secondary control and refrain from any interaction. Thus, seemingly ignoring the other person's needs can mean empathic awareness of this need and an agentic choice of secondary control by intentionally refraining from helping.

In our own studies we have observed a high degree of uncertainty and intrapersonal conflict when Japanese children observe another person in distress (Trommsdorff 1995). The children become quite distressed and tend to focus their empathic attention onto the other person without engaging in active help. This is in contrast to German children, who actually help in the case of empathy, presumably on the basis of their higher primary control beliefs (and belief in self-efficacy). Accordingly, cultural differences in control orientations as an aspect of agency have implications for various aspects of behaviour, including coping and interpersonal behaviour.

Shared Agency

Control beliefs are part of the cultural belief system. Culture can provide different meanings for control based on independent versus interdependent actions. Such cultural meaning systems extend to another aspect of agency, which has traditionally been ignored by ideocentric Western psychology. Control can be shared by the group rather than deriving primarily from the self, a concept known as 'shared agency' (Bratman 1999). Bandura (2001) has similarly dealt with the notion of group-based agency, which he calls 'collective efficacy'. Shared agency derives from being part of a group. This conceptualization is interpersonal but at the same time it is individually experienced, as can be seen by the culture-specific meaning of duty (see Miller 1984; 2003). In individualistic cultures which value independence, the concept of duty is regarded as a constraint on freedom and agency while in socio-centered, interdependent contexts duty is conceived as a part of one's very personal experience. A similar perspective is taken by studies on the Japanese self and agency (Azuma 1986; Lebra 1994). Agency in Japanese culture is based on the interpersonal relationships which constitute the functional unit of the self. Thus, agency is characterized by experiencing interpersonal relationships and shared agency in the in-group (Markus and Kitayama 2002). Accordingly, agency can be experienced as having its source not in the self but in others – in the group, or in the relationship with others. This notion of 'shared agency' has not yet been sufficiently recognized by mainstream psychology even though it is a phenomenon observed in many non-Western parts of the world.

Implications for a Culture-informed Approach to Agency

Cross-cultural and culture-psychological studies on the aspects of agency highlight the fact that the mainstream psychological approach cannot account for the cultural diversity of conceptualizations of agency. However, the problem is to conceptualize cultural differences in a psychologically acceptable way by using psychological constructs which go beyond economic or political descriptions. While the description of cultures along single psychological dimensions such as individualism and collectivism implies oversimplification, such approaches may be a useful start, as studies by Hofstede (2001) or Inglehart (1997) have shown.

Cultural patterns. The dimension of individualism-collectivism is one important axis (beside other dimensions such as uncertainty avoidance, masculinity-femininity, power distance, and long-time perspective) along which cultures have been differentiated (Hofstede 2001). While the cultural level of value orientation has to be differentiated from the individual level of personal beliefs (Triandis 1995), the psychological characteristics of individuals from these cultures have been described in close relation to the theorizing of Markus and Kitayama (1991). In individualistic societies independence is encouraged, and in collectivistic societies, interdependence is highly valued. In both cultural contexts, independence and personal boundaries are valued very differently. According to Schwartz (1994), two aspects are related to the individualism-collectivism dimension. One aspect emphasizes the relative importance of personal versus common interests. The other aspect focuses on autonomy (independence) versus embeddedness (interdependence) of the individual in relation to the group. Accordingly, Kagitcibasi (1996), criticizing the one-dimensionality of the concept of autonomy, differentiates between two cultural models: the model of relatedness refers to interdependent relations, and the model of separateness refers to independency where autonomy and personal boundaries are more important. Kagitcibasi also suggests an integration of these two models as part of social change.

From this discussion one can conclude that culture cannot be ignored in the study of human agency. We do not support a deterministic view on culture but rather assume moderating effects of culture. First, cultures are not homogenous entities. There are significant individual differences in behaviour which sometimes amount more to intracultural than intercultural differences. There is a wide variance in the numbers of people who identify with prominent cultural values (Trommsdorff and Friedlmeier 2004). Also, differences in value preferences between people from different generations have been demonstrated in several studies (e.g., Inglehart 1997; Trommsdorff, Mayer and Albert 2004) even though in more traditional societies, these differences are not very pronounced (Trommsdorff in press). Secondly, cultures do not directly influence behaviour, since cultures consist of multiple, more or less consistently, interrelated subsystems on various levels of the society which interact with each other and with the individual behaviour (for the ecological model of development see Trommsdorff 2007; Trommsdorff and Dasen 2001).

To summarize, we cannot assume that cultures are stable entities which determine individual behaviour. Instead, the question rather arises in which ways and by which processes culture influences human agency. Therefore, we have to take into account processes which affect the relationship between culture and individual behaviour (Trommsdorff and Friedlmeier 2004). A prominent view on such processes is the notion of the 'developmental niche' (Super and Harkness 1993), which consists of parental naïve developmental theories, developmental goals, beliefs and behaviour. The developmental niche can be regarded as a culture-specific context which provides the basis for agency and its development.

The difficult question remains as to how culture is transmitted over time from one generation to the other. Here, the questions of socialization processes and intergenerational relations in the transmission of culture become relevant from a cross-cultural perspective (Trommsdorff 2006; Trommsdorff in press). Such studies may demonstrate that people play a role in modifying and altering cultural belief systems and cultural scripts while at the same time serving to preserve certain culture-specificities which may constitute cultural 'models of agency' (Markus and Kitayama 2002). Thus, cultural models are both reflected in and stabilized by individual agency, e.g., by specific conceptualizations of autonomy.

This study has been supported by the Deutsche Forschungsgemeinschaft (DFG) (TR 169/14–1, 2) 'Limits of Intentionality', University of Konstanz. I am grateful to Holly Bunje and Tobias Heikamp for their valuable comments on an earlier version.

Correspondence concerning this article should be addressed to Gisela Trommsdorff, Department of Psychology, Post Box D14, University of Konstanz, Germany. Phone: +49-(0)7531-88-2917. Fax: +49-(0)7531-88-3039. E-mail: Gisela.Trommsdorff@uni-konstanz.de

Chapter 4

Positioned Meaning in Personal Narrative

Stephen C. Leavitt

In August 1986 I sat down with Ahiren, a Bumbita Arapesh man in his late fifties, to talk about his life.[1] He discussed with me what he viewed as his greatest accomplishment – that he had become a powerful yam grower. Our conversation occurred against the backdrop of troubled times for the Arapesh people of the East Sepik Province of Papua New Guinea. In the previous year an apocalyptic Christian revival movement swept through the area, and in their haste to show a commitment to a belief in the imminent arrival of Jesus, some Bumbita had neglected to plant their crops of yams. Now, in August, it was clear that the harvest would be small, and rumours spread that the coming months would bring something unheard of, a shortage of food.

In that interview, Ahiren related a series of stories about his accomplishments in growing yams, about his having taken the place of his father, and about his impatience with the ignorance, as he saw it, that caused people to give up their farming in pursuit of unrealistic religious goals. The story he told me was, on the one hand, a well-rehearsed version of his life, framed by enduring cultural values, and one he undoubtedly used as a measure of his identity and self-worth. At the same time, the story, as told to me on that day, took its shape from a series of contingencies related to the particulars of the moment. He was telling me about yam growing at a time when people were uncertain about the future, looking towards development and modernization, abandoning tradition with its unknown consequences. He was asserting his skills as an older man, with a younger generation of men coming up behind him, men better educated than he and more sophisticated about modern times. And he was telling it to a young anthropologist who had looked up to him as a father, and who would soon be leaving to bring the Bumbita Arapesh back to America.

I shall use this interview here as the centrepiece in illustrating a method of interpreting personal narratives, or life stories, that seeks to bring out what I have called the 'experience' that lies behind human adherence to a particular cultural system. The term 'experience' has a long history in anthropology, and I myself use it deliberately in part to tie into a series of debates in the field. Over the years anthropologists have staked out a wide range of positions on just how best to describe people's experience, but there has been a fairly consistent view of what the study of experience is – it is the attempt to appreciate the feelings, sensations, thoughts and perspectives of individual people in a way that aims as much as possible to transcend the differences between us. It is, at its roots, an attempt at empathy.

The problem comes in figuring out how best to achieve that. The challenge of apprehending experience has taken researchers directly into debates in anthropological theory over the role of shared cultural values in defining human life. For many years one argument hinged on how important individual psychology and human nature should be in guiding our understanding of cultural experience. Did people's deep commitments to cultural ideas arise from generic psychological needs, or was the cultural impress itself so powerful as to ensure such commitments? Over time the terms of the debate shifted to allow for the many contingencies in life that might affect experience at a given point in time. Researchers recognized that the 'centripetal' forces of culture producing consistency and coherence must contend with 'centrifugal' forces that propel conflict and change (Bakhtin 1981). Culture could no longer be seen as the central unifying force in people's lives. Similar developments affected views of the role of psychology. Attention to the role of a stable 'personality' in cultural life gave way to an interest in how individuals actively construct self-concepts and identities in pursuing their ends. It could no longer be a given that culture simply supplied the meanings that guided people's lives. Rather, anthropologists described people engaged in meaning-making activities that depended on local circumstance. For those interested in experience, such trends have opened up a whole new set of research questions that do not rely on abstract 'top-down' models of psychology and culture.

But even a nuanced appreciation of local cultural constructs is not sufficient for the study of experience. Over the last decade, interpretivists' and ethnopsychologists' earlier insistence on finding 'experience' in the domain of cultural categories has been challenged from many directions within the fields of psychology and anthropology (e.g., Ewing 1990, 1997; Gregg 1991; Paul 1990; Wikan 1995). Even with these critiques, there is as yet, in my view, not enough effort to reintegrate into our understanding of experience the central recognition that our subjects are beings with motives and commitments, that they have a stake in presenting their experience – to themselves and to others – in a particular way. I therefore outline an analytical strategy that begins with the idea that all personal narratives be approached as rhetorical arguments, with a set of asserted meanings juxtaposed against an equally relevant set of suppressed meanings. The key to understanding the presentation of experience lies in appreciating the terms and conditions of the narrator's argument. Experience, if you will, lies in the space between what is asserted and what is denied.

The challenge for such an analysis comes in identifying what it is that shapes the argument presented by the narrator. There may be straightforward political objectives, for example, that guide the rhetoric, and these may have little to do with a narrator's deepest feelings. Or the narrator may be preoccupied with presenting a distinct version of events to the particular person eliciting the narration. The social context of the elicitation itself, in that case, defines the terms of the presentation. Such considerations have been the subject of ample research, from the attempts to identify the political dimension to personal narrative (see especially Rosenwald and Ochberg 1992) to approaches that emphasize the constitutive role of social context in self-presentation (e.g. Gergen 1990). But there have been fewer attempts to look at narrators' arguments from the perspective of possible psychological motives. While psychological anthropologists acknowledge that people are motivated to understand their lives in certain way, their emphasis tends to be on what makes the narrator's life meaningful. This emphasis on creative meaning-making is in itself important, but it often has the consequence of shifting attention away from motives. Often there is the assumption that if a culture provides a coherent set of meanings, the result is inherently beneficial for the individual.

While it is essential to consider the value of meaning systems available to individuals, there is nevertheless a danger in diverting attention from the psychic motives underlying individual experience. Hollan points out that it is easy to 'get carried away by our fascination with cultural logic' and assume that a given cultural meaning system will necessarily yield a more satisfying perspective (1994: 75). While a system of meanings may provide some satisfaction simply through its coherence, it may, ultimately, offer little true emotional relief. It depends on what the meanings are. To take a well-known example from psychological anthropology, it is entirely possible that depression occurs so often in the West not because of a lack of coherent meanings anchoring suffering, as Obeyesekere has argued (1985: 135), but instead because a very strong cultural system defines feelings of depression as based on fundamental moral failing (see Leavitt 1995). A coherent meaning system such as that might only make one feel worse. If one were able to think about feelings of depression differently, by defining them, say, as the product of an inherited disposition entirely unrelated to one's moral worth, then the experiential impact of the system of meaning could be quite different. Therefore, to describe how cultural understandings affect experience, one must go beyond arguments that focus mainly on the relative coherence of a system of meanings.

One way to do that is to reintroduce psychological conflict, but in a way that avoids relying too heavily on an elaborate superstructure of psychological theory. The approach I take here adopts the simple premise that psychological conflict is central. People construct personal narrative not only to make their experience meaningful but also to alleviate or allay anxiety, to manage psychological conflict. Personal narratives use local cultural schemas to build meaningful versions of narrators' lives, but they do so in a way that is psychologically appealing. A psychologically 'appealing' account is one that protects the speaker from anxiety-provoking alternatives. To understand the psychological basis for the account, the analyst must look for defensive moves in the rhetoric of personal narrative, moves that suppress alternative meanings to foreground favoured ones. Personal narratives are thus built through a rhetorical action that strives to create a plausible interpretation fortified against suppressed alternative renderings. It is an action that is both defensive and creative.

In the case of Ahiren, the subject whose material illustrates my methods here, we have a man who, at a time in his life when he should be at the peak of his influence and prestige in the community, chooses to argue to me that he should be regarded as a great yam grower. We have to look at what is at stake for him in making that claim at that particular time. It is, on the one hand, a claim about the value of traditional categories for appraising self-worth at a time when much of the 'traditional' was being undermined by people's reaching out for new ideas in the name of progress and development. At the same time, it is a claim about his identification with his own father, himself a renowned yam grower. In Bumbita, identification with one's parents is framed in dramatically literal terms, where children, as they mature, literally sap the strength of their parents in making their life achievements. Ahiren's argument about taking his father's place works from within these premises. And, as it turns out, Ahiren harbours within him the suspicion that he had in fact indirectly killed his own father by himself pushing for changes in the name of progress. Ahiren's story is thus an elaborate version of himself, built around his most centrally powerful and compelling psychological issues. He tells a story that builds an argument about himself for himself, and understanding the terms of that argument is key to appreciating his experience.

Compelling Cultural Scenarios in Personal Narrative

Ahiren's story is at its root an account of his life as a yam grower. Before we can understand how he constructs his argument, we need to have an appreciation for the activity of yam growing in the context both of Bumbita traditional culture and trends for development in the area. Then we must look at how all this plays out for Ahiren specifically, with his distinct social status and unique personal history. Key here is the insight from cognitive theory that people understand things through interconnected scenarios built from culturally defined 'schemas' (see Strauss and Quinn 1997, D'Andrade 2005). Ahiren's account of himself as a yam grower draws on models he has for the activity and their implications for identity. Included in this are various scenarios in his mind for how successful or unsuccessful yam growing activities play out in people's lives. Describing cultural scenarios here is particularly important because of the nature of the data we are considering. Ahiren's account is a narrative, and as such it is built of a series of stories about himself and others engaged in yam growing. His use of the stories depends on an understanding of what points they make when considered as whole stories. It would not suffice for me to break them down into a series of concepts, for example, because their meaning depends on how they look as whole scenarios.

Equally important is an account of what makes these stories compelling for him. I shall ultimately make the case that Ahiren's account is compelling because he has a psychological stake in constructing the stories in a particular way. Part of this means looking at how he fits these stories into an overall sense of self. Cognitive anthropologists have pointed out that people hold beliefs with different levels of conviction. Some beliefs may be widely held but hold little 'directive force' over people's actions. To get a sense of how Ahiren puts together his argument, we need to know not only what he believes about yam growing but also how it comes to be that some of these beliefs are so compelling as to actively shape the story he tells. Strauss (1992) has outlined a useful typology of levels of cognitive salience for schemas, from the commonly held, relatively weak, context-bounded general bromides about a topic to more deeply compelling schemas that bear on the individual's most central activities and values. The systems of meanings individuals follow may have distinctly personal components, what Strauss calls 'personal semantic networks' (PSN), that package shared schemas according to a unique personal history and speak in multidimensional ways to one's sense of who one is. People's PSNs may overlap considerably based on shared life experiences (Strauss 2005; see also D'Andrade 1992; Quinn 1992). Such research directs us to see in the emotional force of 'yam growing' activities a link not only to a wide range of issues in Bumbita interpersonal life but also to some of the particulars in Ahiren's own personal history.

For the sake of clarity, I shall limit my presentation to scenarios that emerge directly from the Bumbita activity of growing yams. As it turns out, though, in talking about yams we quickly get drawn into some core features of Bumbita masculine identity. Donald Tuzin (1972) laid much of the groundwork in a paper on yam symbolism among the Ilahita Arapesh, Bumbita's neighbours to the east, and much of what he says there applies equally well for Bumbita. Tuzin describes the many ways yams and yam growing activities have become a focus of symbolic elaboration. There are over sixty local names for different types of yams. They are distinguished by shape, smell and texture. There are two species of yams cultivated in the area, referred to locally as 'short' and 'long' yams. Short yams are the daily staple, cultivated by both men and women, and they grow up to two feet long and can weigh several pounds each. Long yams, by contrast, are cultivated only by men; they are an important valuable in local exchange and the focus of dramatic displays. Locals recognize two types of long yams, 'male' and 'female', distinguished by their shape. Male yams are shaped like long poles and can grow up to six feet in length. Female yams have several appendages and can weigh up to two hundred pounds each. Both types are the focus of serious competition among men.

Tuzin argues for three dimensions to yam symbolism. First, the yam is a symbol of masculine prowess. Yams are the central objects of competitive exchange among men. Each man has a life-long exchange partner designated by his kinship position, and throughout their adult lives these two men compete by giving gifts of yams to each other, each trying to show greater power at yam growing. At the time of the yam harvest, each man arranges his best specimens of long yams into displays for all to admire. He paints the yams, adorns them with ritual decorations similar to those placed on young men as they are initiated into men's religious societies. Certain techniques for growing long yams are kept completely secret from women, thus adding to the mystification of male activities. Underlying the entire process is the belief that a successful yam grower must have close spiritual ties to ancestral spirits who influence the yams' growth. Men apply inherited magical concoctions to the crop in an effort to encourage the spirits' participation. Thus, for a man, prowess at growing yams speaks to his efficacy as a male and his ability to enlist the help of the deceased members of his patriline. As Ahiren talks about his yam growing abilities, he must also be thinking about his masculine identity and his ties to his deceased father.

The conceptual link between yam growing and relations with family becomes more insistent once we recognize the second symbolic valence Tuzin describes: the yam symbolizes the human body. I have mentioned the genders of yams and the way they are decorated to resemble initiates. In fact, men use careful secret cultivation techniques with sharpened bamboo poles to encourage the yams to grow into specific body shapes, to have 'two legs', for example, by digging holes in those shapes before filling them with humus and planting on the surface. But even more influential to the body/yam association is the way the Bumbita conceptualize human growth and inheritance as a conceptual cousin to yam growth. Yams grow through the process of 'vegetative propagation', whereby an entire seed yam is planted and the new generation of yams grows out from the top of the seed yam, drawing nourishment from it. The original seed yam rots. In the Bumbita view so it is with humans. As a man and woman produce their progeny, and as the children grow from the nourishing efforts of their parents, they literally sap the strength from the parents, causing them to age. Aging thus comes from raising children – growing food, preparing food, and serving food all work literally to transfer life force from parent to child. When I protested that those few with no children seemed to be aging as well, informants inevitably returned with something like: 'Oh, but [that man] looks young for his age; he would be so much older if he had had children!' It is a literal process that draws on fundamental beliefs about the transfer of spiritual substance from generation to generation.

One implication is that yams have families and lines of descent just as humans do. When a man marries, his wife's family donates some of their yams to his stock so that future generations of yams may share in the substance of both sides. Such views hold deep emotional consequences.

Once, an older man, in the process of burning off debris for a new garden, accidentally burned down the yam house of another man, destroying the entire harvested crop. When word spread we all rushed to the scene in the garden. I could hear the perpetrator off in the long grass, crying and crying over what he had done. We all filed solemnly past the burned remains of the yams, and women sang dirges of mourning. I thought this must be a serious event, that the owner of the yams might risk hunger because of no crop. But no, people assured me, that was not it at all; many would contribute to rebuilding his stock, and he would have plenty to eat. The tragedy, they all agreed, and the reason for the deepest of regrets from the perpetrator, was the fact that an entire line of yams had been destroyed. It was a family that had died, and there was no way to bring it back. Thus, the association of the yam with the human body has far reaching consequences for tying one's activities of yam growing to the deeply significant process of maintaining a family and building a legacy for one's line. It is an example of a symbolic association that cognitive theorists have argued has much 'directive force' because of its close ties to one's sense of self.

Ahiren himself draws on these themes in his interview with me. I turned on the tape recorder as we were discussing the effects of the Christian revival on people's crops. Anticipating the arrival of Jesus, some had not planted their gardens. He then went on to say that people generally were behind in their gardens that year, including himself. So I asked him:

SL: Right. So now I want you to think of yourself, are you yourself strong to plant, or do you yourself feel as though you don't want to?

A: I could plant a lot but just my wife, she is not motivated ['smat'] to work.

SL: Really?

A: Yeah, before yes, she would work, carry the yams, but now no way, because you know how it is when you have plenty of children, and she has raised them, and all the grandchildren are big already and holding knives and swiping with them, cutting things, and it's finished. They've cut my blood now, I am the grandfather, I can't be smart to do any work now. A man who doesn't have many children, he could, do the work, carry the yams. Now she can't, she tries but she can't, and I too her husband am not happy to plant plenty of food; I've left it.

In broaching the topic of his own recent malaise in growing yams, Ahiren turns to the culturally framed explanation, that his children's growth has caused him to age, leaving him unable to work hard. He starts with a distinction between himself and his wife, rallying to say he could still do it, but then ends up shifting so as to include himself as one who can no longer grow a lot because of his children's own growth. To anticipate my later argument, I want to note that this shift is significant, for it alerts us to the fact that he has decided to direct the discussion of yam growing to his own condition. He could easily have simply blamed his wife, as he started to do, but he decides against it. His words reveal that he is weighing options at the beginning of the discussion. Should he take a line emphasizing his continuing strength or should he shift to the reasons for his current lack of strength? In choosing the latter tack and invoking the process of a new generation's usurping the strength of their elders, he frames the discussion to be one about the process of aging. The cultural scenario he applies, based on what yam growing symbolizes, makes the idea of a younger generation's strength a central theme. Only by understanding the local cultural significance of growing yams can we appreciate the emotional power behind that theme.

Yam growing carries with it another host of associations related to development that Tuzin does not discuss in his article. Growing yams has become associated with tradition, in contrast to coffee growing, which bears the mark of progress. Every Bumbita man must make a choice about how to spend his time – should it be in growing yams or in cultivating coffee as a cash crop? This association frames the relations between generations in a different way from what was presented above. Younger people are more interested in coffee growing, recognizing the importance of money. Ahiren draws on this theme as well in the opening moments of the interview. As he muses about the current lack of crops, he frames the issues as follows:

A: [There have been changes in how we work;] now there is no competition with it – it is just for looking after your stomach, for business [i.e. coffee], the government. Before, yes, we would look after our food, plant plenty of food, look after our gardens. Now there are too many types of work so they can't think of their food, no. They'll just plant a few ... They won't sit well and think of their food.

SL: And what do you think of this, is it good?

A: No, no way. I too feel it's not good, so all of the time I tell them, I say, 'This work you've pulled in, you'll work hard, win at business, and then later you'll just sit on money? And later you'll just leave our food to go down, planting just a little? Growing a lot? Forget it, you can sit only on your money!' Now there are too many types of work.

In Ahiren's view there has been a shift away from an interest in growing food, with new activities such as coffee and government occupying people's attention. His sarcastic questions to the people convey that he thinks they are chasing dreams, where one can just 'sit on' money. To him, it is a pipe dream. Since we had started our discussion by talking about the Christian revival, it could be that his reference to sitting on money includes here the revivalists' hopes for receiving wealth from Jesus upon his arrival. In any case, he is invoking here a then-versus-now scenario, placing traditional yam growing and prudence on one side, with new activities and exaggerated expectations on the other.

His contrast between tradition and the current scene becomes more focused when he then goes on to lay out an idealized vision of how things once were, in an evocation of what one of my graduate professors, David Jordan, dubbed the 'Paleo-terrific'. It is significant for my later argument that he launches into this vision after first describing how school has prevented his youngest daughter from helping out with traditional yam cultivation. He says:

A: It's because before I sent her to another work, to school, so now she won't carry the yams…

SL: So school itself did it?

A: No, it's because she's at another work, she just sits there at the house and works only at writing [chuckles]. It's, [pause] like it's okay. True, if she didn't have school and stayed with us, I would have taught her how to do this work.

It is striking that he frames her unwillingness as the result of being put to 'another work', not because of her schooling specifically. With this distinction he is identifying the relevant categories. There is the traditional yam growing past set in contrast to the present with its 'many kinds of work'. The schooling is simply an example of the ways that people are turning their attention to the aims of progress. To complete the picture, he immediately turns to a description of how things once were:

A: Before, no. We had only one work. There weren't many. Grandparents, parents, children, grandchildren, they all just thought of food; there weren't many types of work. They would cut the bush and plant, and then what would they do? Just sit and relax. Just build the house to put food in, and then what? Harvest and put the food in. Now there is too much; we can't think about this work and do it all the time.

Ahiren presents here an idealized version of a simple life. His series of rhetorical questions ('And then what would they do?') underscores how predictable and ordered things were. The yam growing is the core activity. As he launches into the description he makes a deliberate enunciation of the generations, listing them off so as to emphasize that central to all of this is an ordered transition from generation to generation, all doing the same 'work'. Implied by this emphasis is the contrast with the current relations between generations, where younger people are encouraging all sorts of different 'works' in the name of progress. With yam growing we have tradition, simplicity, continuity and an ordered existence. With progress we have its opposites.

I shall close this discussion of the underlying cultural themes behind yam growing by describing one of the stories Ahiren recounted that day. The story illustrates how all of these themes contribute to give it its meaning. Ahiren tells of the time he defeated some men by giving them a huge gift of yams, as an answer to an insult they had sent his way. The story falls into a standard genre of tales emphasizing masculine prowess through demonstrations of powerful yam growing, but Ahiren gives it a slant that foregrounds the other themes, related to his own personal identity, that he had decided to discuss with me on that day. In other words, he offers a glimpse of the 'personal semantic networks' (Strauss 1992) behind his view of himself as a yam grower, and these play a central role in the overall argument he makes about his position in life.

The story goes as follows. Ahiren's youngest daughter took a boyfriend in a neighbouring village. Her relationship irritated Ahiren because it meant that the last of his daughters might well marry outside the village, and he had wanted to be sure that at least one married nearby, so that she would be available to take care of him in his old age. He let it be known that he was opposed to her marrying away. In response, the boyfriend and some other young men took the tack of trying to persuade the girl to stay with them by intimating that in their village she would at least have food, and would not go hungry. Word got back to Ahiren about what was said, and he was incensed. He told his wife he would be asking her to send them a message. He went to his yam house and brought out the smallest, most shrivelled yam he could find. He continues, 'So then I stayed behind. I told my wife to take this "symbol" ['piksa'] and go up: "You go up and give it to those men. They'll have to offer an exchange. Will they exchange it or not?" They had insulted me.' With that gesture he had thrown down a challenge. The gift of the shrivelled yam to his rivals sent the sarcastic message, 'Here is all the food I can produce – can you compete with this or not?' The symbolic act ensured that there would have to be a yam competition between the parties at the harvest some months hence. Ahiren called together all of his clan brothers and told them that they would have to harvest well, that they would be making a huge gift of yams to avenge the insult. Each party would measure their harvest, both the numbers of small yams and the size of long yams, and then give them all to the other party. The side that gave the most would be considered the victors.

To this point this story has all the standard features of masculine competition through yam exchange, features that contributed to Tuzin's arguing that the 'penis' was an unconscious third symbolic valence to the yam for the Arapesh. Like the penis, the long yam is a physical manifestation of masculine prowess, and Tuzin cites many examples to show how that equation makes sense in Arapesh iconography. But then Ahiren, in continuing the story, shifts its focus somewhat so that the story is no longer only about demonstrating his masculine prowess – it becomes, for him, a story demonstrating his assured place as a senior man of his generation. In the story he conveys to me that he interpreted the men's insult not only as a taunt over masculine power, but also as an attempt to usurp the rightful place of the senior generation. Here is how he describes the exchange of messages that went back and forth in anticipation of the competitive exchange at harvest time:

A: [Those men] said, '[After the display of harvested yams], we'll take down our yam bin and we'll heap up the food and give it to him [Ahiren], give him some yams and try him out, try our "father".' I too, I said to them, 'You want to compete with me and give food to my daughter. I am not her husband that you'd compete with me! True, if she were married to a man and then left him to come to you, yes, then you could compete with him. But not me, I am like the "father" [i.e. of the senior generation].'

Ahiren's comments to his competitors, as least as he related them to me, aim to shift the grounds of the dispute. He indicates that their insult did not offend him because of its implications about his masculinity; rather, he was offended because they deigned to treat a 'father', a member of the senior generation, as a rival. For him, as he conveys it, this was tantamount to sacrilege. His victory over them would not, he argues, be mainly about his prowess as a man through yam growing; it would instead be a demonstration of his secured position as a senior man holding authority, commanding respect and, well, yes, showing masculine power.

In closing the story he brings the point home. He described how his own subsequent harvest indeed proved to equal his previous mark, that when they exchanged the yams, matching yam for yam, his rivals' gift had been completely matched, and Ahiren had been able to give them a surplus besides. His gift had surpassed theirs. He describes his speech on that day:

A: We exchanged the food and went back. None [of theirs] was left. I told them, 'You all didn't break my house. You all are just playing. You can't beat me with anything. There is no way. You want to compete with your brothers, that's okay. They're all just youngsters. I, however, am a senior man, you all can't beat me, there's no way.'

Almost in a whisper, he continued to me, 'Now this mark is there. My picture is there. Later, when in the afternoon we go, I'll pull it out so you can see it.' With this story, then, Ahiren builds an identity for himself around his position as an established yam grower of the senior generation. It relies on a set of scenarios fashioned from local cultural ideas about yams as manifesting masculine power tied to the strength that comes from maturing to a senior position, from taking a position as a 'father' in the community. The scenarios offer options for him – he can emphasize his own personal prowess as a man, for example, or his role in defending tradition against progress. Or he can emphasize his position of seniority. To tailor the themes to personal issues important to him on that day, he draws on all three scenarios, but in the end he chooses to emphasize the third option: by vanquishing his daughter's boyfriend he affirms his place as a potent senior man. To convey Ahiren's experience, then, it is necessary not only to outline the relevant local ideas but also to describe how they play out in the personal narrative itself. There are unique symbolic valences associated with his own personal semantic networks, and these must be drawn from the narrative.

Suppressed Meanings

To understand the experience behind the narratives told, it is important to remember just what a narrative is: it is a performance enacted through storied forms to convey a particular point or evoke a particular reaction. Its rhetoric is fundamental to its meaning. To borrow an analogy from tonal music, a musical performance deploys the structure of the musical piece in a controlled way; it remains true to the notes but uses them in a way unique to the setting and performer (see Gregg 1991: 60). To 'experience' the piece, then, audience and performers alike must attend to the way those notes are played. With a recent interest in narrative, cognitive psychologists and psychological anthropologists have in fact begun to build the case for seeing the motivational, or intentional, dimension to narrative as central to its meaning. One of the most influential cognitive theorists, Jerome Bruner, makes the point: 'Self as we encounter it in ourselves and in others ... is "storied", or narrative, in structure.' Analysis of self-representation, then, must consider its narrative nature, how it builds 'scenes' emphasizing 'agents' deployed in 'action' using 'instruments' to achieve 'goals' (Bruner 1994: 43). In my view, this feature of intentionality in self-representation means that understanding experience conveyed through narratives involves focusing directly on the argument implied by the story. The narrator presents a version of self by arguing that it be construed in a particular way.

But we cannot make the mistake of saying that the self presented in the narrative is in fact the self as experienced by the narrator. Human experience is motivated, emerging from a fundamental potential for conflict, so any version of self I present in narrative emerges as a choice from a host of alternative versions that at that point in time I choose to suppress. Narratives, in building arguments, must inevitably have things they are arguing against. It is these things, what I have called 'suppressed meanings' (Leavitt 1995, 2005), that play a central role in our apprehending the experience.

One consequence of an approach that considers suppressed meanings in narrative is a need to address psychological motives. As Ochs and Capps put it, '[P]sychological disorders such as post-traumatic stress, depression, and anxiety involve silencing would-be narratives that deviate from the dominant story by which one lives' (1996: 32–33). The idea that psychological conflict may underlie an embraced set of ideas has, of course, been a long-standing cornerstone for psychoanalytic anthropology. What the more recent interest in narrative offers is a potential for exploring psychological process without having to resort to the sweeping psychological generalizations that hampered studies focused on belief systems as such. Rather than making claims about general psychological dispositions, a focus on narrative can address the more immediate psychological motives that drive the immediate story. Just as earlier psychoanalytic theorists looked to 'inconsistencies' in cultural belief systems as telling evidence for psychological motives, researchers focusing on narrative can look for complications in the story that Susan Chase has said comprise a 'narrative difficulty' (1995: 15). She argues that behind any asserted story of self is a 'submerged' story that may point to a 'feeling of inadequacy'. She argues that '[as] interviewers we need to attend to submerged stories ... and invite their telling' (1995: 17).

My key point here is that submerged, or suppressed, stories can in fact reemerge, redefined as psychologically compelling alternative versions of Self. Ahiren's story is, on the one hand, a claim about the value of traditional categories for appraising self-worth at a time when much of the 'traditional' was being undermined by people's reaching out for new ideas in the name of progress and development. He has to suppress an imagined sense that his lack of education compromises his status as a leader in the modern era. His life story sets up a dynamic of tradition versus modernity, and he places himself squarely on the side of tradition. But in another context, that of making a claim about his identification with his own father, himself a renowned yam grower, Ahiren manages, through some complicated reasoning, to portray *himself* at the forefront of the move to modernize. He takes an assertion of who he is not and in a different context redefines himself in a way that includes the alternative 'innovator' identity.

Such a dynamic of self-construction, where a person is able to redefine a 'me/not me' dichotomy into a deeper sense of self, has been studied by psychologist Gary Gregg, who argues in his book *Self-Representation* (1991) that both asserted and suppressed meanings contribute to what one's self-representation is. What is suppressed in one context may become the basis for an assertion in another. To an extent, one may infer suppressed meanings simply from the shape of what is asserted. For example, if I try to argue to a colleague that I am a dedicated researcher, one may infer that I may be harbouring doubts about my commitment to research, or at least that I am concerned that my colleague may think so. Simply by making the argument I raise the issue of its opposite. At another point in time, however, I might well make a very different argument, also drawing on the same themes. I might argue, for example, that any lack of commitment to research is a choice I have made, a choice to devote myself to another worthy pursuit, say, to teaching. Gregg's point is that people's representations of self are the product of a coherent set of contrasts containing metaphors of 'self' and 'not-self' – the self-representation is not a choice of one over the other, or even a negation of one against the other; rather, it is the product of contextual mobilization of these polar themes. There are fundamental ambivalences underlying our views of who we are. A rejected 'not-self' idea can become an important ingredient to the self when it is incorporated into a larger story of, say, moral transformation (Gregg 1991: 89). In my example above, there is a tension between asserting commitment to research and admitting any lack of commitment (which might imply laziness). Both poles are part of the self-image, and the not-Me pole may become redefined as Me when I argue that a lack of research indicates a noble dedication in its own right, in this case to teaching. What is defined as not-Me in one context becomes a self-justification in another.

All this suggests that in Ahiren's story we should look for ways that suppressed meanings reemerge redefined. Ahiren's narrative asserts his success as a yam grower as evidence of his taking on the legacy of his father. At the same time, he admits to a weakening in his own resolve to plant yams – his own crops are not what they once were. He attributes this to his children's growth. They are now sapping his own strength just as he had his father's. But underlying this whole scenario is an alternative version that has less to do with the noble Bumbita tradition of passing spiritual strength across generations. All around him people are choosing alternative activities to yam growing in the name of modernizing through development. Some are dedicating themselves to cash cropping. Others place their hopes in the return of Jesus. As an older, uneducated man, Ahiren suspects he is being left behind. Ahiren's story of his prowess as a yam grower is laced with suppressed insecurities over the relevance of a yam grower in the modern world.

In a remarkable way, Ahiren's narrative addresses his insecurities over being behind the times by transforming them into a larger fable about his deep emotional ties with his father. Instead of just asserting his continued prowess as a yam grower in the face of modernization, or instead of admitting he is a traditionalist whose time has come and gone, Ahiren reconfigures his story so that he himself becomes a protagonist for modernization. He argues that he, in fact, killed his own father by himself asserting, as a younger man, the importance of embracing modernization. The version of himself he presents in that context is that of a forward-looking younger man arguing against the older man's recalcitrance. And, in the story, his father dies as a result.

It is a story that draws on the same generational themes we saw in his earlier story of the yam competition, only this time Ahiren himself is the young upstart. What makes Ahiren's story remarkable is that while his claim that he killed his father conveys the deepest feelings of guilt and remorse, he does not define it simply as an occasion for renouncing development and returning to tradition. Rather, what Ahiren does is draw upon the cultural themes of passing spiritual power from generation to generation to find solace in his own success as a yam grower. By proving that he can grow yams he can find reassurance that his father's spirit still supports him. Behind that support is a tacit acceptance that Ahiren's choice to defend modernization had some merit. In this way, Ahiren can build a story about himself that redefines his insecurities into a bedrock commitment to a view of himself as built on the passing of spiritual strength from generation to generation while at the same time placing himself in the name of progress.

The generally accepted cause of Ahiren's father's death involved a confrontation between the Christians and the traditionalists. Religious cult secrets in a church service had been revealed to a woman relative so the local sorcerers vowed to kill her but, as is common with sorcery killings, they missed and killed Ahiren's father instead. When I asked Ahiren about it, though, he provided an entirely different interpretation, one focusing on Ahiren's own role in promoting development. It is the story of a younger man's pushing for development, the father's resistance and subsequent suicide, and a final absolution for the son in embracing the themes of tradition.

I began by asking him how his father had died. He said that his father had avoided eating pork throughout his adult life. The power of his magic rendered him singularly susceptible to illness from pork. One afternoon Ahiren's father received a gift of pork from another villager. The meat came from a pig that the men had secretly killed and cooked in the bush. Ahiren describes the incident,

A: They carried [the meat] off and gave it to him in the late afternoon, at dusk. I too I sat there and I didn't know. He gave it to him and he took it and put it aside. Then later I took my wives and we went inside to sleep. After we were asleep he took it and cooked it on the fire and ate it and went inside. Then this meat brought him down.

Right after that he changed the subject. It appeared to me that he had implied some kind of secretive behavior on the part of his father, and he certainly hadn't explained why a man who had forbidden himself pork throughout his life should suddenly want to eat some. Ahiren had conveniently passed over something. I took the opportunity to press the point. I started the exchange:

SL: So your father was a man to kill pigs so he avoided pork?
A: It was taboo.
SL: So then why did he eat it at this time? It is true that this man gave it to him, but –
A: He went and gave it to him.
SL: –but he himself ate it on his own.
A: [pause] Some of the fault lies with me, because I myself had wanted to hit him. He was angry over his grandson Wirihen. I wanted to send him to school.
SL: I see.
A: But no, his grandfather was angry, he said he didn't want to send his grandson to school. I said, 'It's all right, he can stay. You say that you'll hold him here, but what kind of knowledge are you going to give him?' … So then he was angry with me over his grandson. So I too I wanted to hit him. He was angry like that so his sister's son here, when he gave him this meat, just like that he went and ate it. [barely audible]. So he died. He would die.

As he described this last interchange, he was visibly moved, and he immediately changed the subject again. It was not something he really wanted to talk about, but in his own mind he felt responsible for his father's death. The father's anger at Ahiren's insubordination caused him to kill himself by eating forbidden foods.

By casting the story this way, what Ahiren does is set up a conflict over the very themes I've been discussing, and it needs some satisfactory resolution. He could have presented the story as simply a fable about his own foolishness in defying authority and promoting development over cherished Bumbita cultural values. But he did not, in the end, do that.

Instead, he framed the story as establishing an opportunity for absolution. If he could assert that his father's spirit, after death, continued to support him in all his activities, then he could receive forgiveness, come to occupy a position of authority by identifying with his father, and at the same time maintain a stance of foresightedness over the need for education and development. And indeed, in his account of the death-bed scene, the father allows all that to happen. The father pulled Ahiren close:

> A: [He told me,] 'Sorry, I am going to leave you now, and I don't want you to be a trouble maker ['bikhet']. You must sit quietly and think of your food and your [coffee] business that you have pulled in, that's all, and you sit… It will not be that long until I will see you again.' He told me that. 'If you sit well, you'll last a long time. You will stay and then later you will grow old and then you will pass away, that's good.'

Here the father affirms his interest in Ahiren's life and its choice for progress, and his advice corresponds with the type of man that Ahiren had already become. He confirms that they will be reunited after death. As Ahiren presents it, he has received the reassurance that he needs.

With the story of his father's death, Ahiren has come full circle. In what started as an argument about his yam growing prowess and the propriety of senior men's position, with its suppressed doubts over now being impotent and behind the times, has become an account of a forward-looking and insubordinate younger self that has been absolved and accepted. It is by tracing the themes of his argument, by following his assertions and their suppressed alternatives, that we can begin to appreciate the conditions of his experience as he conveys it. His account is one that mobilizes positioned meanings in different contexts to convey different points – what is suppressed at one point becomes refashioned as the basis for assertion at another.

Ahiren's narrative forces us to recognize that his 'coherent picture' comes from a rhetorical action in integrating suppressed meanings into his narrative in a particular way. The argument comes off, both to us and to himself, as no argument at all; it seems instead a legitimate version of 'reality'. It is in this sense that narrative can constitute a sense of self.

Conclusion

Recent trends in the culturally sensitive interpretation of personal narrative have emphasized its creative, constructed nature. Many contingencies affect the form of the narrative itself, and analysis needs to consider such contingencies. I have argued that viewing personal narrative as an argument, with asserted meanings allied against unexpressed suppressed meanings, is an effective way of addressing the positioned nature of personal narrative. At the same time, though, it is possible to overstate the significance of the context of elicitation. Ahiren's story shows an enduring sense of identity built around a small set of central themes, and this story was not just 'created' for the interview with me. Personal narratives often address deep-seated and enduring anxieties about self-worth and identity, and the subject, over time, becomes committed to a particular version of events in order to justify a chosen course of self-understanding. Some of the schemas that comprise this understanding come from the shared cultural values that make up a community of people. But some are more personal variations, with one issue stressed over another, helping to negotiate an identity that quells persistent anxieties. The task for the investigator is to sort out the relevant themes by paying close attention to the arguments laid out in the narrative itself. It is by recognizing the rhetorical nature of personal narrative that we can begin to get a better grasp of the experience that lies behind the story.

Note

1. This article is dedicated to the memory of the man I here call Ahiren, who died three years after our interviews. I owe much to him.

Chapter 5

Actors and Actions in 'Exotic' Places

Andrew Strathern and Pamela J. Stewart

Introduction

Place making and the experience of place have been long-standing topics of anthropological research, whether we are dealing with temporal continuities or the ruptures of deterritorialization as explored by Arjun Appadurai (1996) among others (Feld and Basso 1996; see also Strathern and Stewart 2004; Stewart and Strathern 2005). The boundaries of places as a focus for the study of identities have been one strand of studies of locality (Cohen 2000). Filling a wider locality with numerous small specifiers of place names is a very prominent feature of naming practices in the New Guinea Highlands societies that we have studied. The names form a historical grid by means of which memories of gardens made and rights to land, the stories of happenings or events, and the localization of sacrifices to spirit entities in the environment can all be maintained in a network of meanings stretched over the landscape at large (see e.g. Stewart and Strathern 2000). In the Introduction to our co-edited book *Landscape, Memory, and History: Anthropological Perspectives* (Stewart and Strathern 2003) we explored aspects of how people perceive the land around them, and how these perceptions are impacted by history and the connections between place, space, identity, nationalism and, importantly, the role of memory in constructing notions of emplacement.

The concept of landscape involves three intersecting domains of experience and perception. One is the domain of personhood or identity (White and Kirkpatrick 1985). Another is the domain of embodied knowledge (Csordas 1994). A third has to do with the relationship between human action and what are perceived to be the characteristics of an environment, which in European thought has often revolved around a supposed dichotomy between nature and culture, although such a dichotomy is by no means cross-culturally universal (Bargatzsky and Kuschel 1994). At the heart of this overall view of landscape is the approach via phenomenology: that we should learn to see landscape as the product of human perception and experience, and as a way in which humans themselves frame that perception and experience as local actors. Landscape in this sense is not an externally given fact: it is rather the result of poiesis, the making of place and self together, what the philosopher

Heidegger calls *Dasein* (see a discussion of this concept in Polt 1999: 29–30, 42–45; and compare Casey 1997).

Personhood and identity enter into landscape because landscape is not only a backdrop to personhood but one of the prime means by which it is created and sustained. People identify with landscapes. Nostalgia is built out of attachment to landscapes and the places they frame, such as a range of mountain tops seen repeatedly from the window of a flat. Personhood and wider senses of identity may be fused together through images of landscape. The power of poetry and emotional expression in general is partly derived from this phenomenon (see examples in Strathern and Stewart 2001: 77–97).

Embodied knowledge is also crucial to the understanding of landscape, because such knowledge is gained by actually traversing the landscape itself. Knowledge gained in this way is a knowledge born of movement and interaction, both as aspects of what Heidegger called 'dwelling' (see Polt 1999: 46–48). This kind of knowledge is aptly portrayed by John Gray in his study of shepherds and farmers in the Scottish Borders (Gray 2000). It is systematized in the work of the New Guinea specialist Paul Sillitoe in his numerous scholarly works on Wola agriculture, ecology and hunting (Sillitoe 1983, 1996, 2003). Embodied knowledge also emerges from practice, doing things, exploration, discovery, and development of skill, practical mastery in Bourdieu's sense of the term (Bourdieu 1977: 88–89). Such skills are not equally distributed; some people develop them more than others (on this issue of skills in general see also the work of Tim Ingold, e.g. Ingold 2000).

The third element of landscape, its overall categorization or classification, is important for the topic of this essay, which is centred on the idea of 'exotic places'. Senses of place exist in relation to senses of other places, and these other places are important for the definition of 'home'. Yet this aspect of place making seems to have been given less consideration than the aspect of 'dwelling', which was persuasively formulated by Heidegger and also underpins Edward Casey's work. 'Home' is sometimes seen as the place where the person is rooted, and in horticultural regimes it is also seen as where people plant things. One term used to express identity in the Mount Hagen area of the Western Highlands of Papua New Guinea belongs to the verb form 'to plant' (*rökli*), and the expression of solidarity or shared identity builds on the idea of what is *mbo*, a root-stock, a cutting. The forest, by contrast, is the wild, or *römi*, area, unplanted by people although of course teeming with plants and other life forms. The spirits of ancestors or dead kin are *mbo tipu*, 'planted spirits', whereas other forest spirits are *tipu römi*, 'wild spirits'. If this is so, knowledge of the one category gains force from knowledge of the other. Yet studies of 'senses of place' have tended not to examine this dimension of 'place making' as much as the dimension of place as 'home'. In the highlands of Papua New Guinea, the ability to go to other places and obtain ritual and practical knowledge there was actually important in the creation of local or 'home-based' influence, and lies at the back of the significance attributed to circulating religious 'cults' and their points of origin, among other things. In this regard also, then, the construction of 'home' depended in some ways on the experience of 'not-home'. Magical spells also encapsulated knowledge coming from 'exotic places', both through language and dialect forms used in them and in their enumeration of distant places seen as imbued with power. In practical terms, exotic places were sources of unusual and prized foods and decorative items and substances, which again were used to contribute to the acquisition of status and influence through performance of skills and valued identities 'at home'.

In this essay, then, and in line with the theoretical dimensions outlined above, we will explore the somewhat less studied role of geographical knowledge that people extend to places outside of their immediate arenas of daily habitation. Our case study will be from the Highlands of Papua New Guinea, where this is an important topic, for several reasons. First, people travelled a lot, insofar as political interrelations allowed this, not just out of curiosity but to establish trading and exchange networks. Second, travel sometimes resulted in intermarriage and migration patterns whereby persons gained a more definite footing outside of their original localities. Third, such ties could be important ecologically, because in cases of crop failure people could go to more favoured areas where they had established ties and could tide themselves over during periods of deprivation. Typically, high-altitude dwellers who could subsist by planting and composting varieties of the sweet potato staple crop in ordinary times might find their plants damaged by frost on some occasions, and they then sought to migrate temporarily to areas where they had connections in lower-altitude areas in order to survive (Wohlt 1978).

Senses of connections to other places and knowledge of what those places could offer as resources were therefore important to many populations in the Highlands region. The more prominent and adventurous members of communities tended to be those who travelled most and gained this kind of valuable knowledge as well as creating the connections needed to make use of it. This travelling ethos was extended further in colonial times from the 1950s onward, when younger people made journeys to coastal areas on labour schemes, or were taken to other parts of the Highlands, from where they returned with a variety of new goods and knowledge about other places and other people within Papua New Guinea. In the 1950s, when shell valuables were still used in ceremonial exchanges in the Hagen area (Western Highlands Province) and the demand for them had risen, migrant workers often brought their wages back home in the form of these shells (see Stewart and Strathern 2002a, and Strathern and Stewart 1999 for a full description of the use of shell valuables and the transformation of their use historically).

Ongka-Kaepa, a notable leader of the Kawelka group in the northern part of the Hagen area, was not among those who went elsewhere as a migrant labourer at these times. He stayed in his own place Mbukl and built up a reputation working with the Australian colonial authorities and as a prominent participant in the burgeoning networks of *moka* exchanges, involving gifts of live pigs, valuable shells, and state-introduced money, that emerged in the years of colonization and pacification prior to Independence in 1975. Ongka was, nevertheless, an intrepid traveller in an area to the north of Mbukl, across the Sepik-Wahgi Divide, where some Melpa-speaking groups from Hagen had extended their residence in precolonial times.

In Ongka's autobiographical account (this was spoken by Ongka on a set of cassette tapes and then translated in its entirety from the Melpa language into English by Andrew Strathern, who wrote the account out and arranged to have it published, Strathern 1979) he described this area, the Jimi Valley, from his own experience and knowledge (Strathern and Stewart 2000a: 79–89). His description is interesting both as an ethnographic vignette and as an example of how an area may be viewed as 'exotic' by those who live far enough away from it not to be fully familiar with it but near enough to be able to travel to it from time to time. Little has been written about New Guineans' own ideas of 'the exotic', although from our fieldwork experience each group historically maintained horizons of knowledge of other areas that could be characterized in this way. In this chapter we take a closer look at how Ongka displayed his own knowledge and views of a cultural area that he thought of in some ways as 'other' than his own. First, however, we make some comparative remarks from two other areas in the Highlands of Papua New Guinea.

Comparisons with the Pangia and Duna Areas

A hundred miles or so to the south of Mbukl lies the Pangia area, inhabited by speakers of the Wiru language, numbering some 20,000 in the 1960s. The area was brought under colonial administration from 1960 onwards, and a rough all-weather road for vehicles was carved by hand labour out of limestone and mud. The settlements linked by this road became a kind of 'central' part of the administrative area, while those beyond it, which were geographically more isolated and remote from the airstrip and district office, took on an 'exotic' character. This characterization applied partly to the settlement areas known as Lawe and Wembu, entry into which required negotiating on foot a narrow and difficult pathway over a mountain ridge and a descent into a much lower valley area.

Lawe became a part of what was called 'last Wiru', the end of Wiru settlement, fringed by a bush area without contiguous populations. Its own population was small, and in 1967 when this group held a pig-killing, their Local Government Councillor announced that they were holding the killing because there was no food for the pigs. This in turn, they explained, at least in part, was as a result of their having to do compulsory road work for the government (on a road that did not reach their own settlement area). To the visitors from elsewhere who came to Lawe, this was an exotic area, but one of near-famine rather than of plenty; and in spite of, or even because of its isolation, it was feeling the most coercive effects of colonial rule.

Further out from the 'loop road', eight hours' walk from the nearest other Wiru place, lay Tangupane, a tiny settlement within reach of the gorge of the Erave river, beyond which speakers of a different language lived. Tangupane was the epitome of 'last Wiru'. It had an unusually high proportion of unmarried men, owing to the fact that its women preferred to marry out and very few women from elsewhere were agreeable to marrying into the group. This was not only a result of its remote location. It was much more because of Tangupane's reputation as the home of a terrifying type of assault sorcery, known as *maũa* or *uro* (although some people made distinctions between these two terms, others conflated them). Visitors were warned to take great care if going to an outside latrine at night. Companions stood guard while a wary expedition of this kind was made. People blocked up any holes in house walls to prevent the *maũa* from 'looking in'.

Maũa sorcerers were said to hunt for victims in groups. Their presence could be told, it was said, by a wave of heat that preceded them. When they found their prey, they were said to tear out the victim's entrails, replacing these with dry pieces of rubbish, placing one of their kidneys in their mouth, and telling the victim to go home and roast it. The narrative continues that the sorcerers would test whether the victim was thoroughly disoriented, by asking them from which direction the sun rose. If the wrong answer was given, the sorcerers knew their work of destroying the person was complete. On returning home, after two or three days, the victim was said to have died from this attack (Stewart and Strathern 1999, 2004).

This image of the exotic as a source of magical danger from physically impoverished and weak low-altitude areas to their more densely populated and prosperous uphill neighbours is common in areas on the fringes of the New Guinea Highlands. Among the Duna speakers of the Aluni Valley in the Southern Highlands Province, comparable images were entertained with regard to the Oksapmin area across the Strickland River to their west, where the 'others' spoke a different language and wore very different forms of clothing and decoration from the Duna. They were feared as the sources of assault sorcery just as the Tangupane *maũa* were in Pangia. Women from Oksapmin whom Duna men married were also feared as potential witches. Yet there was an established trade between the Duna and the Oksapmin people, notably for cowrie shell valuables used in exchanges. The Aluni Valley Duna had indigenous salt and red ochre for face and body decorations which they themselves obtained from trading links to their southeast, and exchanged these with visiting parties from Oksapmin.

Valuables, sorcery, witchcraft, trade goods: all these things, then, might be seen as coming from or belonging to an exotic periphery in these Wiru and Duna images. We may compare these images now with those of Onkga in his account of the Jimi Valley.

Onkga's Ethnography of the Jimi

From the beginning Ongka tells his readers that this area may have been exotic but that he had become quite familiar with it, and it offered many special types of foods that could not be found in his own much higher-altitude area, known as Korkakona by contrast with the Jimi (Ndimi) which was called Koponkona (*kona* means 'place' in the Melpa language). The special foods and products were important reasons why people braved the journey to visit there: 'types of cane, nuts, fruits, cassowary birds, wild pigs, snakes, lizards, and fish' (Strathern and Stewart 2000a: 79).

Ongka's spoken description of how travellers reached the Jimi is every bit as dramatic as the travel writing of Colin Simpson (1962) or Gavin Souter (1963). Souter writes about an expedition on which he went into the Jimi area in 1958:

> I accompanied a patrol into the Bismarcks, a range of mountains whose green rococo folds of rain forest form the northern wall of the Jimi Valley ... We climbed an almost vertical slope of mud and tree roots, crossed a ridge of moss forest at 6,000 feet, and then jolted down beside a hectic, nameless stream ... (Souter 1963: 235)

The government officer and Souter met at this time the Gants people, ' forty little men in grass sporrans and plumes and possum fur, and some of them were so nervous that they held each other's hands for comfort' (ibid.).

Ongka's description is:

> The pathways into this area are steep; the hills stand up in rows before you. You take your wives and children and walk in front. Slashing the overhanging growth away, up into the dripping, slippery, mountain moss, then on the other side you head down a sheer slope, down, down, to the banks of the big Muglpin River. (Strathern and Stewart 2000a: 79)

Ongka's narrative was translated into English from his oral rendition in Melpa (as noted above) spoken continuously as a stream into a tape recorder. It bears the marks of his dramatic rhetorical powers of expression. For him the streams and rivers are not nameless. He delights in specifying the name of the Muglpin River, and moves immediately after this into a local ethnographic mode. Souter's presentation comes as if through the eyes of the colonial patrol officer whom he was accompanying. The 'little men' they encounter are waiting to listen to a speech about government control and the requirement to build roads in their area. (These 'little men' were those also described by Melpa speakers as Ndimi Mongaemb Wamb and feared by them as potential flying witches, *kum koimb wamb*. This category would include the Karam and perhaps some Maring speakers, see, e.g., Riebe 1987.)

Ongka, by contrast, gives us the ethnohistorical picture of the groups living there, and provides a typical origin story of one of the groups, the Maplke, which describes a quarrel between two brothers and the journey of one of them to a place where he is befriended and given a wife by an Epilke tribesman. In return for the wife, he gives his bow and arrows to the Epilke man. The man and his wife founded the three major groups of the Maplke group through their three sons. The Maplke and Epilke have since fought, Ongka notes, and in their ceremonial oratory the Epilke claim that they have been able to beat the Maplke in fighting precisely because the Maplke originally gave them the weapons (bow and arrows) with which to do so: a witty observation, of the kind that actually did enter into speeches in Hagen, playing on the contradictions between hostility and friendship and between autonomy and interdependency. The Maplke are said to owe their entire existence to the gift of an Epilke wife to the wandering brother who became their ancestor. The Epilke say they owe their fighting success against the Maplke to the Maplke man's own original yielding up of weapons in payment for his wife because at the time the man had no female to complement his abilities and raise pigs that could be used to promote social exchange.

At the end of his discussion on the Jimi, Ongka makes a further point, in counterpoint to his description of the enmities that could arise between residential groups in the Jimi itself: 'We never fought with the Jimi people, for we realized that if we did we would have no way of obtaining all those special foods: the nuts, the pandanus fruit, the cassowaries, the wild pig, the fish, greens, and all kinds of bird plumage' (Strathern and Stewart 2000a: 88).

The bulk of his account is devoted to an ethnography of these special food resources, carefully observed in detail from his own experience. In line with all of these accounts, the scenes are vividly and deftly depicted, and their social context is as much a part of the narrative as the environment and the techniques used by people. First, he describes the hunting of wild pigs. Hunters make traps, and if they find a pig in one they call on their relatives and friends to come, and these arrive with a special kind of condiment, the *uya* (*Pangium edule*) nut which they use to spit onto blood sausages made when the pig is killed. Actually, the pigs like these nuts too, and they are used as bait in the traps when they are in season (ibid.: 80). Ongka explains how the ovens are made deep and the meat is steamed for a long time. He does not say why, but the probable reason is to make the strong flesh more tasty and easier to eat.

He moves on to a vignette of how people catch fish. As with the pigs, his emphasis is on the great plenty the forest and streams can offer. Males who go out on fishing expeditions also, he says, observe a taboo on sexual intercourse, and must not eat hibiscus leaves or red pandanus fruit. These restricted foods are slippery, and if the fisher ate them they might make the fish slip out of his traps; or his smell might frighten the fish away; or the fish might die in his traps (p. 81).

Human ingenuity is used to catch cassowaries. Men may bait snares with their own excrement, because cassowaries like things that smell rotten (p. 82). To catch eagles (whose feathers were important for decorations) men suspended wickerwork baskets above their nests, then dropped these on the birds when they returned. They made bird blinds behind which to shoot the white bird of paradise (also valued for its plumage, and not found in the Mbukl area which was Ongka's home) (p. 85).

There is a long account of how the *uya* nut is gathered and processed by men and women working together. They work in teams, steam-cooking and shelling the nuts, then loading them into special netbags and lowering these into pools, there to ferment in storage for two months or more. When they at length take them out, they recook them in bamboos, each person receiving a bamboo as their share. People feast on the nut and its oils along with forest vegetable greens before returning home. Packs of *uya* mush and greens are so much appreciated that, like the oil of pandanus fruits, they can be given in exchange at marriage and in *moka* ceremonial exchange as an 'extra' item to make friendship (p. 98). People from the Mbukl area, Ongka notes, go in numbers to get pandanus fruit and oil in season in the Jimi via their connections (in Ongka's case a daughter of his was married to a Maplke man, hence his recitation of the Maplke origin story).

Ongka's natural history and hunting ethnography ends with this disquisition on the *uya* (*Pangium edule*) nut and the *neka* (pandanus) fruit, abundantly underlining his point about the reasons why people from his area never got into serious conflicts with the Jimi people. Although he does not specify place names in this part of the account, his descriptions are clearly first-hand and locally observed, suffused with his own observation and participation. The overall account is largely impersonal, as though Ongka were an observer from elsewhere – as indeed he was, engaged in 'participant observation'. The chapter closes, however, with a sharply personal narrative. In mid-1965, Ongka was visiting the Jimi. His eldest son at the time, Kuma, decided to visit the Hagen Show, a biennial festival begun by the colonial government, held 30 miles south of Mbukl near the Mt. Hagen township. Kuma had been warned not to go because he would need to pass through the enemy territory of the Minembi group on his way. In the event, he was found drowned in the big Gumants River halfway to Hagen, the cause unknown. Ongka explains in his account how he heard the bad news about his son: 'I was far away in the deep bush of the Jimi cooking *uya* nuts. We were just about to lift the nuts from the oven when the news came to me' (Strathern and Stewart 2000a: 89). The news came to Ongka at a time when he had been in the Jimi for some time and had already received many presents of nuts from earlier cookings, which were now hung on poles for the journey back to Mbukl. A clansman came and gave him an indirect message about what happened: 'Your son has come back very sick', he said (a polite way of breaking the news that his son was dead). Ongka understood the true import of the message; and he comments on how he felt: *Nanga muntmong poköu nitim*, 'my heart caved in'. He told his people to bring the nuts and fish later, and he himself hurried back to arrange the funeral. 'Later I cooked twenty pigs for his funeral feast, I invited everyone to come and divided out the pork to them all' (ibid.: 89).

The emotional intensity of this event is recounted with a spare economy of expression, and in later years Ongka did not dwell on Kuma's death, although it affected him deeply, and his visits to the Jimi were much less frequent after this event. The experience of being in a distant place, far from home, and learning that a close relative, particularly a child, parent or sibling, has died, is something that Hageners have come to know with increasing frequency as a result of travel, education and employment elsewhere. Dreaming of a relative is a bad sign. People who are away from their home places telephone someone back in their area anxiously, if they can, after such a dream. Ongka's own experience was primordial. Engrossed in local exotics of abundance, his adventurous idyll was interrupted by the most unfortunate of news. His thoughts about the Jimi changed after that event.

The Structure of Ongka's Narrative

Ongka's overall narrative has a clear dramatic structure: the Jimi is an exotic place, but by travelling in it he got to know it well. It attracts people by its abundance of rich foods. But it is far away from home. The Korka (mountain) people do not fight the Kopon (Jimi Valley) people; but the Kopon area cannot be home to them either. Perhaps Ongka could have persuaded his son not to go on his own dangerous journey to the south if he himself had not been absent in the north. Kuma's death was filled with local emotional intensities and ambiguities that impacted the lives of many kinsfolk.

As Ongka notes, the Jimi is not perfect. People like to send their daughters to get married there; but 'the place is very hot and sweaty, and there is sickness there too. It is difficult to stay inside the houses, so people cook and sleep outside at night, getting bitten by mosquitoes; they don't make fires inside the houses at all, for if they do it is far too hot' (Strathern and Stewart 2000a: 88). Ongka's Kawelka are mountain people; the Jimi is more like a coastal region. It has its own dangers. Mosquitoes bring malaria, which without medicine can be lethal.

Ongka's story of the Jimi carries some nuances, therefore, about the dangers of the place. What is remarkable is that he leaves out of his discussion the elements of perception that are shared with the 'last Wiru' of Pangia and the Oksapmin area as seen by the Duna: that is, its association with the dangers of witchcraft, sorcery and spirit attack. Central Hageners tended to think of *kum koimb* cannibalistic witchcraft as coming from the Jimi. Another Hagen man whose biography has been published, Ru-Kundil, describes how he was subject to a *kum koimb* attack when he accompanied an agricultural patrol into the Jimi (Strathern and Stewart 2000b: 85). The Kopon area was proverbial in this regard, and to Hageners of the central plains Ongka's own area of Mbukl was also known as Kopon, and witchcraft was said to be prevalent there too, as a result of its proximity to the Jimi.

The Jimi also figures in a number of striking folk-tales as one home of a set of spirit cannibal beings (*Tei wamb nui wamb*); and as the setting for a dramatic tale of a young man who goes to look for his spirit father, the folkloric character known as Ndepana Nikint ('red parakeet of the forest'), a Sky Being, and who returns to Korkakona with the help of his mother, there marrying and settling down (Stewart and Strathern 2002b: 160–63). The Jimi Valley is a home of forest spirits, witches and monsters. Its far western edges are said to be the place where the spirits of people go in dreams in the shape of birds and marsupials (*Mötamb lip pana*, the Mötamb plain where the *lip* plant grows).

All these images were much more a part of popular imagination in the 1960s than Ongka's own information bank about luxury foods and how they were obtained. Perhaps it was characteristic of Ongka that he stressed his own version of 'rationality' and human ingenuity in his account of this most exotic of areas; almost, but not quite, turning it into a safari area for intrepid visitors like himself. The death of Kuma, which he described at the end of the chapter on 'local knowledge', brings this vision abruptly to a halt (Strathern and Stewart 2000a, chapter 12). The ties with home are reactivated, patently and painfully. And Ongka sets off back for Mbukl, where thirty years later again he will bury another of his sons, Namba, making for him a special fenced-off enclosure at the head of the old ceremonial ground where he had his 'big moka' in 1974 (see the film *Ongka's Big Moka. The Kawelka*, Granada Television, Disappearing World Series).

The Theme of 'Exotic Places'

This discussion of indigenous views on exotic places indicates several analytical points. First, as we have argued, people construct knowledge about a range of areas other than their own, defining these by categorical contrasts. Second, such 'other' areas could be important for trade, exchange and luxury goods, hence the development of positive knowledge about them. Third, prominent families took the lead in making ties with such areas, as Ongka's family did in relation to the Jimi. But, fourth, in the broader contours of popular imagination, such places often took on frightening characteristics. Tangupane in the Pangia case was seen as entirely negative: it had no valuable products, and was instead seen as the source-place of dangerous sorcery. Oksapmin, for the Aluni Valley Duna, had valuable trade items, but was also home to assault sorcerers. The Jimi for Hageners had fine food resources, but was a place of sickness, sorcery and witchcraft also.

In studying the category of 'indigenous knowledge' we need to make many distinctions between different types of such knowledge and their bases in specialist experience or popular hearsay and imagination (for thorough discussion and examples, e.g. in the context of development studies, see Bicker, Sillitoe and Pottier 2004 and Sillitoe, Bicker and Pottier 2002; in the sphere of medicine, Lindenbaum and Lock 1993; verbal arts, Finnegan 1992 with references). In his story Ongka stressed his own immediate, lived experience. Hageners who had never gone to the Jimi would be more likely to describe it as a place they feared because of its dangers. The Jimi Valley, like every place, is several different places, depending on how it is perceived, narrated and experienced.

Place and Memory in Landscape

Ongka's account of the Jimi Valley is not about his own place, Mbukl (a name that means 'mountain ridge'). But it is framed from the very beginning as a contrast to Mbukl. The Jimi is the land of the Other in Ongka's narrative, as it is in the narratives of other areas that stress the themes of danger from witchcraft and sickness. The memories that Ongka calls on are relatively recent biographical ones; but he weaves in with these memories accounts of origin stories of groups and general observations of technical procedures in hunting, fishing and food gathering, thereby giving us a composite snapshot of his personal indigenous knowledge. The memories of his own experiences include both what he had seen and done or observed and what he had heard from others, either in the Jimi itself or from relatives and partners from there who visited him in Mbukl. All these sources constitute the composite mosaic that becomes "the Jimi Valley" in his account. What is the significance of such an account in terms of more general ideas about emplacement, landscape and history?

First, narratives generally establish horizons and boundaries of experience that define the expansive agency or extensions of personhood of people. Ongka's knowledge of the Jimi defined an ambit of activities that set him apart from many of his fellow-clansfolk at Mbukl through his greater acquaintance with this area. This claim to extensive geographical knowledge is typical of leaders in the New Guinea Highlands. Knowing other places and dealing with them is the mark of the entrepreneur, a way of singling out the leader, not as an individual separate from his group but as one who can bring back benefits to it in addition to assisting in his own projects. Ongka's journeys to and from the Jimi were acts of trans-placements of his agency, associating him with the ability to know two places and bring them into transactional relationships. By agency here we do not mean simply the allocation of responsibility for action (we could call this imputed agency) but the active ability to make a difference in the world (creative agency). Ongka's self-emplacement in a wider landscape that went beyond his own place and encompassed the wild, fertile and abundant Jimi Valley therefore contributes to the portrait he drew of his own agency, and, thereby, his personhood. The Jimi, external to Mbukl, becomes an internal landscape of reflection and memory for Ongka as he thinks about it: a part of himself as well as being outside of himself.

Second, the Jimi is for Ongka to some extent an area outside of history, if by history we mean the major narratives of exchanges, alternating with hostilities, that took place in the Valley area running east and west from Mbukl, joining the Kawelka to allies among the Tipuka group, and to their enemies among the Minembi. As Ongka says, 'We never fought with the Jimi people.' They were a source of foods and wealth items that fed into exchanges, and networks of individual trading partnerships developed along lines of kinship and marriage. Some men of Ongka's group had grown up in the Jimi after their seniors took refuge there in times of warfare prior to the 1930s. One man, for example, stayed there for a long time because he was 'sorry' for the place where some of his paternal lineage kin died and were buried (A.J. Strathern 1972: 243 note 8). Such individual acts of an affiliative kind, however, are different from the full-scale engagement of groups in ongoing relations with one another. So the Jimi becomes a kind of timeless land, lending itself as a backdrop to folklore and ideas of witchcraft. Such reservoirs of experience as nonhistory are of course not impervious to historical change. Roads cut across the Sepik-Wahgi Divide into the Jimi area opened it up to traffic, and a Catholic mission station was established along with a small airstrip. But such manifestations of change, although known to Ongka, are not made a focus of his account.

Third, if the Jimi has a history, for Ongka it is the mythological history of the founding of its major groups, and early on in his account he gives one such story, which has a classic narrative structure. To recap this here, two brothers quarrelled over the rights to a garden, and they separated. One of them left the area, while the other stayed. The one who left (and became the ancestor of the Maplke group) took just a little pig and his bow and arrows. Feeling hungry, he took a ripe banana from a garden. The garden's owner, a man of the Epilke tribe, offered him more fruit and saw that he ate this carefully, piling up the skins in a heap. (This is an action that people would take to prevent sorcerers from gaining access to their leavings.) The Epilke man accepted him as human and persuaded his sister to marry him. As we have seen earlier, Ongka then 'fast forwards' his story to contemporary times and the competitive speeches made between the Epilke and the Maplke on occasions of exchange of wealth. Onkga thus clearly recognized that in their own area the Epilke and Maplke have a history which becomes memorized as a part of current political relations. In this picture, then, the Jimi is not timeless after all. It has its own rhythmic time, outside of and separate from the historical time of Ongka's own political ambit.

Finally, the way in which memory and place become a part of politics and political history in Ongka's view of the world is very much tied up with political speech making. Such speeches were made on ceremonial grounds where *moka* exchanges (Strathern and Stewart 2000c) took place (*moka pena*). All major discussions would be held at these open, grassy areas fringed by planted trees, cordyline bushes, and other ornamental plants. The laying out of a new ceremonial ground marked a fresh point in history. If an old ground was abandoned and used for gardens, the memory of it was weakened. Leaders who died were placed on high trestles for mourning, situated at their own ceremonial grounds. For each *moka* held at a ground a new *pokla mbo*, or ornamental tub of bushes or trees, might be set up. Above all, people gathered in numbers to hear the speeches made at these grounds. History was thus made in speeches as well as in the material gifts handed over; and gifts could not be transferred without a highly stylized verbal counting out of them, performed by one of the leading men among the dancers. These speeches usually also would locate the occasion clearly by naming the ceremonial ground on which they were made and addressing the major groups among those present. The speeches presented, therefore, a 'wordscape' to go with the visual images of these occasions. They were also the meeting places of individual and collective biographies, where individuals made their marks and at the same time began to merge with their collective worlds.

This same differentiation combined with merging appears also in the account by Ongka of the Jimi Valley around which our chapter here has been organized. The account bears the unmistakable marks of his own personal eloquence and keen powers of observation and humour. But it is presented in an objectivist way, like an ethnography, until, that is, its end, when his personal experience and memory break sharply into the objectivizing structure of the account with the painful story of where he was when he heard of his son's death, and how it made his heart 'cave in'. Characteristic of him, too, is his response to the news of the death. He left at once for home, and arranged the funeral, merging his personal grief with a presentation of pigs cooked and distributed for the occasion.

Ongka's own personal memorial is set into a local landscape of highly charged significance. He died in late April 2003 and was buried at the ceremonial ground Wayake belonging to his clan-section of the Kawelka group, the Klammbo, just beside the entrance gate to the old government agricultural research station at Kuk, not far from the Mount Hagen town (on Kuk see Strathern and Stewart 1998, n.d.). At the time when he gave his self-account, in the 1970s, he had been living at Mbukl close to the Sepik-Wahgi Divide. The bulk of the Kawelka later moved back to an older ancestral territory around the place Kuk, taking advantage of colonial 'peace' to reclaim an area with rich potential for cash cropping and served by a government road. The Kawelka had also yielded a large part of their old ground at Kuk to the Australian Administration for use as a research station. This ground was swampy, and the Administration drained it, revealing its fertile soil.

Subsequently again, in the 1990s, the research station fell into disuse, and the Kawelka and others occupied its houses and began to reoccupy its lands. The whole area became contested between the government and the people as well as among the people themselves. Ongka had earlier expressed a wish that he should be buried back at Mbukl along with his son Namba; but he had reportedly changed his mind and agreed to be buried at Wayake instead, where his grave was marked by a cement monument and symbolically guarded the entranceway to the contested site of Kuk, becoming an icon for the whole Kawelka group, not just his own clan-section within it. With his impending death from old age, we had been approached for contributions to his funeral. We sent, at his request and the request of a younger leader of his group, substantial contributions for his subsequent funeral, as well as other sums for his use prior to his death. We also sent, at his request, enlarged photographs of him for possible use at this gravesite.

The outcome of the efforts to emplace him permanently after his death is that Ongka is still a presence among the Kawelka, his memory sedimented in stone at a place where he had made many a challenging, ingenious and fiery speech. Onkga has himself become a kind of landscape of memory for the Kawelka, anchored to the place Wayake, no longer mobile except through the mobility of his reputation, his 'name'. Ongka has made his mark on the physical, social and political landscapes of two different Kawelka territories during his lifetime. In choosing, finally, to be buried, and so to be commemorated, at Wayake rather than Mbukl, he was acknowledging the fact that the more ancient territory of Kuk was also the land of the future for his group. His final placement in a prominent position within that landscape, at once authentic because of its past and politically relevant because of its future, meant that his kinsfolk incorporated him not only into their memories but also into their anticipations.

Part II
Emplacement and Landscape

Chapter 6

Power, Knowledge and the Organization of Space

Peter Meusburger

Introduction

Since division of labour developed in early human history, there was a need to coordinate, control and integrate the elements of social systems in space. The term 'social system' is used very broadly in system theory. In this paper only goal-oriented social systems (organizations), which provide workplaces, are discussed. Each social system develops its authority and communication structure, its hierarchy of control and decision making, and its relations with other systems. In the spatial dimension the authority structure of a social system is represented by workplaces for highly and lowly skilled persons, by networks of communication, by flows of material, energy and people, and by areas of influence. It was predominantly the vertical division of labour, namely the bifurcation of skills and decision making in a social system, which produced hierarchies of increasing complexity and shaped the disparities of knowledge and power between centres and peripheries. New technologies of storing and transmitting information and of transporting goods have extended the potential spatial range of coordination, control and surveillance and increased the possible size and complexity of organizations.

This paper deals with coalitions between power and knowledge and their manifestation in space; it focuses on the role of centres and peripheries in 'power-geometries' (Massey 1999), and on central–peripheral disparities of knowledge and decision making.

Visibility of Power, Social Differentiation and Demarcation of Space

Terms such as power, authority, dominance, control or influence are not traits of the individual; rather they describe asymmetrical social relations between persons or between parts of institutions. Analysing the spatial dimension of organizations helps to describe power relations and social differentiation. Power creates and organizes social systems in space; it demarcates spaces for different functions; it defines and defends its territories; it provides identities, functions and status to places. Obviously political authority organizes the use of geographical space; and the organization of space evolved through a political process (Gottmann 1980: 18). We should keep in mind that the term 'region' has the same etymological root as the words rex, regulate, regime, regiment, or the German word *regieren* (to rule). In its original meaning, region was a space that was organized, controlled and influenced by a power centre. Authority operates within certain spatial limits, and thus influences certain areas or specific spatial networks.

In order to minimize the number of conflicts in a social system, to maintain social order as well as achieve cohesion and collective action, authority and power relationships must be displayed. Status differences and power relations are made visible by signs, symbols, titles, ceremonials and positioning in space. Foucault (1972, 1980) was by far not the first to study the relations between knowledge, space and control, but since he explicated his concept of disciplinary power, which was based on visibility and surveillance, this issue has been studied extensively in geography and other disciplines. In the horizontal dimension spatiality is expressed by distance, demarcation, exclusion or positioning in relative space; in the vertical dimension by elevation. Elevation of a throne, altar, stage or speaker's platform, seating arrangements in a meeting, or the staging of a ceremony in many cases have all the function of 'status reminders' (Harper 1985: 36).

Positioning in space is one of the most frequently used means to distinguish between ranks of status. As spatiality displays the 'contemporaneity of difference' (Massey 1999: 35) and the gradation of authority and status in social systems, a multiplicity of places is needed to represent social differences. Even in early human history, partitioning of space was used to display gradations of sanctity, authority and status. The word 'temple' has its root in the Greek verb *temnein* which means cutting or marking. The term temple (*temenos*) identified a precinct or demarcated area bounded in some way by lines, hedges or walls. The temple as a holy district was a marked space (Turner 1979: 15). In all types of societies, the varying degrees of importance, power and authority were expressed by separation and demarcation of spaces, as well as by exclusion and elevation Meusburger 2005, 2006.

The processes of conditioning, generalizing and stereotyping ascribe to places positive or negative symbolic meanings which may rouse various emotions, memories and actions. A certain location can serve as a medium for cultural memories and collective identity. It can be a symbol for prestige, reliability, credit-worthiness and power; however, another location or street name may suggest danger, poverty or criminality. In many cultures identity is first of all constituted in relationship to place. In some Asian and Pacific cultures place is even more important for the establishment of social relationships than kinship or religious affiliation. Places are needed to perform or participate in divine, political or secret rituals. One cannot be baptized by radio, fraternize with Freemasons by telephone or receive the Holy Communion by email.

The importance of spatiality for representation of authority and construction of difference has not decreased in recent history; spatiality is more than ever an essential to the production of society (Massey 1999: 28–40). Space is a structure of perception and 'a means of intervention that controls the behaviour of individuals' (Townley 1993; Feldman 1997: 944). The gradation of power and importance in governments, ministries, banks or headquarters of multinational companies is still to a large extent represented by demarcation of space, by restriction in access or exclusion from places. Top managers have their own elevators, their own parking lot and reside on floors with the most beautiful view. The more power an interest group exerts, the more it tends to spatial exclusion, restriction of access and secret initiation rituals. It is again worth noticing that the words 'exclusive' and 'exclude' have the same etymological root.

Focusing on the spatial distribution of power and knowledge in social systems, the term space is understood as relative space and not as container. Relative space is a product of interrelations and interactions. It is never finished and never a closed system; it is always in a process of becoming, always being made (Massey 1999: 28). Also the concept of centre and periphery should not be regarded as a spatially fixed system. Knowledge and power, manifested by centres, are always on the move, always contested and socially constructed.

The different types of relative space can be categorized according to the degree of how much authorities are able to interfere with or control the activities of individual actors and whether their locations are stationary or mobile. Some types of activity patterns are the result of spontaneous decisions and interactions of numerous actors not coordinated by a centre of power. An example of this type would be the main catchment area of a shopping centre. Another type of relative space is organized, coordinated and regulated by the power hierarchy of a social system. The spaces coordinated by public administration, controlled by organized crime, or monopolized by a firm belong to the second type. It goes without saying that between these two extremes there exist a number of intermediate types. The activity patterns of tourists in a national park or in a ski resort are partly due to spontaneous decisions of individuals, but they are also restricted to certain areas and times by the regulations of an authority responsible for the park or resort.

Relations of Power and Knowledge

The mutual dependency and coalition of power and knowledge and the social construction and symbolic meaning of centres can be traced back to early history. In pre-literate societies, the shaman, the sage, the dream interpreter, prophet or, later, the priest were in the centre of power or were representing themselves as the centre of the social system. Their claim, to know better, to be in connection with the gods or ancestors, or to represent God's will on earth, gave them a central and privileged position in their social system (Konrád and Szelényi 1978). In later periods, experts, scientists, interpreters of ideologies and think tanks took over this role.

Why do power and knowledge so frequently build coalitions and why do they seek proximity to each other? Those in the top ranks of power depend in their decision making upon the analytical abilities, knowledge, creativity and advice of experts and professionals. Both instrumental power, which is the capacity to make others do one's will and the ability to control, regulate and dominate others, and associational power, namely the power to do things by acting in concert or using institutional mediation (Agnew 1999), have to rely on experts, scientists, intellectuals, priests, academies or think tanks to achieve and sustain their position (see Figure 6.1).

Experts are persons who, by objective standards and over time, consistently show superior and outstanding performance and competence in typical activities of a particular domain (Gruber 2001). Like oracles in ancient Greece, experts are paid for predicting possible developments and likely consequences of actions, for anticipating potential risks and for giving advice on how to cope with uncertain events. They are supposed to offer more certainty in a complex and risky environment than a layperson is able to do. Experts are needed, on the one hand, for perceiving and understanding certain problems and offering solutions for them; on the other hand, for interpreting signs and patterns of change that are not understood by the majority of people. As expertise is an action-orientated advice, it should be free of errors. Those asking and paying for expertise do not want to play the role of guinea pig.

Lacking expertise and having faulty information make it difficult to maintain a grasp on reality and lead to mistakes in analysing and solving

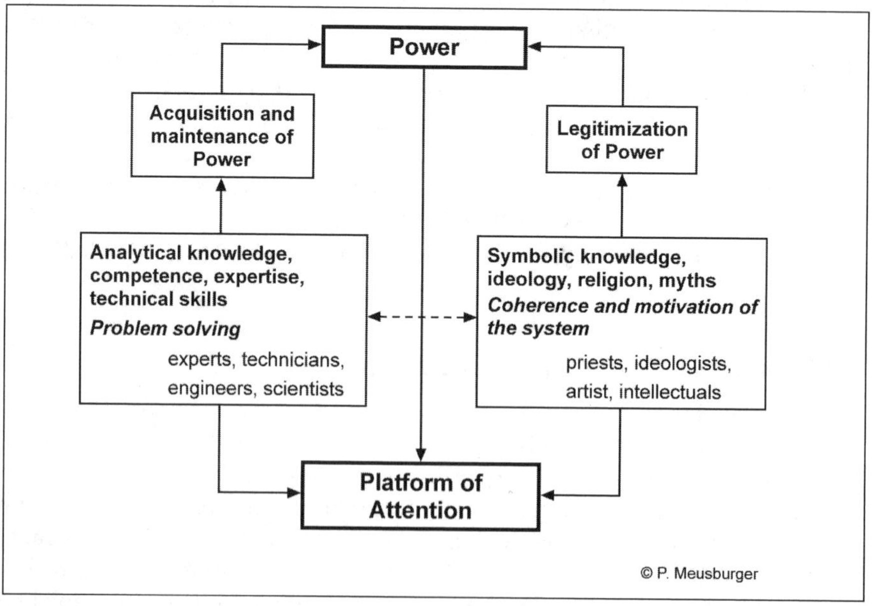

Figure 6.1: Coalition between knowledge and space.

problems. Incompetence and poor decisions waste resources and create unwanted dependencies. They invite questioning of the ruler's authority and prestige, and weaken the cohesion of the social system. Therefore, those in power need the knowledge, the analytical capabilities and the competence of experts, advisers and scientists. To preserve power for significant periods of time in an uncertain and dynamic environment, a social system has to be successful and has to retain its ability to learn and adapt to new situations. A lack of knowledge endangers a social system's capability of self-regulation.

It is not sufficient to obtain power. It has to be legitimized, if it is to endure over significant periods of time as shown in Figure 6.1. Since the legitimacy of rulers, dynasties and governments is socially constructed, it may be fragile and questioned. It is in the interest of those in power that their authority is acknowledged and accepted by their subordinates. The rulers obtain this legitimization of their power mainly from representatives of symbolic knowledge. In earlier times, these were shamans and oracles, later they were priests, intellectuals, editors, ideologists, propaganda departments and artists. An important role of the representatives of symbolic knowledge is to give the social system an identity, to sustain the internal cohesion and motivation of the social system. They set standards and goals, convey values and views of the world, glorify rulers and historic events, demonize political opponents, create myths, found cultural practices and rituals, declare taboos and decide which questions and terms are politically correct and which are not.

Representatives of symbolic knowledge are trained and experienced in the art of influencing, convincing and manipulating people; their task is not to analyse the real situation or to search for truth, but to make moral judgements, to support the value system and cohesion of their own organization, to undermine the cohesion and collective memories of opposing groups. Institutions of knowledge production and media have to create, disseminate and defend the norms, collective memories, symbolism, culture, identities and ideology that hold a social system together. In modern times, their tasks might include: the production of propaganda for the government or a political party, the censoring of media, the manipulation of statistics, the falsification of documents, the construction of new collective memories, or the manipulation of voters and consumers.

As power and knowledge have a rather ambiguous relationship, it is in the interest of those in power to integrate potential competitors, the best analysts and experts, and the most important representatives of symbolic knowledge into 'networks of assent' (Agnew 1999).

Power Decides What Knowledge Is

Knowledge can be used as a means of domination. It is a capacity to control and manipulate physical and human nature (Brown 1993: 154, 164). As experts and the representatives of symbolic knowledge have the power to influence thought, opinion, value system, and action of novices or laymen, those in power try to influence or manipulate our patterns of cognition. Power tries to procure those experts which are conducive to representing and legitimating the kind of reality it wants, while it ignores those experts which do not serve its interests. Power tries to define what should be regarded as knowledge, truth and reality by the members of their system, and they try to control the production and distribution of knowledge. 'In modern societies the ability to facilitate or suppress knowledge is in large part what makes one party more powerful than another' (Flyvbjerg 1998: 36).

In this manipulation of knowledge, places play an important role. Neuropsychological research has shown that the memory of events strongly relates to the memory of places. Monuments, memorials, symbols, graffiti, art objects and architectures distributed in space can form a sort of external information repository. Places, names or monuments can serve as 'mnemonic aids' or as 'spatial anchors for historical traditions' (Foote, Tóth and Árvay 2000: 305). Such places of memory can trigger associations, sensations and memories both consciously and subconsciously. A spatial location of rituals and cultural artefacts contributes to renewing certain associations, thus solidifying them in the conscious and the subconscious. Memorials, monuments, museums, parades, street names and graffiti are designed to guide public attention and collective memory in a certain direction. At the same time, other historic events for which no monuments exist are intended to be erased from the memory of future generations. The news not produced and the monuments not built are just as interesting as those that exist. The plurality of opinions is much reduced when it comes to monuments and memorials in public space.

In democracies cultural hegemony is not necessarily expressed by open suppression or censorship, but by moving the spotlight of public attention on certain issues and letting others remain in the dark. Hegemonic filtering of information uses a number of practices and techniques. One of the most effective ones is to provide platforms of attention for certain experts who serve the interests of those in power and to deny these platforms to others.

To be able to fulfil the role of an expert, or conveyor of salvation and meaning, the experts both of analytical and symbolic knowledge require a platform of attention, which puts them in the spotlight and guarantees their presence in the media. Public attention is a very scarce resource (Franck 1998). Since our memory and information-processing capacities are limited, attention and learning are selective. Information is not processed unless the individual attends to it. Attention, which is a general term for selectivity in perception, determines what is learned and kept in the memory and what is excluded. Judgement of significance is neither impartial nor spatially invariant: it is rather an instrument of executing power. Since attention is an increasingly rare resource in an information society, various sophisticated techniques are developed to attract, manipulate or divert attention and to control the access to these platforms of attention. Spotlighting the attention of the public to certain objects, persons or issues has become a big business and a very powerful device of accumulating resources, of executing power, of manipulating learning processes, value systems, identities, collective memories, social coherence, and, last but not least, consumer behaviour.

Considering today's flood of information, the contents of a message or its usefulness for society are often less important for its wide diffusion than the platform on which it is presented. The locality where new knowledge is proclaimed determines to a large extent the relevance, visibility and credibility of the knowledge claims.

Focusing attention on certain places, products, persons, historical events or issues resembles a kind of gate-keeping mechanism, whose practices include selecting, making a choice, categorizing, promoting, favouring, hiding and excluding. The capability to direct public attention like a spotlight to selected issues, persons, objects or places, and to divert it from spots being left in the dark, belongs in the twenty-first century to the most effective instruments of power.

The coalition between knowledge and power is reinforced by the fact that both – the representatives of scientific knowledge and those of symbolic knowledge – seek for themselves the proximity to power for various reasons. Centrality is the spatial manifestation of authority, importance and prestige. Most experts and intellectuals are fascinated by authority; they want to be concerned with the essential in existence, and strive to exert influence and to be recognized. Network centrality determines access to valued resources and information unavailable to those at the margin (Ibarra and Andrews 1993). Experts, scientists and artists in many cases have to prove their usefulness to power in order to get the resources they need for their work. As centres bestow experts, intellectuals, artists and scientists with existential safety, access to resources and influence, they act like magnets for them. Proximity to power may also increase their chances of influencing important decision makers. Even intellectuals opposing those in power can best play their role as opposition in centres. This was the reason why dissidents were banned to remote places.

Important decision makers, professionals, experts, artists and intellectuals striving to have prestige, influence or success, sooner or later find their place of work in one of the centres or will create for themselves a new centre in their field.

Social Systems in Space

The overwhelming majority of employees do not act as autonomous individual persons but as members of one or more social systems or organizations, which are defined as goal-oriented social systems. Social systems may be classified along four variables as either mobile or stationary; and as either sharing common space with competitors or as demanding exclusive territories. The first type could be labelled as stationary organizations with a claim to exclusive space. It is represented by institutions such as public administration, which by definition is exclusively responsible for a certain administrative area. It is unthinkable that the same territory be administered and ruled by two governments. The second type can be categorized as mobile organizations with a claim to exclusive space. This type consists of organizations such as armies during war or nomadic tribes. They change their areas and range of activity at short notice, but defend their (temporary) space against intruders or adversaries. The third and most frequent type comprises stationary organizations sharing common space with their competitors. Typical examples are economic enterprises, banks or retail shops. They have no exclusive territory of their own, and compete with each other for the same customers. The fourth type consists of mobile organizations sharing common space with competitors, such as organizations of travelling salesmen, or of forwarding agents. All these types of organizations have their centres, their hierarchies of decision-making, routine activities, peripheries and areas of influence.

The key elements of any organization are communication, information processing, knowledge acquisition, asymmetrical power relations, and hierarchies of decision making. They all serve, first of all, to reduce uncertainty. As the so-called span of control of a supervisor is limited, and since a single decision maker does not have the capacity to absorb, process and pass on all the information necessary for the long-term survival of an organization, each social system develops its lines of formal communication, its authority structure, and thus its hierarchy of information processing, control and decision making. Once it reaches a certain size and complexity, no social system can exist without adopting hierarchic structures of communication.

The need to organize, coordinate, control and integrate the elements of a social system in space began when the vertical division of labour increased dramatically after the invention of writing several thousand years ago. The bifurcation of decision making and skills in social systems was the functional basis for the development of cities and created social and spatial disparities of knowledge and power.

Knowledge and Uncertainty – an Evolutionary Approach

The evolution of economy and society is a dynamic process which continually confronts social systems with uncertainty, new challenges, competition and conflicts. Competition and uncertainty of the environment constantly force social systems to acquire new knowledge, to be efficient and productive and to adapt their structures of decision making and production to a changing environment. According to this approach, it is not the 'fittest' who survive, but those who are able to learn and to use new knowledge at an early stage; those who have a head start or competitive advantage in knowledge, skills and information; those who are able to react quickly to new situations and challenges; or those who are able to transform their own environment (Meusburger 2000); 'it is evolution through learning that clearly dominates and accounts for the major variations in human social behaviour' (McClintock 1988).

According to this evolutionary approach, reducing uncertainty is one of the fundamental goals of social systems. A social system can reduce uncertainty by acquiring additional knowledge and new information. New knowledge, skills, experience and information not only lead to new inventions and technologies, which provide the system with a competitive advantage; new knowledge can also influence the orientation, goals, perception and motivation of the members of a social system. A better knowledge of the environment (e.g. the market and the competitors) will help to avoid mistakes and unintended negative results of decisions, and will allow resources to be implemented more effectively. However, additional knowledge, new information and a deeper understanding of mutual dependencies and factors of influence can raise new questions and create new uncertainties that did not exist before. As the need to reduce uncertainty and to acquire new knowledge never ends, the type of knowledge important to survive competition or conflicts will always be a scarce resource. It is important to notice that various types of information, expertise and knowledge cannot be acquired everywhere. They are created in particular contexts and through interaction within certain spaces. The creation and diffusion of new knowledge is place dependent (Livingstone 2002).

Therefore, the crucial questions are: where to locate the scarce knowledge and the high-level decision making in the architecture of a social system; how to organize the internal and external formal structures of communication; how to organize the interaction with the environment; and where to locate the centre of the social system. All these questions have a spatial dimension.

Optimal Architectures of Social Systems

The architecture of a social system is defined as an ordered arrangement of different functions (workplaces with different tasks). It can be described by the number and hierarchical arrangement of units fulfilling line and staff functions, by the distribution of responsibilities and control functions, by the centralization or decentralization of decision making, by the channels of formal communication, and many other attributes. The purposes of a hierarchy are to reduce the flood of information to those elements essential for decision making, to accelerate information processing and decision making, and to facilitate control of performance and evaluation of results (see Figure 6.2).

The architecture of a social system is not a matter of deliberate choice. The optimal architecture of a social system depends rather on the varying degree of uncertainty of the environment, the tasks, age, autonomy (power, resources), size and complexity of an organization. It also depends on the availability of technical means of information processing and transmitting and on the skills of the available labour force. Since most organizations can only cope with a limited degree of incompetence for any extended period of time, they act in their own interest when they fill the key positions of information processing and decision making with highly skilled, competent and experienced people. In particular, those positions and subsystems that are constantly confronted with uncertainty and/or those elements whose decisions have long-lasting consequences for the entire system require special skills and knowledge. In social systems skills and knowledge have a similar function as redundancy in technical systems (Meusburger 2000). They are needed in nonroutine situations; they reduce uncertainty and enhance stability of a social system. Since it is not knowledge or information per se, but an advantage or lead in knowledge, or a prior access to information that brings about success in a competitive society, the skills, experience, training and knowledge needed by the highest decision makers of a system will always be scarce and expensive commodities.

So the first crucial question is where to locate scarce knowledge and skills within a social system. In which circumstances is the centralization of knowledge and decision making more efficient, and in which cases the decentralization? Which functions of the social system are dependent on the potential of face-to-face contacts, a certain type of action setting or context, and which are not? In bureaucratic systems and in systems dealing with simple tasks and a stable environment – meaning with low degrees of uncertainty – decision making, problem solving, research, development

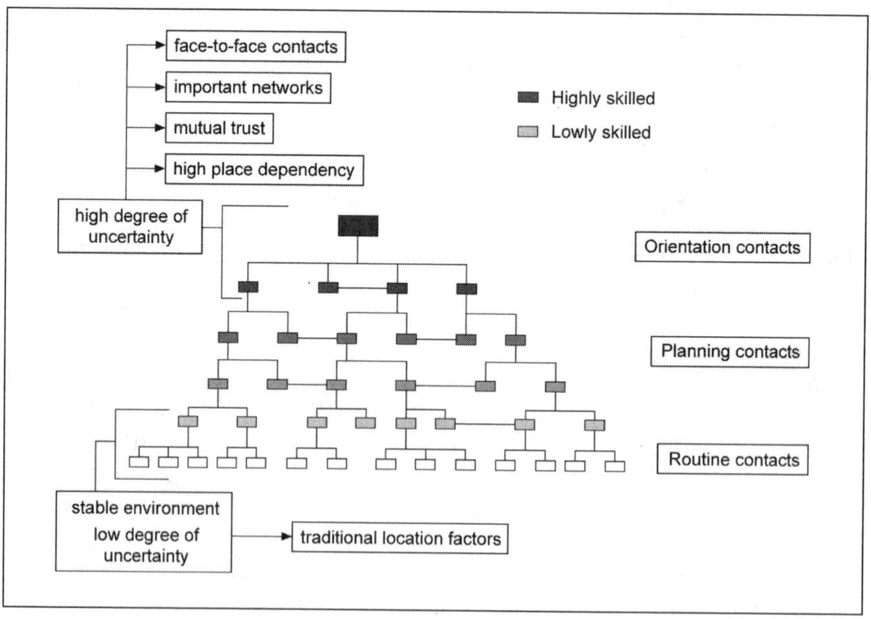

Figure 6.2: Architecture of organizations, degree of uncertainty, face-to-face contacts.

and planning will shift to the upper levels of the system's hierarchy, and consequently, the lower levels will predominantly keep routine activities and workplaces for the low-skilled (see Figure 6.2). In systems confronted with a complex task and a dynamic environment and dealing with constantly changing, unpredictable, one-time transactions, decentralization of competence and authority within the system is more effective (Meusburger 1998). For a more detailed discussion of various types of organizations, environments and means of coordination see Mintzberg (1979).

The second question is, where to locate scarce knowledge and skills in the spatial dimension. Low-skilled routine activities for which there are plans, rules and regulations call for very few face-to-face contacts with other organizations. They are less place dependent and can be situated in a great number of smaller cities or at the rural periphery, as long as the traditional location factors such as transportation costs, wage costs, or the location of consumer markets make it feasible. Only low-skilled routine activities in production and administration show a trend towards dispersion and decentralization.

Each social system has its hierarchy of decision making and information processing, its centre and its periphery. In the view of organization theory, the centre of a social system is the place where the highest authority of a social system is located and where the most important decisions are made. The centre is the nodal point of interaction and communication, from where the elements of a social or spatial system are ruled, coordinated and controlled (Strassoldo 1980). It collects and distributes resources, sets the rules, norms and standards for the members of the system. It controls, legitimates and accumulates knowledge and creates memories and identities. The centre is the place where important decision makers of different realms (politics, economy, science, media and culture) meet, communicate and interact. The main functional reason for the spatial concentration of knowledge and power and for the spatial clustering of high-ranking decision makers in a small number of centres is the dependency of the top management on high-level face-to-face contacts and on trustful relations with important decision makers of other organizations.

Theoretically, each centre of a social system could be located at a different place. The main question is why so many centres of social systems choose the same location. The harder it is for decisions to be governed by guidelines and plans, the greater the uncertainty about the consequences of an important decision, the future development and the correctness of the methods and objectives, and the higher the competition a social system has to face, the more necessary it is to have frequent and spontaneous face-to-face contacts with knowledgeable and well-informed decision makers and highly skilled specialists of other organizations. Top managers and other high-level decision makers have to observe innovative and successful competitors; they have to get access to crucial information as early as possible, and have to establish trustful relations to the leaders of other social systems.

The higher the professional status in an organization is, the greater the amount of external face-to-face contacts to decision makers of other organizations. Also the amount of contacts involving planning and orientation will rise accordingly, whereas the proportion of routine contacts and indirect contacts will decrease. The most important reassurance against the uncertainties of business life is prompt knowledge of technical innovations, of social changes, of possible future developments and of likely political changes. This kind of soft information is not presented in business reports, press conferences, or data bases but is first revealed by rumours, nonverbal communication in informal meetings, and in small pieces of information which can only be put together by experienced and attentive observers.

The need of the highest management levels for trustful face-to-face contacts and valuable networks involving leading decision makers in government, research, finance, consulting, public relations or industry can only be fulfilled by a few large centres, offering a large diversity and density of knowledgeable specialists and prominent decision makers. Another important reason for the concentration of power and knowledge lies in the role of trust. The role of an expert involves being trustworthy, accountable and credible. Credibility is the most important asset of an expert. Trust in the expertise and knowledge of professionals and experts is an indispensable prerequisite for coping with the increasing complexity of society, and it is the crucial strategy to deal with an uncertain, unpredictable, and uncontrollable future (Hardwig 1991; Jones 2001; Sztompka 1999, 2001).

Complex social systems would suffer severe damages, if they could not trust experts. Social relations between expert and layperson or between experts of different domains would not function without a certain level of trust. Trust is derived both from continuous face-to-face contacts and from belonging to certain networks and institutions. Trust cannot be established by telecommunication, for the acquisition of trust needs frequent and spontaneous face-to-face-contacts, rituals and symbolic acts.

Centres offer greater chances for new networks and creative associations of diverse knowledge bases; they permit earlier access to important information; and they contain more varied knowledge sources and more possibilities for high-level face-to-face contacts than smaller cities or peripheral locations. Observing the competitors and imitating successful decision makers from other systems is one of the most efficient methods of learning.

Social Attributes of Centres and Peripheries

There is no doubt that the functional approach of organization theory can explain central-peripheral disparities only to a certain extent, and that the symbolic meaning of places should not be underestimated. The symbolic meaning of a centre is as old as its functional role. According to Eliade (1969) and Turner (1979: 9–10), in many religions the centre was identical to a sacred place where the gods or ancestors dwelt or revealed themselves.

The sacred place was the point of communication between heaven and earth. The most important function of the shaman was his relation with the supernatural beings and with the ancestors. In Asian cultures the shaman was the one who conducted the dead person's soul to the underworld.

The translation of the name 'Babylon' is 'gate of the gods'. The connection with heaven and with the gods was initiated from this centre. The meaning of the sacred place as the gate to (divine) power was soon transferred to centres in general. Global cities are no longer the 'gates of the gods', but still the seats of the most powerful institutions and companies.

Apart from its functional role, a centre is associated with social attributes such as power, authority, dominance, prestige, control, attractiveness, distribution of resources and influence. Centrality is the spatial representation of authority. The centre of a social system is a point of reference and orientation. From the centre the rest of the social system is guided, coordinated or valued. The centre provides meaning and direction for the system's actions and it is the anchor that gives security (Turner 1979: 33). The centre gives the system its identity. Being in or near the centre has important psychological significance. Proximity to the centre (either genealogically, or via rank, or through close social relations to powerful decision-makers) offers security, priority of access to resources and privileges. It indicates importance, competence and trustworthiness.

The counterpart of a centre is the system's periphery. Centre and periphery display asymmetric power relations in space and resemble an uneven spatial distribution of power and knowledge. However, they are not regarded as a dichotomy, but rather as a gradient of power, influence and attractiveness. The periphery symbolizes dependency, marginality, alienation, weakness, backwardness and subordination (Bailly 1998). Innovations come late and investments tend to have a predominantly exploitative and colonial character, with limited positive spin-off effects (Strassoldo 1980). Most of the workplaces offered at the periphery are low-skilled and lowly paid routine activities, coordinated and controlled by external decision makers. The periphery constantly loses the largest part of its highly educated people to centres.

In social psychology, those at the periphery, at the edge or margin are called outsiders or *Randgruppen*; they have little influence, are disadvantaged and excluded from major political, cultural and economic processes, and enjoy less prestige. The terms centre and periphery also represent different action settings and *Erfahrungsräume* (experiental spaces), which produce different types of knowledge and different identities.

Conclusion

A centre is the spatial representation of authority. Power relations and communication structures of organizations have a spatial dimension. Centres and peripheries carry a number of social attributes and serve as points of reference and orientation. In centres, the divine or secular powers give meaning and direction to the system's actions; from the centre all else is orientated, guided, understood or valued; it is the anchor that gives security against danger from outside (Turner 1979: 19–20, 33).

A big advantage of organization theory and the concept of centre and periphery lies in the fact that both are applicable on any scale, from the small social system of a primary group, to the urban hierarchy of a nation-state and the system of world cities. Most small social systems are elements of larger and more complex social systems. Moving between scales, almost every centre can be another centre's periphery, and a complex system may be the element of another larger system.

On the micro scale of a small social system a position in or near the centre or periphery has important psychological significance for all subjects involved. Proximity to the centre offers security, priority of access to resources, and privileges, it indicates importance, competence and trustworthiness. On the meso scale a centre or a periphery offers a spatial context that influences the production of knowledge, the career chances and the activity patterns of people. Moving up to the macro scale of spatial systems, the multiplicity of spatial contexts in a centre or periphery increases. The high potential of valuable face-to-face contacts of a national capital or world city affects only a small part of the functional elites. World cities contain also an internal periphery, represented by low-skilled workers and marginal groups.

Chapter 7

On the Constitution of Space and the Construction of Places: Java's Magic Axis

Werner Hennings

Recently there have been laments about the 'disappearance of space' (Virilio 1996), about the 'implosion of space' as a result of the 'critical mass' being exceeded (Baudrillard 1995), horrific images of 'non-places' (Auge 1994) 'atopia' having spread (Willke 2001); space as a medium of perception has been denied any relevance as a meaningful dimension of its own (Luhmann 1997) and yet here we are almost simultaneously experiencing a renaissance of scientific discussion on the topic of space at conferences, in articles, textbooks, in theory.

Keeping in mind this reverse development in publications, some years ago I accepted a guest professorship at Gadjah Mada University in Yogyakarta, Java, Indonesia. The subject of my lectures was in a completely different context (sustainability), and my work so intense that I rarely left the campus except for an occasional visit to the Kraton, the Jalan Malioboro with its magnificent historical buildings, shopping arcades, and the bazaar as well as the world famous shrines Borobudur and Prambanan – the things you have to have seen as a stranger before going home again. For my last evening, it was recommended that I attend a typical Javanese event, probably because it was suddenly realized that as well as sustainability I also had anthropological interests. Accompanied by three Javanese colleagues, I therefore drove to the 'South Sea', as the Indian Ocean is also called here, to Parangtritis and the Queen of the South Hotel, a heavenly resort high up on the cliffs above the sea, where we had a delicious dinner in an open palatial building, about which our driver commented with great respect and in awe of the luxury: 'Only the Sultan can usually afford this.'

Having eaten, we looked down from the edge of the cliffs to the beach and a sea of lights made up of torches and candles in the midst of masses of people who had gathered there (as my colleague Dhesta Titi Raharjana explained, who at the time was conducting anthropological fieldwork there), in honour of Ratu Kidul, the 'Queen of the South'.

From here we then went down among the crowd to the 'night market' with its various locations and functions and the meditations for Ratu Kidul – a happening and contexts which Dhesta explained to me for hours, although in the face of their exoticism and esotericism I understood practically nothing.

It was only when looking at the map the next morning, because I wanted to see again where this curious hustle and bustle had taken place, that the geographer in me all of a sudden discovered spatial relationships between the village of Parangtritis on the Indian Ocean, the city of Yogyakarta and the Merapi volcano, relationships I found confirmed on the city map and which I wanted to investigate more closely on the occasion of my next guest professorship the following year.

The investigation of 'Java's magical axis' is therefore really a by-product of completely different scientific activities and has been substantially initiated and favoured by coincidences.

Spaces, Places, Scenes[1]

Landscapes and spaces do not exist simply as such; they have been constituted over the course of a more or less lengthy social process of construction (Löw 2001: 18). In this sense, they may also be considered as the spatial 'archive of a society', where its culture and experience are reflected. 'Landscapes (and spaces) are texts written and read by individuals and groups' (Knox and Marston 2001: 276), i.e. integrated in a social process.

The process of constitution of spatial texts has an active and a passive side: as in the linguistic process of communication a sender transmits linguistic signs to a receiver, thus actors position in places specific material and social goods (coding) containing a message which can be understood by others (decoding). The active part of the process of constitution, the 'spacing', consists of the arrangement, the creation of a spatial structure; the passive part, the 'achievement of synthesis', encompasses the interpretation and reception of the already existing spatial order and structure.

As a rule, 'spacing' is done by individuals and social groups with the necessary material and social resources (wealth, power) and capable of writing down the spatial order they have created in a social book of rules as well as of depositing and anchoring them in social institutions (see Meusburger in this volume). Spatial structures are therefore to be understood as social structures and they reflect the social hierarchy (rule). On the other hand, those social groups who do not have the material, social and cultural capital for 'spacing' are being socialized through the 'achievement of synthesis' via the spatial structures as well (Löw 2001: 226).

Excursus: Place/Space and Message/Code

The terms 'location' and 'space' used here seem to approximately represent the conceptual context which Hirsch (1995: 4) sees as placed between two poles in tabular and dichotomous form:

'place'	⟷	'space'
'foreground activity'	⟷	'background potentiality'
'inside'	⟷	'outside'
'image'	⟷	'representation'

In order to understand what is meant by these dichotomies with respect to the terms 'place' and 'space', I go back to the comparison used above concerning communication theory in general and linguistics in particular, since there the concepts and the system have been clearly and without contradiction defined and comprehensibly explained by means of de Saussure's *Cours de Linguistique Générale* and Chomsky's generative transformation grammar.

'Place' and 'space' are to be understood in the same way as in linguistics 'parole' and 'langue' (de Saussure) or 'performance' and 'competence' in Chomsky's terminology. While the terms 'parole/performance' comprise all the linguistic signs originating in the actual act of talking and writing, the terms 'langue/competence' are used for the relevant units and rules of combination on which the linguistic system is based and from which oral (acts of talking) and written texts are generated. Formulated according to communication theory, it is thus about the distinction between 'message' (parole/performance) on the one hand and 'code' (langue/competence) on the other hand. Only the 'message' is accessible for direct observation and empirical recording because it alone is articulated in the (according to Chomsky) 'surface structure' while the 'code' initially remains hidden to the non-native speaker due to its being embedded in the 'deep structure', and can only be analytically accessed with the help of numerous small 'messages' (actual texts). If I am not mistaken, Hirsch means by 'place' the very same thing that in linguistics is called 'parole/performance' (= 'foreground activity'), 'message' (= 'image'), and 'surface structure' (= 'inside'), while 'space' corresponds to 'langue/competence' (= 'background potentiality'), 'code' (= 'representation'), and 'deep structure' (= 'outside').

When applying linguistic theory to a theory of space, places as specific 'spatial texts' can only be generated (coded) and understood (decoded) on the basis of the knowledge of the respective spatial system with its relevant units and rules. This spatial system (*Raum* in Löw's terminology, 'space' in the terminology used by Hirsch) is always influenced by society and culture, i.e. spatial communication ('spacing' and 'achievement of synthesis') is primarily possible among those belonging to a sociocultural group with a common history, while outsiders first have to find an analytic access – as is the case for foreign languages – to the 'code', i.e. the respective socioculturally and specifically determined spatial system with its relevant units and rules in order to understand places, spatial orders and arrangements.

Through 'spacing' and 'achievement of synthesis' purely topographically marked places are in a first step socially identified as a 'relational order and arrangement of social goods and people in places' (Hirsch 1995: 224). In the social perception, these places, meaning the spatial structures and arrangements, develop specific 'symbolic effects', the 'atmospheric quality' of a space (ibid.: 229). Due to the atmospheric quality, people feel good or turned-off in places, i.e. in specific spatial orders, they feel at home or like strangers, and they react to the atmospheric quality of places with approval or rejection. That is why places – with the help of symbolic effects and atmospheric qualities – create attraction, identity and feelings of belonging (inclusion), yet they can also cause organized exclusions via the spatial order (ibid.: 228).

The result of the atmospheric quality of places can then be an 'emotional sense of place' (Knox and Marston 2001: 284; cf. also fieldwork reports in Feld and Basso 1996) right up to 'topophilia' (Löw 2001: 290), a strong bond between people and places, the sum of human feelings through the connecting of personal experience with the symbolic meanings of a specific place.

An emotional sense of place develops as the result of a specific process of the spatial perception which fixes on 'elements of cognitive images' (ibid.: 286) with the help of which information from the environment is filtered and selected because it determines imagination and memory.

Typical basic elements of cognitive images are:

- 'paths', i.e. the trajectories of encounters, e.g. roads;
- 'boundaries', i.e. barriers of delimitation, e.g. coasts, mountains, rivers, walls;
- 'districts', i.e. areas with characteristic features which people can enter and leave mentally and/or in reality, e.g. business districts, government districts, ethnically or culturally determined districts;
- 'knots', i.e. strategic points of orientation and encounter, e.g. crossings and squares; and finally
- 'landmarks', i.e. physical-geographical points of reference, e.g. specific forms of surfaces.

If I understand Strauss and Quinn (1997) correctly, there is a lot to be said for the fact that the perceptions I call 'cognitive images' correspond to what they define as 'cognitive schemas' or 'cultural schemas': 'not distinct things but rather collections of elements that work together to process information at a given time' (Strauss and Quinn 1997: 49). The term 'schema' is used in this sense 'to refer to generic knowledge of any sort, from parts to wholes, simple to complex, concrete to abstract' (ibid.). With the help of such 'schemas' we can 'reconstruct our memories of past events, determine the meanings we impart to ongoing experience, and give us expectations for the future' (ibid.).

According to Strauss and Quinn, on the basis of the 'connectionist theory' models can explain how 'schemas' generate meaning, meaning in the sense of mental states 'which are shaped by the learner's specific life experiences and are sensitive to activity in a particular context … often similar from person to person, context to context, and one period of time to another' (ibid.: 50). Such models imply

> many simple processing units that work like neurons … arranged in layers. Sensory neurons are activated by particular features of the world (e.g. in the context of space and place outstanding spatial features like paths, barriers, knots, landmarks, maybe additionally accompanied by features of color, smell,

> noise etc. [and particular social groups, W.H.]); motor neurons are connected to muscle and gland cells, sending the signals to lead to behavior and feelings; and interneurons lie between, combining signals from, in some cases, over a hundred thousand other neurons and learning to react to complex combinations of features. In each layer beyond the first, all a neuron does is receive excitatory or inhibitory signals from other neurons, combine them, and, if it is activated past its threshold, send excitatory or inhibitory signals on to other neurons (or muscle or gland cells). No single neuron knows much, but thousands of them working in parallel produce intelligent action. (Strauss and Quinn 1997: 51).

According to 'connectionist models', 'schemas' (or the 'cognitive images' here used as a basis) are typically learned in everyday life through repeated participation in all kinds of social interactions, which can also be embedded in a certain spatial context with specific features (i.e. typical basic elements such as roads, barriers, districts, crossings, piazzas and landmarks) and which produce 'a gradual strengthening of the weight of association among the units' (ibid.: 52).

On the basis of such an understanding of 'cognitive images' as comparable with 'cognitive schemas', we are better able to comprehend what happens in the process of 'achievement of synthesis' and 'spacing' and at the same time understand the motivation of the actors in the spacing process to change nature into landscape and to create space and spatial order from territory since with the help of symbolic charging the cognitive images can to a high degree be used to create and stabilize social power and rule.

It is above all the symbolic effects and the atmospheric qualities of places that motivate those who have the necessary material, social and cultural resources (wealth, power) to turn places into 'scenes', to establish spatial orders that tell 'stories' (narrative spaces) and in their symbolic effect convey an atmosphere of amazement, of fascination, and of cultural and social support, of belonging and identity while, at the same time, the existing power and rule relationships can be reproduced and legitimized.

The following is about actually investigating, in the case of a topographically clearly demarcated and defined place, 'which symbolic, emotional, and subjective meanings (this) specific spatial detail' has for the society and how the specific 'socio-cultural meaning' of this spatially relative symbolic construction originated and continues to reproduce itself (Werlen 2000: 348ff.). The spatial example is in Central Java (Indonesia), more precisely in the vicinity of approximately 30 km from the city of Yogyakarta.

'Desolate Places'

As far as terminology goes, we distinguish between *places*, *nonplaces* and *scenes*. Accordingly, we call *places* any historically grown and culturally determined location, whereas *nonplaces* are defined as 'transit spaces' which are characterized by speed, change and lack of attachment (Augé 1994). *Places* as well as *nonplaces* turn into *scenes* if they are staged as 'narrative spaces' with the help of 'special effects' and dramaturgic tricks. In this sense the two *places* considered here are at first visually 'desolate', at least for the non-Javanese viewer, anything but spectacular and far from being able to be considered attractive and 'atmospherically' charged.

First there is the village of Kaliurang, a mere 30 km northeast of Yogyakarta, on the southern slope of the Merapi volcano, 900 m above sea level, 2,000 m below the peak, at the dividing line between agriculture and tropical mountain forest. Mostly in the clouds or in mist, the place admits only little sunshine and gives the visitor the impression of being grey and cold. The lack of hospitability is reinforced by two additional adversities of nature. A little above the place the weather becomes unpredictable; the view can be close to zero within minutes if fog develops. The unpredictability of the weather is accompanied by the unpredictability of the volcano. Its crater issues a permanent cloud of smoke. Streams of lava, showers of ash, pillars of fire and avalanches of slag are a threat to any life in the vicinity.

The second place is called Parangtritis, situated less than 30 km southwest of Yogyakarta right on the Indian Ocean, with a wide flat beach of dirty grey sand. In the morning mist reinforced by the evaporation steadily raising off the sea, only the outlines of buildings beyond the beach can be seen, without definite contours, seemingly without structure, colourless, flat. On the beach and in the village nobody can be seen, and there are no fishing boats at sea either – it all looks lifeless. Before one even arrives at the place, large signs on the beach warn against swimming: as proof, large-format statistics are printed on posters showing how many people have drowned here in the past years. The inconspicuous, grey, cold, even life-threatening impression of the place only became relative once it came alive with many friendly Javanese and the hustle and bustle of numerous vendors with trays and other small-scale entrepreneurs.

'Desolate places', the European viewer would say, if not for the many small hotels in Kaliurang that testify to the masses of visitors to the place before the destructive eruption of the Merapi in 1994, or to the more than one million people who come to Parangtritis every year and so change the place into a *scene*. The *dramaturgy* of the places explains their attraction and their 'special effects' are activated simultaneously in distinct productions on two stages that show a drama, on the one hand of nature, and on the other, of Javanese culture.

Stagings of Nature

Seen from the point of view of the natural sciences, the two places form an integrated whole: namely a scene of the highest tectonic activities. From this perspective both places are focal points of a geotectonic weak zone: here, south of the Indonesian islands, the oceanic lithosphere of the Indo-Australian shelf slides beneath the Oceanic lithosphere of the Eurasian shelf and forms an active zone of subduction. In this process so-called island arcs are created – in this case the Indonesian islands from Sumatra to the Moluccas – whose core consists of a chain of highly active volcanoes, subduction volcanoes, for the most part explosive: the famous/infamous 'ring of fire'(Schwarzkopf and Schminke 2000: 4).

One of these explosive giants, the Merapi, dominates the area investigated here, in particular the village of Kaliurang. In the case of a major explosive eruption, the Merapi would be as destructive for the city of Yogyakarta as was Vesuvius 2,000 years ago for Pompeii. Due to the volcano and its activities the place becomes, with regard to geology, one of the most interesting *scenes* worldwide for volcanologists. Experts list the Merapi as 'one of the world's most active and most dangerous volcanoes' because it shows the most eruptions in historical times (about seventy since 1548) (Schwarzkopf and Schminke 2000: 10) and represents a permanent and deadly danger for the city and for the region.

Indirectly, the second place of the investigation is also influenced by the Merapi: Parangtritis is situated not only close to the subduction zone between the Indo-Australian and the Eurasian shelf but also in the estuary of the two rivers Progo and Opak, which collect the rainfalls west and east of the Merapi and transport them together with the sediments to their base of erosion in the Indian Ocean near Parangtritis. The sediments have not only created the fertile fluvial soils of the province of Yogyakarta but extend far beyond the coastline into the Indian Ocean where they have overlaid the reef protecting the coast with sand. The result of this process is a flat coast reaching for several kilometres out into the sea that is full of sandbanks and shallows which constantly changes their location and shape. The coast at Parangtritis thus presents itself as a bay where the waves of the Indian Ocean, unprotected by an offshore reef, sometimes rise to a surprising height of several metres and where being in the water becomes a deadly risk every day due to the incalculable swell and undertows. Should there be a seaquake in the subduction zone like the one in December 2004 off Sumatra, gigantic tsunamis would have unimpeded access all the way to Yogyakarta and would be an equal potential danger for the city as the explosive force of the Merapi volcano. The shelf tectonics potentially has the city and the 3.5 million people of the region firmly in its grip with its destructive force, from both sides, from above and from below, from the volcano and from the sea.

Stagings of Culture

Spatial Scenario

If Kaliurang and Parangtritis are already impressive scenes from a natural science point of view, from the perspective of the humanities they are a masterpiece of the Javanese art of staging, a perfect combination of mythology, religion, magic, architecture, spatial arrangement, and politics in the framework of topography, morphology, and geology provided by nature. The cultural construction of Javanese reality between volcano and sea spatially reveals itself as shown in Figure. 7.1.

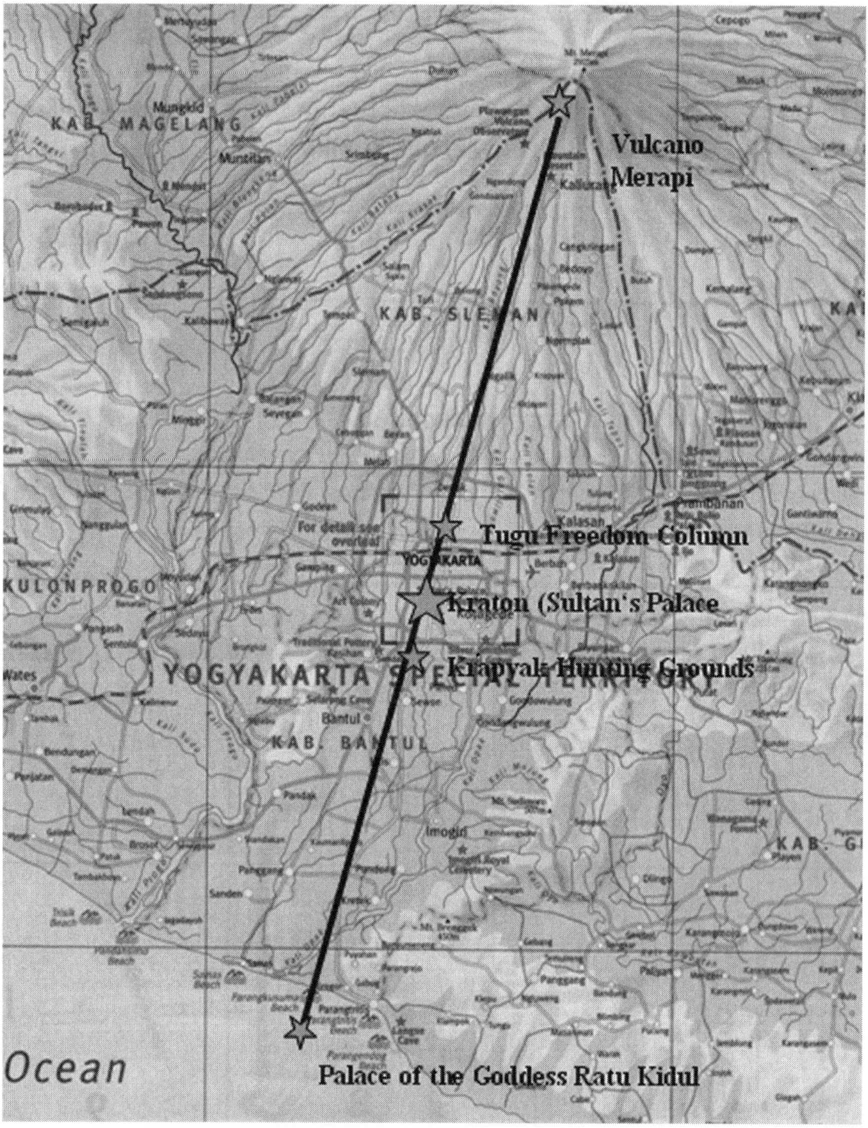

Figure 7.1: Total view of the magic axis with its five so-called 'cardinal points'.

If we connect the top of the Merapi in a straight line with the village quarter Parangkusumo of Parangtritis (almost 60 km), the city of Yogyakarta, more precisely: the Kraton (palace of the Sultan) lies exactly in the middle. On the same line of this imaginary axis are the most important institutions of Central Java (see Figure 7.2): bordering the Kraton to the north and the south are the two squares Alun-alun Utar and Alun-alun Selatan, with the ancient lofty and sacred banyan trees in their centres. Approximately 3 km south of the Kraton lies the former hunting reserve of the Sultan, the 'animal park'(Adukidang Krapyak); a good 3 km north of the Kraton stands the Tugu Monument, the column of independence. In between runs the central shopping mile of the region, Jalan Malioboro, where nearly all inner-city traffic is squeezed through and where, in addition to the traditional bazaar and the most modern shopping malls, the most important and powerful buildings of the colonial and postcolonial administration have also been placed (the Dutch fortress Vredenburg, the residence of the Dutch governor, later the first seat of the Indonesian government and today a government guesthouse), the central post office, the central station, the large mosque, and the Taman Sari, the country seat of the Sultan and his harem which because of its pools and baths was also known as the Water Palace. Finally, the route from the Kraton to the Merapi leads past the most renowned university in Indonesia beside the National University, the Gadjah Mada (white elephant) University. On the way from the Kraton to Parangtritis at Imogiri are the most important burial grounds of Javanese culture, established since the sixteenth century, the graves of the Sultans of the legendary Mataram Reign and the later Sultans of Yogyakarta and Surakarta (Solo). The magic axis is flanked by two first-rate historical cult places: in the northwest the Buddhist Borobudur temple, constructed in the shape of a pyramid during the eighth and ninth centuries and in the east the Hindu temple complex Prambanan, built in the ninth century. In this way, the magic axis between volcano and sea unites – as if strung onto a staff – the essential stations and institutions of Javanese history of the past 1,200 years. In its construction, the axis corresponds to spaces Foucault calls 'utopia', a 'placement which stands in an immediate ... analogy to the real space of a society. Perfectionization of the society ...' (Foucault 1990:10).

For the filtering, selection and lasting determination of the spatial information in the process of human perception, the 'spatial scenario' in Yogyakarta uses all the essential basic elements of cognitive images according to Knox and Marston:

- as clear 'landmarks' the Merapi volcano and the sea at Parangkusumo;
- as the axis between the two the 'path' of encounter particularly within Yogyakarta, the Jalan Malioboro;
- along this 'path' a series of 'knots,' sites of encounter and orientation: the Kraton as the traditional government district, the Jalan Malioboro as the business district, in between the colonial government district, the sacred districts at the Merapi and near the sea in Parangkusumo.

The spatial construction of the magic axis is not last a stroke of genius because the chain of various 'knots' is a connection and overlapping of profane and sacred places to which the people equally make pilgrimages in order to both demonstrate their faith and their devotion to the ruler.

In this 'territoriality' the individuals and social groups become anchored in places and cultures; here they are given a feeling of safety, confidence and wholeness, which on the other hand makes it easy for the rulers to have control over people and resources. To this day the Sultan of Yogyakarta is *the* charismatic figure and uncontested authority in Central Java, although he has more of a subordinate power position in the formal political hierarchy of modern Indonesia. In this respect the spatial structures seem to pave the way for and support social communication and rule.

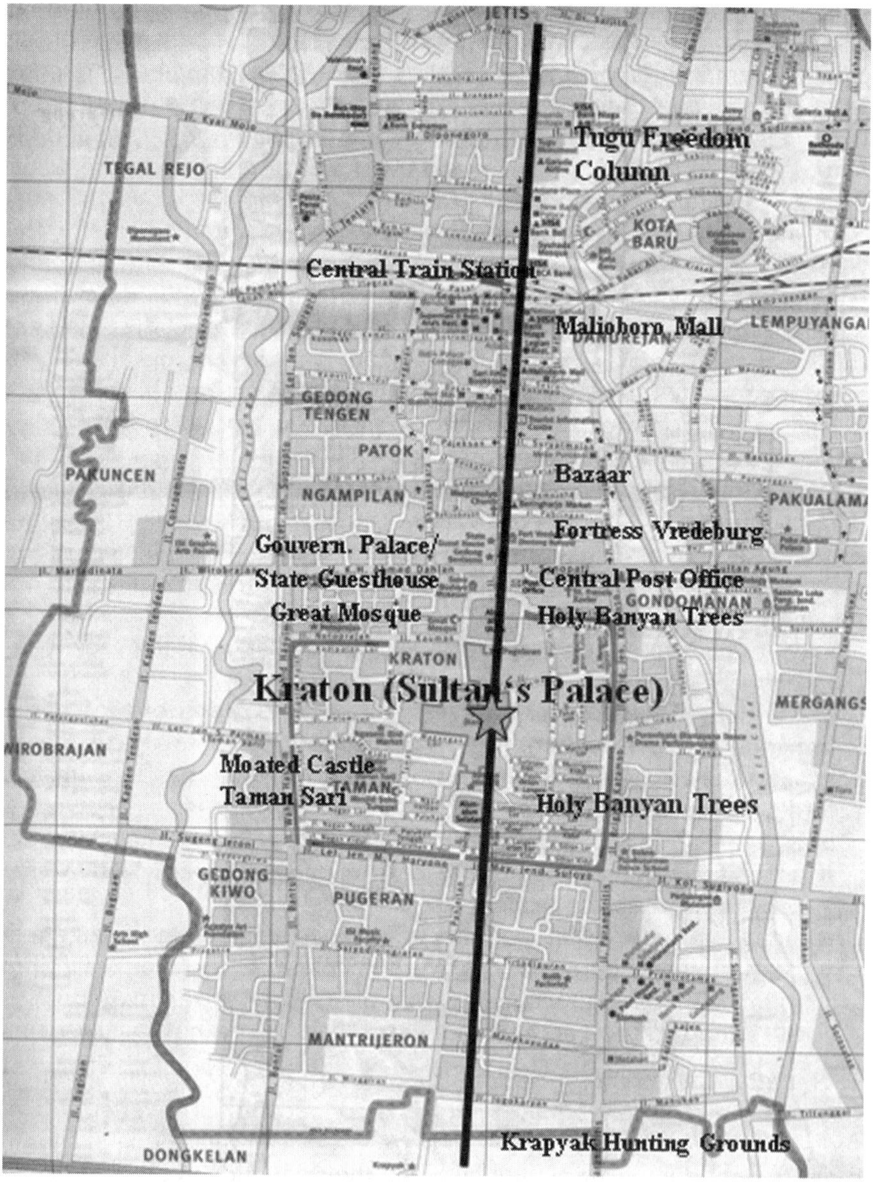

Figure 7.2: Town section of the magic axis with original magic points (Tugu, Kraton, both squares north and south of the Kraton with the holy Banyans) and the attempts at continuation of the following rulers (Dutch colonialists, Islam, infrastructure of modernity, institutions of the market).

Mythological Scenario

The end points (or starting points) of the magic axis itself reach back far into the prehistoric time, which today is only handed down to us as legend and mythology. To my knowledge, nothing is written down concerning the history of the axis so oral history remains the only source. To this day, the early Javanese myths have remained present and alive throughout all cultures and religions in Java, among the rulers as well as among the people. Animistic traditions and the various consecutive cultures of Java (Buddhism, Hinduism, Islam) have again and again had a mutual influence and in the course of this formed symbiotic links, a phenomenon Geertz calls 'syncretism' and to which Magnis-Suseno gives the label 'cooption', i.e. assumption and embedding. Schlehe gives a more precise definition: 'The overall realm of the respective modern time is integrated into a traditional understanding of reality as a new experiential element' (1998: 39).

Maybe what this result means for Javanese everyday life deserves to be emphasized again: despite all the tendencies of an increase of Islamic influences, 'traditional beliefs ... form a framework for dealing with ... the elements of modern time which determine life in reality' (Schlehe 1998: 39). There is a marked and strong Javanese identity beyond Islam and it is conserved in the spiritual traditions.

The writings of the *Babad Tanah Jawi* are the sources for the mythology of the scenes around Yogyakarta. In addition to Ancient and Middle Javanese myths of origin they also list those of the Sultans of Mataram. The essence is always to legitimatize a mythical man and a mythical woman as the founders of the respective dynasty as being descendants of one or several gods. The following account of the myth of origin of the Mataram reign is the *Babad Tanah Jawi* version as summed up by Schlehe (ibid.: 49f., 112f.) and supplemented by an oral version.

Meditating in magical places, the founder of the reign Panembahan Senopati Ingalaga (1575–1601) was searching for divine inspiration and power in order to strengthen his rule. His meditation was so deep and strong that the Merapi trembled and spat fire. The sea was boiling and seething. The commotion made Ratu Kidul, goddess and queen of the South Sea and ruler over all Javanese spirits, appear in front of Senopati. The two fell in love and spent three days and nights in Ratu Kidul's crystal palace in the Indian Ocean. In saying farewell, she promised him help and support for all situations and times and as a gift and promise of love handed him an egg he was to eat. But the mistrustful Sultan gave the egg to a servant who, having eaten it, at once turned into a demonic giant. As ruler or minister in the reign of the Merapi and with the name Kyai Sapu Jagad, the latter since then supervises the army of spirits and the state of nature in the Merapi and in addition is responsible for guarding the Mataram reign from the Merapi, in particular for protecting the city of Yogyakarta and the Kraton against streams of lava.

Both places have been given magic power, in Parangtritis the Ocean Palace of the goddess Ratu Kidul ('walls of gold and silver, the courtyard plastered with jewels, magnificent gardens', ibid.: 50) and in the Merapi the Fire Palace, seat of the demon Kyai Sapu Jagad, with its entrance guarded by evil, restless spirits and demons (Prasetyo 1998: 235). The cultural construction of the spatial reality between the Indian Ocean and Merapi can as a summary also be interpreted dichotomatically:

sea/mountain	divine/demonic
deep/high	good/evil
water/fire	female/male

Exactly in the centre of these oppositions, spatially as well as mythologically, stands the Kraton, seat of the Sultan, mediator between the divine and the demonic, between rulers and subordinates. The Sultan's permanent task is to find a balance between the oppositions in nature and those in society: Javanese culture as an expression of harmony, balance and conflict avoidance.

To do justice to this task, the Sultan formed an alliance with the metaphysical forces of nature: the demon in the volcano and the goddess in the sea. It is institutionalized that every Sultan of Yogyakarta is married to the goddess Ratu Kidul and the visible sign of this union are the three thrones standing in the Kraton: one for the Sultan, one for his mortal wife, and one for the goddess.

Places and Scenes of Magic Attraction

As described above, the places Parangtritis on the Indian Ocean and Kaliurang at the Merapi are, because of their physical-geographical features, of so little attraction that none of the numerous European, American or Australian tourists visiting Yogyakarta and the temples of Borobudur and Prambanan ever strays here. Yet the Javanese come in hoards: year after year more than a million of them.

The attraction of the superficially inconspicuous places must thus be looked for in specific Javanese characteristics: the *kejawen*, an ideology dominated by esoteric knowledge and a spiritual basic attitude.[2] The *kejawen* assumes a cosmos, 'where everything that exists forms a whole', where 'all the phenomena and all the people [take] their proper places ... in an ordered cosmos in harmonic balance.' The esoteric practices include, especially among the elite, the 'refined forms of meditation and asceticism', among the common people in contrast 'the simple belief in supernatural worlds and forces and in spirits': the goal of all the esoteric practices is the 'control of the emotions, ... enrapture and ... calming down of the person' – harmony in the 'correspondence between microcosm and macrocosm ...' (Schlehe 1998: 36f.).

The appearance of the visitors to the beach of, for instance, Parangtritis to a large extent accords with this ideal Javanese basic attitude (Figure. 7. 3). All activities are marked by tranquility, calm, modesty and steadiness: the majority of the people stand on the beach, most of them in small groups; many walk a few steps into the waves while fully clothed, sit or lie down, let the waves wash over them, look out to sea. People stroll, go for a ride in small horse-drawn carriages or on horseback, fly kites, play ball. Refreshments and snacks are consumed, preferably from small food stalls (*warung*) directly on the beach – a quiet, balanced, peaceful picture exuding contentment, joy of life and harmony and where loud and shrill sounds have no place.

Cult Stage: '*Labuhan*' Rituals and Meditations

Spatially separated and away from the normal streams of visitors, by the sea as well as on the Merapi, are the places for the 'pilgrims', visitors with a strictly ceremonial purpose and with explicitly spiritual activities. As signs of the sacredness and consecration of the place, in both locations are found sacrificial stones and small temples for the rituals.

Once a year the ceremonies reach a peak. The *labuhan* ritual is to guarantee the Sultan, his reign and his people protection, to calm the fury of nature (and of the spirits), and to bring wealth and happiness to the people of Java.

Labuhan is derived from Javanese *labuh* meaning to throw away, to drift. Thrown into the sea or into the volcano respectively on the day after the coronation of the current Sultan are specific offerings for Ratu Kidul and the demon Kyai Sapu Jagad. In addition to various consumer items (from batik cloths, silk, lace, belts, hair ribbons, powder, perfume and cigarettes to small bottles and money) and, above all, fingernail clippings and clippings

Figure 7.3: Javanese harmony in the neighbourhood of the Queen of the South.

from the hair of the Sultan as well as flowers from his household. Giving these objects to the sea and the volcano is the high point of ceremonial preparations lasting three days, which are started on the day of the sacrificial ritual with prayers at the *cepuri*, the sacred stones (Figure. 7.4), then continued with a procession to the beach or the top of the volcano, and concluded with meditations for several hours.

Scenes and Markets

It goes without saying that the places Parangtritis and (before it was closed off to tourists) also Kaliurang are highly charged due to their stagings, not only with regard to cult but also to economics. The region of Bantul in the province of Yogyakarta achieves by far its highest tax income right here. In 1999 more than 700 shops, restaurants and accommodation for visitors were counted, an increase of almost 100 per cent in 12 years.

In Parangkusumo clear spatial zoning exists. Business is conducted mainly during the night and, so to speak, illegally yet tolerated informally and on the move on the so-called night market, the *pasar malam*. A visitor passes the following 'business districts' on his way to the beach. In front of the entrance to the prayer site (*cepuri*) with the sacred stone are the stalls with blossoms and traditional beverages and a few scattered *warung*. The passage between the prayer site and the dunes, which lies completely in darkness, is above all the contact zone for prostitution, in part also the location for vendors with their trays. The adjacent stretch of dunes towards the sea is the spiritual business district lit by gas and petroleum lamps: the

Figure 7.4: Labuhan ceremony at the holy stones.

site where traditional medicine is sold and various kinds of gambling goes on. Beyond this zone, again not lit, is where prostitution actually takes place. On the beach itself it is possible to get a snack even at night. The small hotels and restaurants grouped around the prayer site play only a marginal role while the illegal shops in the dark zone are frequented the most; at least, they have the most visitors and the crush is worst (Figure 7.5).

Figure 7.5: Scene of the night market.

People in the Place in their Everyday Life: Local Actors

It has already been mentioned that there is a marked and strong Javanese identity beyond Islam which is conserved in spiritual traditions (see also Geertz 1960, 1990). How is this spiritual tradition expressed in the everyday life of the local actors?

Schlehe, who in her comprehensive and differentiating empirical work of 1998 looked into the 'spirit politics in Javanese everyday life', distinguishes between spatial and social aspects of spiritualism. Thus, for example, she e.g. describes as sacred sites 'everyday' magical-potent places as 'natural parts of the landscape' (1998: 165–75) in open spaces – not temple complexes of the ruling dynasties with their closed spaces but sea, beach, sand dunes, caves, hills, springs and rivers as 'topography of magical-mystical power' (ibid.: 165). As a second spatial-sacred category she lists 'graves and shrines' related to venerated spirit beings and ancient kings and situated decentrally (pp. 175–80). Her third category of spiritual-sacred spaces derives from the necessity of modern times to also offer the spiritual tradition as a 'touristic performance' in hotels, such as the regular 'Saturday Night Queen Party' in the Queen of the South Hotel in Parangtritis, or the room exclusively and permanently reserved for Ratu Kidul at the Samudra Beach Hotel in Pelabuhanrata (West Java), and even in Sanur on Bali at the Grand Bali Beach with an exclusive reservation of room no. 327 and bungalow no. 2401 for Ratu Kidul (pp. 180–83).

In everyday life, all these sacred places serve the pilgrims as resting place, meditation site and accommodation where, depending on their concern, they spend 'a few hours or one night, sometimes three, some of them seven, and a very few forty days' (p.165).

Geertz sees the daily spiritualism in the form of pilgrimage as an increasing example of the revival of a neotraditionalist 'Neo-Javanism': 'The very rapidly expanding pilgrimage pattern (exploding is perhaps not too strong a term)' (1990: 90). According to Schlehe, visiting the sacred places is a personal matter, 'based on individual wishes' (Schlehe 1998: 184), in order to receive, for example, through

> contact with the supernatural spiritual strength and welfare and to attain inner tranquility and peace. Some hope for a divine revelation or very generally for protection and blessings from Ratu Kidul ... There is a clear tendency to seek for help in case of actual profane problems ... [Much of it] relates to health, but mostly above all to livelihood. About half of the pilgrims in Parangkusumo are market women and female vendors. They are asking for success in business. But numerous women with small and larger businesses also come there. Younger women are sometimes looking for husbands, mothers pray for their children, be it for their health or for success in school exams, female singers pray to keep their voice, prostitutes look for customers and for protection against being arrested by the police. People looking for work and students worry about professional positions, civil servants, policemen, and members of the military hope for a career, candidates for the office of mayor want to turn the election in

> their favor. On the other hand, many people also wish for wealth coming to them in other ways than via their professions and business. Until it was prohibited in the early 1990s under pressure from the Islamic side, the government lottery was a central hope. A clue from the world of the spirits with regard to the winning number was what one wished for'. (ibid.: 187f)

With such deep and widespread roots in the spiritualism of the Javanese masses it is understandable that there is a series of almost institutionalized specialists who, if not as priests, still function as indispensable mediators to the world of the spirits: guardians of sacred sites (*juru kunci*), 'wise old people and paranormals (*orang paranormal*)' and 'possessed persons, spirit mediums, and saviors' (pp. 208–22).

The whole extent of the spread of animism and spiritualism in Java becomes manifest if one looks at the representation of the Ratu Kidul myths in the various forms of art: ceremonial and folklorized court dances (*bedhoyo* dances), gamelan orchestra concerts, modern painting and photography, legend and fairytale books not only for children (Cerita Rakyat, folk tales), press reports (in the column: 'Believe it or not'), comics, folk theatre (Kretopak, which is also broadcast by radio and television) and, finally, numerous films (production per year approximately 100). Java's magic axis – with its starting points Parangkusumo on the South Sea and Kaliurang at the Merapi volcano, as well as with the Kraton and the Jalan Malioboro in its middle – to this day holds extraordinary attraction for the majority of the Javanese because it symbolizes in their everyday life the desired individual and social harmony of the accordance of spatial microcosm and philosophical macrocosm: the balance between the natural and the social forces in the sense of the *kejawen*, with all phenomena and all people taking their proper places in a harmonious whole.

The people right in the place are thus living the spatial relationships. The spatial order *is* the mirror image of the social order: all the places anchored in it are familiar to the people and they are aware of what these represent symbolically. In my opinion, this awareness of the philosophical background of the magic axis is so clear and unambiguous that the Javanese neither talk nor write about it. As far as I know, there are no more than written allusions which, however, mean little to a stranger. Neither do local tourist brochures and travel guides mention the axis; here the tourist spots are listed only as individual objects removed from context. If one asks pointed questions, however, the people are in most cases able to name the cardinal places of the axis yet lack the knowledge of the philosophical contexts.

The Proximity and the Distance or Inclusion and Exclusion

With respect to system theory, Java's magic axis turns out to be an apt example and well suited to give a dimension of meaning to space as a medium of perception (contrary to Luhmann 1997: 30, fn. 24).

When discussing the question whether space should be included in a theory of society as its own dimension of meaning, Stichweh (2001) distinguishes two categories:

- 'spatial differences in the environment of the society which are given causal meaning for social systems as spatial differences' (ibid.: 191); and
- 'spatial differences which the society generates with the help of objects and sites, distance and proximity as guiding distinctions and hence as inner-societal distinctions' (ibid.: 193).

The spatial construction principle of Java's magic axis fulfils the conditions for both categories. First of all, the spatial differences in the environment of the society have to be mentioned – differences caused by nature without the actors of the society having established markers or placements; differences, however, which as such are causally decisive for elementary societal processes of constitution. Differences in this sense are at the base of Java's magic axis through the specific constellations of the shelf tectonics in this region of the world, since it is here precisely that the Indo-Australian shelf slides beneath the Eurasian shelf. In connection with the processes of subduction at the edges of the shelf, the Indonesian 'ring of fire' on the one hand contains the Merapi volcano, and on the other hand the subduction zone itself is a permanent source of seaquakes of the highest intensity. Now these two spatial differences in the environment of the society present not only a constant source of danger for society, they are at the same time the decisive factor for the location's establishing and developing an advanced culture which – with these features and with this prosperity (population density: 1,000 inhabitants per square kilometre in Yogyakarta Province) – was certainly only possible because the alluvial cone built up between the Merapi and the sea and the repeatedly occurring rain of ash from the Merapi created probably the most fertile soil in the world. These are spatial differences in the environment of the society with a causal significance for the social systems: on the one hand, causal because decisive for the determination of the site and fundamental for the economic and social wealth, but, on the other hand, also causal as a permanent existential threat to the social system as a whole.

On the basis of these literally existentially significant spatial differences in the environment of the society, the society itself then constructed spatial differences and thereby generated guiding distinctions and society-immanent differences with the help of the spatial differences of proximity and distance. The proximity to, more precisely, the complete congruence of individual topographic points or markings with the placement of specific material and social goods in the space relating to nature and society creates a straight line or axis with topographically and socially precisely defined starting or end points and a just as clearly defined middle. The spatial order of the individual placements on this axis can be read as a reflection of the social structures: Java's magic axis 'tells' about the origin, the descent, the power and rule relations, and the principles of Javanese society and culture through the guiding distinctions of *distance* and *proximity*, about who and what belongs to this society and culture. Via the symbolic effect of the spatial order, atmospheric perceptions are generated which reproduce, perpetuate and legitimize not only the existing power and rule relationships but in the end Javanese culture as a whole.

Although the increasing integration of Indonesia into the world market has led to a 'Westernization' of society, to be clearly seen in the vertical and horizontal mobility, urbanization, tourism, higher standard of living, consumerism, and the development of a mass culture, on the other hand experts of the situation, among them also Geertz (1990) emphasize, 'that we can continue speaking more about a Javanisation of Indonesia which is based on the culture of the two Middle Javanese principalities of Yogyakarta and Surakarta' (Schlehe 1998: 38). The same is true for the increasing attraction of Islam in Indonesia and worldwide. 'In the religious sphere indigenous animism and mysticism remain the most significant counterweights to the strength of Islam in Indonesia' (Stange 1986: 76f.).

Summary

Thus on Java's magic axis, in a process of the spacing lasting for centuries, various social forces and groups have established placements of material and social goods and thus constituted a space which in its 'relational order' has linked places with specific 'symbolic effects' that in their 'atmospheric perception' condense into a 'narrative space'; a space which recites the 'great tale' of Java in the language of geography.

The spatial order of the magic axis is seen by all the people in Java as something that represents them symbolically and that radiates for them atmospherically – as a unique spatial and social construction of places which magically attract them, where they feel at home, develop a feeling of cultural belonging, and are able to proudly demarcate themselves against strangers. In the reception and interpretation (achievement of synthesis) of the spatial order of the magic axis of Java, the Javanese partake of routines, i.e. in their regular social practices of their everyday basic existential functions – all present in condensed form in the context of the axis right in the place where people live, work, buy, are educated and part of the traffic in masses – of the great experience of a stable development of identity: to be like all other Javanese and at the same time like no other society and culture: unique.

Java's magic axis is therefore a prime example for what Augé calls *place*:

> Places have at least three characteristics in common. They see themselves as identical, relational, and historic. The outlay …, the rules of residency, the zones …, the altars, the public squares, the division of the territory always correspond to a totality of possibilities …, whose content is connotated spatially as well as socially, … is constitutive for the individual identity. (1994:. 64f.)

It is a place where all of society is spatially concentrated and structurally connected, and the 'symbolic, emotional, and subjective meaning of the culture of [this] specific spatial section of the world' (Werlen 2000: 348) unfolds.

The space into which the axis is fitted is, on a reduced scale, a cosmos of itself, clearly demarcated by natural features and relatively sheltered by the surrounding mountains and the coast in the south (without ports).

Java's magic axis thus encompasses a spatial ensemble, an entity of physical-geographical forces with an ever-present threat (tectonics of plates, volcanism, seaquakes), on the one hand, and the cultural answer to this threat (*kejawen*, control, harmony), on the other. In its construction the axis therefore corresponds to a space as 'utopia' – to a 'placement that entertains with the real space of the society a relationship of immediate analogy: 'Perfectionization of the society' (Foucault 1990: 10).

At the beginning and at the end of the axis (volcano and sea: in the volcano lithosphere is produced and in the sea it disappears again) stand the *forces of nature* which justify and legitimize everything: the forces of the gods and the demons which at the same time symbolically stand for the *differences and the dangers* of uncontrolled and intemperate nature (water/fire; inscrutably deep/high; good/evil; female/male; divine/demonic). Precisely in the centre of these primeval forces (spatially) is the Kraton, the palace of the Sultan whose (social and cultural) task and function it is to work towards a balance between the otherwise uncontrolled forces of nature and of society. As mediator between the natural primeval powers, between gods and demons, and between rulers and subordinates, the Kraton is the spatial centre and the Sultan the social embodiment of Javanese culture: the place and incarnation of *harmony, balance* and conflict avoidance. The ritual expression of the strategy of conflict avoidance is the regularly conducted *labuhan* ceremonies at the Merapi and in Parangkusumo by the sea with the sacrificial offerings of the Sultan which are rendered to the volcano and the sea.

According to information from the Kraton there are five cardinal placements ('condiciones sine quibus non') on the magic axis, from north to south in sequence: Gunung Merapi (place of the demons), Tugu (place of independence), Kraton (place of harmony and cultural centre), Pangung Krapyak (place of hunting), and Segara Kidul (place of the goddess).

In addition, further placements were made on the magic axis, by other social powers and other social systems, with the aim of finding a connection, to be included in the existing social system, and to be seen as belonging to it: the *residence* of the Dutch governor and the fortress Vredeburg situated opposite, legitimization of foreign rule in the context of traditional structures of rule; the great mosque on the Alun-alun Utara, the spatial and symbolic attempt by Islam to find a religious and cultural connection to pre-Islamic Java; the buildings of the central post office and of the central station, both the decisive guiding technologies of the modern infrastructural systems at the end of the nineteenth and the beginning of the twentieth century; as well as the Jalan Malioboro, the central traffic axis of motorized individual traffic, the leading infrastructure technology of today; the traditional bazaar and the modern shopping malls, inclusion and exclusion of the market forces and the economy; and finally in the north of the city and lying to the east of the axis the university campus, thus also spatially and socially integrating science into the traditional Javanese social order.

Off the magic axis but central in the space between the Merapi and the sea are the places of the religions that came between animism and Islam, the temples Borobudur (Buddhist) and Prambanan (Hindu).

Their significance in the spatial and mystic order cannot be deciphered for the time being and has to be left to future research.

Notes

1. For the following conceptual definitions and theoretical deliberations I essentially refer to the publication by Martina Löw (2000) on 'spatial sociology', as well as to the chapter, 'Interpretations of Place and Space', of the textbook by Paul Knox and Sallie Marston, *Places and Regions in Global Context – Human Geography* (1998).
2. The literature about the *kejawen* is extremely extensive. Repeatedly cited are: Geertz (1990), Weiss (1977), Mulder (1992), Magnis-Suseno (1989), Koentjaraningrat (1985), and Keeler (1987). The account given here follows Schlehe (1998).

Chapter 8

Elementary Methodological Tools for a Recursive Approach to Human-Environmental Relations

Katja Neves-Graça

1 Introduction

Students often ask whether anthropologists who investigate human-environmental issues must develop a distinct methodological approach and, if so, how might this be done? Answers to methodological questions are often simultaneously simple and complex, and this one is no exception. The methods we commonly use within our discipline are highly pertinent to environmental anthropologists. Generally, we rely on classic qualitative anthropological tools to tackle environmental phenomena. Just like any other anthropologist, we do so within the specific ethnographic settings of our research project, our research goals, and in relation to the theoretical framework that informs the investigation. In turn, as with any field of anthropological investigation, a successful account of human-environmental relations requires that we fully understand the specificities of this subject of inquiry, which methods might best serve the unique purposes of the research project, and how reliance on these tools relates to the goals that guide our efforts.

During the past three decades one of the most important developments in environmental anthropology has been a paradigmatic shift in how we conceptualize our subject of inquiry. This has changed the way we implement the research methodologies we have at our disposal. While individually we may identify with specific theoretical scaffoldings such as constructivism, phenomenology, Marxism, postmodernism or critical realism, an increasing number of environmental anthropologists seem to operate within a paradigm that refuses to allow research tools to dissect environmental phenomena into artificially discrete units (see Brosius 1999; Kottak 1999; Little 1999 for relatively recent reviews of the literature; see also West 2005 for a discussion of more recent developments). Notwithstanding theoretical preferences, contemporary studies explore, for example, the mutuality of human-environmental phenomena (e.g. Gunderson and Holling 2002), the inextricability of environmental perception, practice and cultural knowledge (e.g. Ingold 2000), the interaction of emotion and cognition in environmental knowledge and practice (e.g. Milton 2002), and the study of environmental global connections, governance and subjectivity (e.g. Carrier 2004; Tsing 2005; Agrawal 2005).

Within this paradigmatic setting there has been a renewed interest in accounting for human-environmental phenomena from a holistic ecological perspective (see for example Berkes 1999; Biersack 1999). Not surprisingly then, earlier contributions by Gregory Bateson have become extremely relevant (Bateson 1958, 1972, 1978, 1991; Bateson and Bateson 1987). In effect, Gregory Bateson was one of the first – and rare – anthropologists to fully explore the epistemological, ontological and theoretical implications of a holistic understanding of human-ecological phenomena (see also Rappaport 1979). Unfortunately, the recognition of Bateson's major contributions has been a slow process (Nuckolls 1995; Harries-Jones 2005). So also has been the 'transduction' of his epistemological insight into anthropological research methodology. For this reason, the present chapter constitutes but an inchoate and diffident step in this direction.

In this necessarily brief chapter – the topic in itself could easily be the theme of an entire book – I will provide a short overview of how one can use some of the most conventional anthropological methods within the framework of a holistic epistemology premised on notions of relation and holism. The intention here is not to provide an overview of the proliferous anthropological literature on environmental issues (see Brosius 1999; Kottak 1999; Little 1999; West 2005), nor is it to pursue an extensive analysis of all existing anthropological methods. Rather, relying on examples from my own field research in the Azores-Portugal (1998–2003), I will sketch a small practical tool-kit for those who intend to begin to investigate for the first time human-environmental relations from a recursive perspective.

This is meant as an uncomplicated starting point, which should subsequently be complemented with additional readings about the various theoretical approaches and methodological strategies that anthropologists, environmental or not, have developed since the inception of the discipline. You should not be misled into thinking that existing alternative and/or complementary methodological approaches to human-environmental relations are less relevant merely because reviewing them all falls beyond the scope of this paper.

2 From Dualisms Overshadowing Holism to Batesonian Epistemology

I begin this section by discussing some of the ways in which dualistic theoretical premises and research goals can obscure the understanding of human-environmental issues – which, as we will see below, Bateson sought to overcome with a new epistemology of holism. These reflections guide the development of the methodological approach I propose in section 3. As stated in the introduction to this chapter, the present summary is by no means exhaustive, nor is it representative of the numerous anthropological studies that have focused on human-environmental relations. Rather, my intention is to present a simplified substantive example of how anthropologists reflexively develop their methodological tools. I hope that by understanding this process clearly, students will be in a better position to envision their own research strategies within the specific context in which they may find themselves.

Although it is quite apparent that dualistic epistemology is scientifically reductionist and politically dubious, this acknowledgement raises some problems that environmental anthropologists are still grappling with. In effect, it seems that it is easier to identify and critique existing dichotomies and their relations to power/knowledge matrixes, than it is to overcome them by means of correcting our methodological schemes and/or by engaging in more reflexive forms of writing. I will address this contention in what follows by pointing to three essential dualisms that we still try hard to avoid when investigating human-environmental relations. As you will see, these selected examples of dichotomous thinking in environmental anthropology are: culture versus nature; scientific versus traditional knowledge and the implicit contrast between global discourses versus local practices; objective history versus popular beliefs.

Environmental anthropologists can no longer accept views that reduce ecosystems and environments to culturally constructed categories which presume the external existence of an independent 'Nature'. One of the main premises of a Batesonian-derived approach is that the ecological environment of any given sociocultural group includes both cultural and biophysical elements (Rappaport 1979; Harries-Jones 1995; Berleant 1997; Berkes 1999; Rappaport 1999), and that the reverse is also true: a culture is also constituted by both cultural and biophysical elements. The view I follow is that cultural practices are, over time, mutually reciprocal parts of a wider phenomenon – environmental activity – so that culture transforms the environment and the environment transforms cultural practices (what the above authors would consider as 'co-evolutionary adaptation'). It is in this sense that human-environmental relations can be described as recursive.

It is true, nonetheless, that accounting for the reciprocity that characterizes this relation has been a major tension within environmental anthropology. The study of how it is that biophysical elements become cultural, and vice versa, was pursued dominantly on the basis of a Cartesian dualism that conceptually separated culture as mental schemata from nature as raw experience. The history of environmental anthropology is marked by a favouring of either ideational culture or experiential nature to explain human-environmental relationships. This is often deterministic. Within this framework, the former is currently privileged as the locus of enquiry and analysis, such that accounts of knowledge obtained through the senses by means of practical engagement with environments have been obscured during the last three decades. The goal then is to reveal the specificities of knowledge and forms of knowing which co-develop with the ecosystems of which societies and cultures are part, and which societies and cultures continuously transform. We will see in section 3 that amongst the many methods that we have at our disposal, life histories and story telling can be used towards understanding the mutual causality of biophysical and cultural elements through time.

For environmental anthropologists it is certainly important to explore relationships amongst globally dominant discourses about the environment and ecology, and locally developed understandings and practices. One way to do this is by scrutinizing different groups of people attempting to define and implement alternative 'ecological' projects. In this context one notices that often international discourses on the environment follow a type of scientism that removes ecosystems from their contextual uniqueness and abstracts them into universal, context-removed models. On the other hand, our 'own' informants describe their 'native' knowledge as inextricably local, based solely on direct environmental experience and on local oral histories.

Nonetheless, one should also be warned against assuming that there is a clear difference between locally developed (indigenous) knowledge of the environment versus scientific accounts and global economic practices. I have shown elsewhere that even though such clear distinctions might be a reflection of what our informants choose to tell us, different notions of ecology often come to exist and are linked to one another through communicative interactions. Often, those intervening from what seem to be radically different perspectives and interests learn from one another, which further blurs the distinctions between 'types' of knowledge and discursive settings (Neves-Graça 2004, 2006).

Moreover, if we consider that ecological systems as well as cultural understandings are simultaneously space specific and globally connected, accounting for the articulation between global environmental projects and their local manifestation/implementation becomes crucial. Even though this critique has already been made within the literature, the problem remains of finding ways to productively link scientific globalized knowledge with local specific knowledge such that neither one is privileged over the other (see Berkes and Folke 2000).[1] This, I suggest amongst other possible strategies, can be pursued by means of an adaptation of Foucauldian-based scrutiny of the relations that take between texts/discourses and contexts as well as between the practices of differently situated social agents who rely on distinct knowledge reference points.

One of the basic tenets of the Batesonian-inspired framework (see the references mentioned above), is that human-environmental relations are characterized by mutual causality – they co-construct one another. In this way these relations tend to form an interconnected pattern that unfolds through time between what is normally called the 'human' and the 'environmental'. This means that human-environmental relations are recursively linked (see Harries-Jones 1995 for a thorough philosophical analysis of Gregory Bateson's 'recursive vision', and Bateson 2005 [originally printed in 1972, 1991] for the contextualization and explanation of some Batesonian concepts by Gregory Bateson's oldest daughter, Catherine Bateson), such that knowledge, forms of knowing, and experiences of past human-environmental practices are constantly being reembedded in new contexts that are themselves simultaneously sociocultural, historical and ecological. Hence, the ontological condition of 'human-environmental relations' is irreducibly interactive and dynamic. It is marked by great temporal and spatial complexity.

Having just said that there is no clear a priori existential or phenomenological boundary between the 'human' and the 'environmental', however, I must note that distinct cultures at specific historical junctures do in fact assign varying degrees of co-dependence, autonomy, or independence between the terms of the relation.[2] So do we – researchers – particularly when we turn them into heuristic devices. Heuristic devices in themselves are not problematic. I think that without them it would hardly be possible to conduct research and/or engage in ethnographic writing. The problem arises when the anthropologist confuses the heuristic for that which s/he is studying or representing – a wrongful procedure we now agree to call 'objectification'.[3]

In order to avoid this type of error, methodological approaches to human-environmental themes should seek to unveil the mutual causality that takes place among the co-developing systems that form the relation I am heuristically labeling human-environmental. The goal is to avoid falsely establishing a dichotomy between them by presenting the 'human' and the 'environmental' as if they constituted structural-functional essences. It goes without saying that another implicit objective is to avoid the less common but equally pernicious blunder of coalescing the 'human' into the 'environmental' – or vice versa.

In this context, the holistic epistemology envisioned by Gregory Bateson can inform the development of an effective methodological approach. It allows us to overcome the dangers of falling into a Cartesian dualism that conceptually and substantively separates culture as mental schemata from nature as raw sensual experience. A holistic epistemology brings forth the reciprocity of relationships between social parts and ecological wholes – as in the narrow example provided in this chapter, for instance, by bringing together the anthropological study of environmental dwelling or practice, perceptions of the environment, and of processes of sociocultural construction of 'nature' (for a wider encompassing example of these applications see Neves-Graça 2002).

Bateson's contribution was mostly conceptual since he did not develop any ready-to-use theory for purposes of anthropological analysis. For this reason, his ideas can easily be distorted or applied erroneously when directly used in models dealing with the cultural construction of 'nature'. This clarification is important. As Rappaport once stated, while theories and models explain phenomena and 'how things work', conceptual frameworks 'help to think about phenomena, to order material, revealing patterns' (Rappaport 1999: 256).

Bateson was wary of anthropological accounts that reduced the complexity of multilevel phenomena to a single level of explanation where, for example, the metaphor of social power became the single explanatory factor for a given cultural configuration. This he labeled a form of upward reductionism and one of the great myths of social science. Bateson was equally suspicious of anthropological categories that created clear borders across several ethnographic units of analysis, *ab initio*, such as between History, life history, environmental practice, perception, or discourse. Alternatively, Bateson saw these phenomena as mutually constitutive and thus not separable into distinct categories, which is to say that they all affect one another time after time (i.e., recursively).

Nevertheless, it is sometimes difficult to escape the deeply ingrained habit in our society of treating history as a self-evident linear sequence of events that unfolds in a progressive manner such that the 'past' determines the present (and the future) in clearly predictable ways. The problems of such assumptions are well known, though certainly worth recapitulating so that we can reflexively correct them. First, this understanding of history objectifies the past as de facto truth which refuses to take into account the 'histories of the people without history' (Wolf 1982) or, in other words, that most written history reflects the viewpoints of particular elites. Secondly, the objectification of 'history' fails to acknowledge that people relate to 'past' events from 'present' contexts, which means that they necessarily reinterpret these events subjectively in terms of their current social-cultural positionality. Thirdly, any essentialist view of history becomes oblivious to the fact that differently situated social agents often struggle to define the meaning and veracity of historical events, and that they do so as a strategy to make sense of their contemporary predicaments as well as to legitimize political, cultural and economic goals. In section 3, I suggest that archival research can be used to reveal the processes by which the past is reembedded in different historical contexts. This constitutes one possible method to begin revealing the temporal complexities of human-environmental dynamics.

3 Sampling Research Methods

In this section I present three short examples of how environmental anthropologists can adopt some of the most commonly used anthropological methods in order to avoid the pitfalls identified above. I subscribe to the contention that the main problems of modernist anthropology reside in the epistemological groundings of our research methods and in the political implications of our strategies of representation – not in the research tools per se. Therefore, any of the methods presented in this section can be used successfully towards overcoming the research challenges and dangers discussed so far (Delamont 2003; Law 2004; Steinmetz 2005). For instance, by focusing on life histories I bring forth the mutual causality of biophysical and cultural elements through time. This allows me to avoid archaic nature-culture dichotomies. I could equally have chosen to demonstrate that life histories and story telling reveal the constant reembedding of the past within new contexts, thus avoiding linear or progressive accounts of history. In fact, I have applied the various methodological approaches discussed below to a multitude of related analytical purposes (Neves-Graça 2002). It is needless to remind you that we have many more methodological tools at our disposal, and that their efficacy depends on how we adapt them to specific research settings and analytical objectives (see for example Denzin and Lincon 1998; Jorgensen 1998; Schensul, Schensul and LeCompte 1999; Crumley 2001; May 2001; DeWalt and DeWalt 2002).

Life Histories and Story Telling

Anthropologists have for long relied on collecting verbal accounts from research 'informants' in order to reconstruct their societies' pasts. Folklorists and early cultural anthropologists first used this method to catalogue fast-disappearing Native American cultures, as well as to build a written record of these people's rituals, artefacts and traditions. These accounts were based on recollections of lived events, on knowledge that was transmitted by those who had experienced them first hand, or on myths of origin. For the most part, however, it was assumed that because most of these societies had no written records, they had no history. We must understand that this prejudice was typical of a positivist framework according to which supposedly objective historical truths can only be extracted out of documentary sources, and even then only with proper scientific methods developed by historians.

These assumptions were critically scrutinized during the postmodern turn in the social sciences. As a consequence, they are no longer acceptable (see for example Said 1979; Fabian 1983; Clifford and Marcus 1986; Marcus and Fischer 1986). First, the post-modern critique revealed that the social sciences – anthropology in particular – had systematically denied coevality (Fabian 1983) to their subjects of inquiry by presenting 'them' as if 'they' were suspended in a perpetual present and within clearly bounded spaces. Insofar as anthropologists tended to conflate history with 'tradition', which in turn was construed as an unchanging monolith, this critique extended to most orthodox anthropological research endeavours that aimed to explore people's pasts. Secondly, these scholars showed that representing our research participants as a temporally suspended 'primitive Other' not only promoted the erroneous impression that these people had no agency, but also legitimized numerous forms of colonial rule and violence. Finally, these authors demonstrated that there are no objective fundamental historical truths. Instead, there is an interaction between dominant historical discourses promoted by elites and the alternative historical views and realities of those who live at the margins of power (see for example Fabian 1990; Dirks 1992; Keck 1998; Wassmann 1998).

It is from within this context that anthropologists developed an interest in collecting life histories from research participants. The use of oral histories as a reflexive research tool was first developed and used by feminist and postcolonial researchers (e.g. Crapanzano 1984; Behar 1990; Linde 1993; Peacock and Holland 1993; Ochs and Capps 1996). Their objective was to reveal the alternative histories of women and colonized peoples that remained at the margins of social-scientific endeavour and of official histories. These scholars meant not only to give voice to subaltern and abject subjects but also, in so doing, to dismantle the notion that there exist fundamental historic truths that can be represented by means of unquestionable objectivity. It soon became evident that the silences of marginal people could speak as loudly as did the official textual histories of literate elites. Searching within the fissures of these margins scholars began to unveil entire universes of parallel histories (e.g. Berreman 1981; Guha 1982; Spivak 1988; Asad 1991; Karkelar 1991; Comaroff and Comaroff 1992; Butalia 1998). Thus feminist and postcolonial scholarship set the grounds for rethinking the presumed objectivity and factual basis of History. Instead, it became evident that all histories reflect their authors' positionalities as well as the subjectivity of those who interpret them.

In the aftermath of these contributions, we should be warned not to reify 'life histories' under positivist auspices. We should also keep in mind that, just like official elite-produced histories, these accounts reveal but glimpses of a person's life and these too are produced at particular sociohistorical, political and emotional junctures. The life histories that our research collaborators share with us are not all-encompassing unicursal biographies. They are fragmented narratives that are put together in specific settings (see Leavitt in this volume; Markahm 2005). Often, life histories constitute highly elaborate forms of self-construction and/or self-presentation (Ochs and Capps 1996), or become discourses that legitimize a person's predicament and actions. Sometimes life histories rely on fiction or metaphor to convey 'truths' (Gilgun 2004), whereas on other occasions they entail hidden knowledge and/or habitual practices that the narrator may not be aware of.[4] As many have argued, we must resist the danger of romanticizing and idealizing life histories as the voicing of authentic, stable, agent-laden, self-conscious and uniform selves (see Carry 1999).

Life histories are highly reflexive narratives that are part and parcel of the contingencies of a person's social and cultural life. It is the deep reflexive nature of life histories vis à vis the complex nonlinear dynamics of people's lives that I find most useful in the context of environmental anthropology. Although, within the postmodern predicament I allude to above, I have relied on life histories as a means to reveal the hidden histories of my informants, in this section I will focus on their potential to bring forth the recursiveness of human-environmental relations. To be sure, collecting life histories we can study people's reflections on the relation between their biographies and the sociocultural, environmental processes of which they are part.

There are various techniques for doing this. These techniques range from asking the informant to freely construct the story of their lives (or that of their ancestors, as is the case below), to asking people to provide details about their lives at particular historical junctures which the anthropologist has identified as being relevant to his/her project, to various forms of directed interviewing. It even includes taking notes from participant observation as people casually engage in such narration (see the following sources for more detailed information on relevant methodological techniques: Agar 1980; Denzin 1989; Van Maanen 1995; Bernard 2000). In my experience, combining these various possible approaches yields the most revealing results as it allows the researcher to analyze different narration strategies in relation to different settings. More importantly, in terms of the topic of this article, it allows the researcher to observe how the narrator herself conceives of her relations with the environment within changing social, cultural and ecological environments. An example from my fieldwork experience amongst former whale hunters in the Azores illustrates this argument.

In Lajes do Pico, in the Azores, men hunted whales for about 100 years, between the late 1800s and 1983. Their life histories are fascinating. Obviously I could not interview members of the first generations of whalers about their life histories since I began fieldwork in 1998. Still, following the theoretical analytical logic presented so far in this subsection, I was able to re-'construct' the invisible and marginal life histories of these hunters. This entailed collecting letters they wrote to family and friends, gathering stories they transmitted to their descendants, and on rare fortunate occasions listening to old tape-recorded interviews. As with any other historical account, the views I present here are necessarily incomplete and permeated with my own subjective understandings of what it might have been like to be a whaler during the early stages of whale hunting in Lajes do Pico.

Throughout the history of whale hunting in the Azores, whalers navigated fragile sailing canoes from which they harpooned these gigantic mammals by hand. During the first two decades of hunting, whalers also relied on these canoes for towing the whales back to shore. In early accounts, whale hunters portrayed their lives as being characterized by extreme hardship. They described the experiences of hunting whales as imbued with great fear and unspeakable sacrifice. Most whale hunter life histories of this period are constructed around stories of storms and the dangers of the ocean. These whalers described sperm whales as frightening creatures, and the processing of these cetaceans as a most unpleasant job that came with a hellish stench. A great deal of narrative emphasis was placed on the anxiety that their wives, mothers and children lived through as they waited for the safe return of the whalers.

As mentioned above, I obtained these stories second and often third hand, in the form that they were relayed by the sons, grandsons or great-grandsons of the first whale hunters of the Azores. Though each story is unique to a whale hunter and his family, they all seem to share in presenting whale hunting as an activity that people only pursued because of dire necessity. Many of the whalers I interviewed – then in their seventies and eighties – remember their fathers telling them that their grandfathers had hoped the profits of whale hunting would provide their descendants with a better life, preferably with the means to pursue an education that would lead them to a living that did not depend on hunting whales.

Clearly, in the late 1800s the people of Lajes had little experience and/or knowledge of their oceanic surroundings. With the exception of a few fishermen, most people never set foot on a boat in their entire lives. Whales and the ocean were a mystery to most Lajence. As Catholics, most people in Lajes were familiar with the biblical story of Jonah whom God had a whale swallow as punishment for his arrogance. This story is even the theme of tile panels in some Azorean churches. It is possible to hypothesize then, that the first men who hunted whales off of Lajes saw and experienced the biophysical world of these cetaceans through the lenses of scary mythical stories and fear of the unknown. As seen above, accounts of the practices of whale hunting during this period were intertwined with these views as well.

By the mid twentieth century, especially during the Second World War, whale hunting had become a rather lucrative activity. Also, from the 1920s whale hunters relied on the support of motorized boats to pull their canoes closer to the whales, as well as to tow the whales back to shore. These boats were equipped with radios so that communication with on-shore stations permitted fast responses in case of accident. In many interviews I conducted, those who did not hunt whales described these radio communications as a great form of entertainment and a way to follow the exciting adventures of whale hunting.

Whalers who hunted within this period tell their life histories in a very different manner from that of their ancestors. They talk about whale hunting as a source of great personal joy and financial security. They described whales as magnificent creatures of enormous beauty, and presented the oceanic environment as relatively safe for those who understand the signs that indicate oncoming storms or other unfavourable conditions. These whalers even qualified the smell of processing sperm whale oil as a wonderful scent, that of money. They displayed a far greater level of confidence in their ability to offset the risks of whale hunting due to the knowledge that they had acquired as young practitioners under the 'safety' of guidance provided by older, highly experienced whalers. Most still wished a better education for their sons but were very proud of them when they joined a whale hunting company.

One thing is certain, as successive generations of whalers learned from one another, and as they practiced whale hunting, their perceptions of the biophysical world of sperm whales changed and, therefore, so did their practices. This was no longer a world of mythical fear-inducing creatures and storms, but rather an increasingly familiar environment that could actually be appreciated for its beauty. These changed perceptions became integral to whaler stories and life histories which were shared with relatives and friends. Thus, in time, they also became part of the Lajence cultural worldview. As this happened, a new generation of whaler hunters learned this art with greater confidence, which, in turn, became part of their interactions with sperm whales.

Finally, by the time that whale-derived products were no longer in demand and whale hunting began to decline, the most recent generation of whale hunters saw themselves as heroes, who kept alive a 'tradition' that rendered their village's culture unique. In telling their life histories, these whalers constructed their sense of identity as inextricably linked to whale hunting, to knowing whales at the level of greatly identifying with them, and to being part of the entire oceanic environment of Lajes. As I argued in a recent publication, these whalers developed an aesthetic awareness of their surroundings whereby a sense of self co-developed with knowing and interacting with sperm whales (Neves-Graça 2005). They were extremely proud of being whale hunters, and were hailed by the rest of the population as the village's most special members.

It is important to add that many generations of Lajence inhabitants – whether they belonged to a whale hunting family or not – spent most of their lives listening to all the above stories, such that eventually these stories became points of reference within their own lives. Many people who did not experience whale hunting, or who spent little time in the ocean, came to appreciate whales and the oceanic environment of Lajes by these means. Indeed, this is detectable in how *they* tell their own biographies.

In actuality, I have gathered data that covers roughly five to six generations, so the example I just evoked is a much simplified illustration of how life histories and related story telling can shed light on the recursiveness of human–environmental relations. One thing these narratives reveal is that as people experience and narrate the environmental systems of which they are part, particular forms of knowing these systems emerge. For example, early whale hunters knew the ocean of Lajes and whale hunting as dangerous and fear-inducing. As these activities persisted through time, they became more familiar to the village's inhabitants, and as their socioeconomic context changed we notice that alternative ways of knowing the environment began to emerge as well. Hence whalers began to focus on the beauty of the ocean and of sperm whales, and eventually even came to deeply identify with both. When in the 1990s whale watching was introduced in Lajes do Pico, these several views of the oceanic environment and sperm whales again were important points of reference for reflexively conceptualizing and developing this new economic activity.

In short, as changing stories of whale hunting – told in the form of life history narration – became part of Lajence culture, the biophysical world of the ocean and of sperm whales was transduced into the cultural world of the inhabitants of Lajes do Pico. Eventually, it became the main point of reference for the construction of the village's cultural identity. In turn, as the social and cultural context of whale hunting changed, so did the whale hunters' perception of the village's oceanic environment and of their practices of hunting whales.

Historical-Archival Sources

As mentioned above, during the 1960s and the 1970s anthropology was severely, though justly, criticized for its tendency to represent cultures and societies as if they were suspended in a perpetual present (see Bird-David 2004 for a recent critique of Eurocentric biases in anthropological approaches to the past and history), as if one could understand societies and cultures without addressing their inescapably processual nature. As a response to this serious limitation, anthropologists working in the 1980s and the 1990s shifted their attention and efforts towards collecting data about the histories of the societies and cultures that they studied. In fact, during this time historical anthropology emerged as a new and exciting field of research (in addition to the references listed in the previous section see also Rosaldo 1980; Biersack 1991; Hastrup 1992; Gulliver and Silverman 1995; Borofsky 2000; Silverman 2001; Axel 2002 for examples of anthropological approaches to sociocultural, political and economic histories).

Within the framework of feminist and postcolonial critiques of objectified understandings of history, anthropologists now explore historical-archival sources very cautiously. We know that these sources might have been produced by elites with specific political economic interests in mind. Often, historical archives do not include data about those whose social position was less prominent, or who were illiterate. In turn, keeping records itself has resulted in 'biases' as people make decisions about the types of information that are worthy of being stored for future use. For example, many cultures have a preference for storing data that concerns economic transactions, while others might prefer to keep records of genealogies. Hence, archival sources are always incomplete evidence of the past and, moreover, they reflect the selection criteria of those who were in a position to gather and keep this information. Consequently, the knowledge we obtain from historical-archival data is always provisional (Mathur 2000).

Another dilemma emerges in the context of historical-archival research. Sometimes, anthropologists find that archival data contradicts the collectively negotiated memories of our research participants (see Hagen 1999 for an interesting ethnography of strategic constructions of collective memories and how they can be analyzed). It would be wrong, of course, to think that archival sources are truer than social cultural memories (Hoskins 1987). Indeed, such fallacy is typical of the types of reductionist positivist thinking that originally kept anthropologists away from pursuing historical analysis. What is interesting to us is the array of processes by means of which public memories and archival documents show different interpretations of events which, in turn, allow us to begin to understand the sociocultural positionings of these interpretations. Moreover, this allows us to scrutinize the unfolding of interactions between discourses and the sociopolitical-economic orders (discursive formations) with which they are associated (Foucault 1994). Notwithstanding many limitations and concerns, archives are still highly productive sites for anthropological research. Archival research like, for example, that conducted by critical theorists, can reveal entirely 'new' perspectives on historical processes by unveiling the power relations that lay implicit within them (e.g. Foucault 1972, 1978; van Dijk 1995; Briggs 1996a; Rodman 2004). An example from my research in the Azores illustrates these contentions (see Hodder 1998; Manning and Cullum-Sawn 1998; Rodman 2003; McCulloch 2004 for information on how to conduct documentary research).

In Lajes do Pico there is an annual religious festivity during which the whalers' ritual takes place. This festivity is very important for all Lajence people, and visitors gather from around the island, even from other Azorean islands and the Portuguese mainland. Collective memory has it that the ritual was first performed because the relatives of whalers who had been caught in a storm spontaneously invoked the Lady of Lourdes. The story tells us that as the whalers struggled to sail back to shore, the population who waited for them on shore prayed to the Lady of Lourdes for their safe return. The population promised that if she saved the whalers they would hold a special religious celebration as a gesture of gratitude.

While conducting archival research, however, I discovered a letter from the person who at the time was the head priest of the village's parish. The events he recounts are somewhat different. In his letter the priest recalls that he had been very worried about the Lajence people's lack of religious faith, and with their absence from church. In an attempt to recruit the Lajence population into religious Catholic practice, he had ordered an image of the Lady of Lourdes from France. In the letter there is clear indication that as a strategy to increase people's faith in catholic saints it was he who suggested that the local population ask for the saint's favourable intervention on that 'historical' day.

It would be wrong for an anthropologist to argue that one story is truer than the other. For instance, that the archival source is of higher value than the collective memory of Lajes or vice versa. In fact, both reflect distinct perspectives on an event (that of a spiritual leader, and that of the population who waited for the whalers' safe return). It must be noted that the version of events held by the collective memory was perpetuated partly through the annual sermons of Catholic priests during the whalers' rituals.

Whether the idea of invoking the saint's help initially came from the village's priest or from the population, whether or not it was the result of successive objectifications during the annual sermons, the fact is that the idea itself was adopted by the Lajence and incorporated into their cultural world along with related practices. What is most interesting in this discussion is that both collective memory and archival sources demonstrate that whaler engagements with the biophysical environment of Lajes became part of the sociocultural world of Lajes in the form of its main annual religious celebration. On the other hand, whalers felt less alienated from the rest of the village – especially while at sea – by carrying paper images of the saint on their canoes. In this way, the cultural world of Lajes also became part of whaler practices at sea.

Archival resources can be informative in many additional forms. One I wish to point out is that, used in complementarity with life histories and story telling, they add great depth to our understanding of the processual nature of human-environmental relations. As I mentioned above, oral history tells us that the first whalers who hunted sperm whales in Lajes did so out of great economic necessity, to secure the survival of their families. Archival data not only confirms this, but also shows the dimension of these statements, as I explain next.

In Lajes, as in basically all Azorean villages, people bought their 'groceries' at small local shops. Since many Azorean families had practiced subsistence agriculture for centuries, they bought mostly those things that they could not produce themselves, such as coffee, sugar, shoes, clothing, oil etc. Sometimes these goods were bought in exchange for other goods such as eggs, or corn, but often people had to pay their expenses in cash. The owner of the store kept a record of these expenses throughout the year, and expected cash payments to be made once or twice a year. Comparing this source of information with data on people's wages (e.g. by analyzing the municipality's records of taxation), it is obvious that most of the population of Lajes lived in a very precarious economic condition. It is evident that for many families, during the first decades of Azorean whale hunting, the whalers' cash income was the only way they could afford basic goods such shoes for their children. The archive does indeed add to the anthropological analysis of the reasons why people decided to pursue this activity.

Another interesting aspect of doing archival research is that it sometimes reveals aspects of local history that people have decided to 'forget'. As mentioned above, whale hunters of the Second World War period described themselves as involved in a very profitable activity. Again, archival resources do confirm this assertion. I analyzed the annual accounting records of many whale hunting companies to come to this conclusion. Interestingly – although by no means surprisingly – the owners of the factories who processed the whales made by far the greatest profit. From a Marxist perspective one might even wonder whether the whalers were exploited by risking their lives for the benefit of those who were in a position to make capital investments. I put this question to many former whale hunters. Though some made allegations that whale factory accountants had at times cheated the whalers out of their 'real' shares, most whalers believed that they were paid well and justly for their work.

Nonetheless, the question that the archive reveals begs further research before it can be answered.

Finally, by means of archival research I could confirm beyond doubt that the introduction of whale hunting in Lajes was part and parcel of historical processes that we would currently label as 'global'. The archive shows that the first Lajence whaling company was founded by a contract that was celebrated between a former Pico immigrant captain who had lived in the U.S.A., and the Dabney families, who developed a major import-export business. Captain Anselmo da Silveira had learned to hunt whales in one of the many American whaling ships that ported in the Azores, and had been able to finance his share of the firm with money he made as an immigrant in the States. On the other hand, during the entire time that whales were hunted in Lajes do Pico, the records show that most of the whale-derived products were sold to European markets especially the U.K. and Germany.

Hence, even though the people of Lajes see 'their tradition' of whale hunting as the main marker of Lajence culture and worldview (and indeed, as I mentioned above, human-whale interactions are a core point of reference for the inhabitants of Lajes), historical archival research reveals the enormous extent to which the history of Lajes do Pico is inseparable from the histories of many other countries. In this sense, the recursive process of human-environmental relations in Lajes is partly embedded within the dynamics of international economics.

This confirms my argument earlier in this paper to the effect 'that ecological systems as well as cultural understandings are simultaneously space specific and globally connected'. As we will see in the subsection that follows, the same is true of 'global' environmental discourses, 'local' ecological understandings, and whale hunting as well as whale watching practices.

Discourses and Practices

The term 'discourse' – and its use within the context of the social sciences – has been discussed in an endless series of articles, books and web-based discussion groups. While it has been developed the furthest within fields that are concerned with linguistics, discourse analysis has been a precious tool for social and cultural anthropologists as well (Blommaert and Bulcaen 2000). Unfortunately, this is not the place to pursue a review of this extensive literature nor is it the context to thoroughly scrutinize the notion of 'discourse' analysis (for a brief introduction to 'critical discourse analysis' consult Foucault 1972; Lowe and Willis 1980; Bourdieu 1991; Fairclought and Wodak 1995; Wodak and Meyer 2001; Weis and Wodak 2003). Instead, I will provide an abridged operational definition of the narrow terms in which I evoke the concept. My main goal is to explain how discourse analysis may shed light on the co-development of 'global' and 'local' forms of knowing, as well as on the intersection of 'global' and 'local' practices (see Agrawal 2005 and Tsing 2005 for two recent outstanding analyses of environmentalism and globalization). For the sake of conciseness, I will rely on Lindsay Prior's definition:

> a representation should be understood not as a true and accurate reflection of some aspect of an external world, but as something to be explained and accounted for through the discursive rules and themes that predominate in a particular socio-historical context. The task of the researcher is to disentangle the rules of association by means of which the representation is structured, the genealogy of the various elements contained in the text … and the image of 'reality' which the text projects. (2004: 324–25)

Following Foucault, Prior adds that 'discourse not only restricts and limits what can and cannot be said about the phenomena within its domain, it also empowers (and disempowers) certain agents to speak on this or that question of fact'. As I, too, have argued before (Neves-Graça 2004, 2006), this empowerment leaves some social agents in a more credible and, therefore, authoritative position to make statements about the world. An example of our contemporary world is the degree of higher validity that is placed in statements that scientists make – so long as these statements are produced in accordance to the 'discursive rules and themes that predominate' our times (see Briggs 2005 for a parallel discussion of methodological and theoretical applications of discourse analysis in the context of medical anthropology).

Thus, it should not come as a surprise that when whale watching was introduced in Lajes in the aftermath of the prohibition of whale hunting, mainstream Western science became the frame of reference for defining what constituted sound ecological practices for this new form of eco-tourism. Indeed, it is easily argued that currently scientific texts constitute the dominant way of representing and producing the world. As such, scientific texts and related scientific performances (i.e. conference presentations) follow the discursive rules and themes that predominate within a liberal, Western, modern to postmodern sociocultural context. These rules and themes are quite familiar to us, so I will refer only to those that are more directly related to the substantive example I weave through my analysis.

One of the central rules in the production of scientific texts is that they must convey knowledge that is universally applicable, which is, therefore, presented in its 'abstract' form. For instance, scientific articles that explain the social behaviour of sperm whales are supposed to be true of all sperm whales, and not of the specific groups that were studied. Indeed, this fits the criterion of generalization that is central to the validity of scientific findings.

In the case at hand, after the introduction of whale watching in Lajes do Pico, former whale hunters challenged this basic premise of scientific discourse. This happened when the Azorean government decided to regulate the growing business of whale watching and different views came into play concerning what might or might not be harmful to the whales. One group of whale hunters worked for a company that had been deeply influenced by 'deep ecologists',[5] who believed that whale watching had the potential to disrupt the whales. They were very cautious in adopting this form of eco-tourism. Most other companies, however, relied on work by university-based scientists whose research with sperm whales in many other areas of the globe (e.g. New Zealand and Hawai'i) indicated that whale watching did not pose any serious long-term environmental threats to whales.

Alternatively, a group of former whale hunters from Lajes do Pico contended that the Lajence ecosystem was quite unique, especially because it was a nursery area for resident sperm whales. These whalers argued that this insight was the result of a hundred years of knowledge accumulated through whale hunting, and that therefore it was akin to what environmental anthropologists might call 'traditional ecological knowledge'. It was by thus implicitly creating the rules and themes of their own discourse that the whalers of Lajes do Pico challenged the validity of abstract universally applicable scientific texts (see Neves-Graça 2004, 2006 for a detailed ethnographic analysis of this interaction).

In the process, the whalers and their supporters constructed and reified a divide between global scientific knowledge and local traditional ecological knowledge. In the remainder of this section, I will engage in the analysis of 'the genealogy of the various elements contained' in these discourses to demonstrate that both have relations to the specific Lajence context, and to globalized discourses.

It is true that scientific findings from scientists working in New Zealand and in Hawai'i became reference sources for commercial whale watchers as well as for the legislation proposed by the Azorean government to regulate this activity. In this sense, these discourses fit the category of abstract universal science. However, following the genealogy of these texts, I was able to discover that the earlier scientific work of Robert Clark and Jonathan Gordon had enormous influence on the texts. Interestingly, these two researchers had developed their work from data they collected amongst whale hunters in Lajes do Pico during the whale hunting period. It turns out that Azorean sperm whales were prominent – though invisible – characters within the genealogy of 'universal abstract scientific discourse'. It was quite obvious as well that Robert Clark and Jonathan Gordon had learned a great deal from Lajence whale hunters during the 1950s and 1970s. Finally, I must point out that Clark's and Gordon's learning about sperm whales also took place at their home universities and within the wider context of the scholarly world. As abstract as they may indeed become, global scientific discourses are produced in specific locales by mobile and variously socially and culturally situated agents.

On the other hand, it is also virtually impossible to present the knowledge of former whale hunters from Lajes do Pico as truly local and/or traditional. As stated above, Azoreans first learned about hunting whales in the American whale ships that ported in the Azores to obtain supplies. Also, these whalers must have learned about 'scientific' views of sperm whales through their interactions with scientists such as Gordon and Clark.

Finally, recent generations of Azorean whale hunters were quite familiar with the discourses of anti-whaling environmental groups. For instance, in the 1980s environmentalists had accused the whalers of Pico of being murderers of cetacea. In engaging defensively with environmentalists, the whalers became familiar with environmental discourses. These same whalers coopted these environmental discourses in the later 1990s to present themselves as 'real ecologists'. They argued that by hunting whales for a hundred years they had developed a deep understanding of the mammals and of their environments and that, therefore, they were in a better position than anyone else (including scientists) to make valid statements about what constituted sound ecological practices for whale watching.

This example reveals, I hope, the practical usefulness of discourse analysis and more specifically of focusing on the genealogy learning practices through which discourses are (re)produced. This shows that rather than there existing a dichotomy between global and local knowledge (and associated practices), the 'two' are interrelated processes of mutual construction.

4 Concluding Remark: Methodological Tools as Part Perspectives of a Recursive Relational Whole

In this paper I have focused on three dichotomies that environmental anthropologists can easily run into. These constitute sets of biases that present culture in opposition to nature, scientific knowledge or discourse in contrast to local knowledge or practice, and present objective historical accounts of human-environmental relations against subjective interpretations of history. There are many other possible fallacies within this field of anthropological scrutiny, though many have been identified and corrected throughout the history of our discipline. My goal was to explore three of anthropology's standard methodological tools in the context of findings from my fieldwork in the Azores, in order to discuss how we might use them reflexively to overcome some of the pitfalls I had identified in the beginning of this chapter. My wish is that this might assist the reflections of those students who plan to engage with environmental issues for the first time. If there is one message I hope to have got across, it is that each methodological tool provides us with a view of only part of the relational whole we are meant to explain.

I believe that we can never provide completely holistic views of phenomena we study – amongst other reasons because we are constrained by our own positionality and historical predicament. As Khun would say, we think and write within paradigmatic contexts of which we are most often not aware. Still, I am convinced that the more we apply different tools to the various themes of our work, and the more we remain open to possible alternative perspectives we might be misrepresenting, the less likely we are to produce the kinds of monolithic-reductionist ethnographic work that became unacceptable in the aftermath of the postmodern turn. Indeed, one of the most precious lessons to be learned from the so-called postmodern crisis is that even though we cannot fully represent the wide relational unities we study, we should also not misrepresent them by taking short cuts in the form of careless and ill-thought methodology. We must strive for holistic accounts while knowing and working with the limitations of our endeavours.

Notes

1. Following this line of reasoning, one should be wary of casting accounts of local knowledge as pristine harmonious relations with nature in opposition to scientific knowledge construed as the result of a conjunction between academic laboratory production and capitalist pursuits. That is, we should not reify a polarity, between an imagined pristine, harmonious relation with nature and a science supposedly subjugated to capitalism whereby humans become alienated from 'nature'.
2. To the point that for some cultures the human and the environmental are completely separate entities – in fact, these cultural demarcations of boundaries constitute the central themes of environmental anthropology.
3. Matters become even worse when this misunderstanding becomes the ground on which future relations are built between such objectified terms. We can find our own anthropological example in the construction of a falsely concrete distinction between 'Us' and 'Other'. While this anthropological dualism emerged within the settings of European colonialism, it also contributed to the maintenance and justification of further exploitative forms of colonial rule. Similarly, it is easily argued that human practices that exploit and degrade the ecosystems of which they are part are frequently fuelled by the false belief that somehow the environment is an externality of the human world (i.e. exists outside the human world).
4. For instance, memories are quite literally living processes. Human memory is not a hardwired data base wherein the brain stores unprocessed and unchanging information. While memories may appear to be fixed and permanent, retrieving them entails a neurological process of (re)production which is often socially and culturally mediated. Arguably, we researchers are active participants in the creation of the context wherein this mediation takes place.
5. Deep ecologists follow a branch of environmental thinking and activism that was first introduced by the Norwegian philosopher Arnold Naess. They criticize most other forms of environmentalism as being 'shallow' for following anthropocentric views and promoting mainly human interests of conservation.

Chapter 9

Tempestuous Landscapes: Persons, Places and Memory in Two Vanuatu Hurricanes

Margaret C. *Rodman*

One premise of this volume is that everyday behaviour and decision making reveal how differently placed Pacific actors create their identities in relation to memories of place and conceptions of belonging. Extraordinary experiences, especially remembered life-threatening ones, can highlight processes integral to the everyday, and to identity formation. This chapter contributes to understanding persons, places and memory through discussion of two hurricanes in Vanuatu, one in 1939, the other in 1985. Each storm transformed the landscape, left survivors homeless and hungry, and in each storm one person died. A settler's memoirs provide an account of the November 1939 storm that struck low-lying Pakea Island. Multiple accounts of the hurricane that took Ambae islanders by surprise, in January 1985, are compiled from field notes that William Rodman and I recorded later that year. At a theoretical level, I have urged anthropologists to consider how different actors construct, contest and ground experience and identity in place (Rodman 1992, 2001). Narratives about the two storms provide an opportunity to reflect on how these processes occur when wind and water turn taken-for-granted landscapes into chaos.

Interest in entwined relationships of personhood and place has a long history in Pacific studies.[1] Debate continues on the complex ways in which place, personhood, emotions and the self are psychologically, socially and culturally constructed in the Pacific islands.[2] To what extent are emotions, persons and selves products of culture? How does place, in this case, tempestuous Pacific island environments, shape identities? Much attention has been paid to exploring answers to these questions in Melanesian contexts but what about the culturally European residents who have called the islands home for generations?

The two hurricanes at the heart of this chapter hit islands in an archipelago in the southwest Pacific known from 1906 to 1980 as the Anglo-French Condominium of the New Hebrides and later as the Republic of Vanuatu. Called 'hurricanes' in the southwest Pacific, such storms are called cyclones and typhoons elsewhere.[3] The hurricanes of 1939 and 1985 shattered worlds that were different in crucial ways. One storm hit a high, volcanic island populated with scattered hamlets of indigenous families. The other overwhelmed a low-lying atoll on which a settler family lived with ni-Vanuatu (indigenous Melanesian) employees. Nearly half a century separated the two events, but the impact and response were similar. People had no warning of the hurricanes; they were on their own, reliant on local leadership and local resources. Survivors' narratives point to what is culturally distinctive about these experiences as well as to what is shared. These are particularly evident in terms of displacement and emplacement, community, gender and identity.

Pakea Island, 25 November 1939

Violet Whitford was eight years old when she experienced the 1939 hurricane. She remembered it vividly for the rest of her life. Her mother told the story of the storm many times, and Violet remembered that, too. In 1998, when I met her, she was no longer Violet Whitford but Vi Bowhay (her married name). She had left Vanuatu, and lived in Sydney all her adult life. By then, her memories and her mother's narrative had become so entangled that it was hard to say where one left off and the other began. At the urging of her daughter, Anne, Vi had begun writing a memoir of her childhood, but she had stopped with an account of the 1939 hurricane. Vi read it aloud to me, until she began to weep as she reached the part where she had written: 'Anger, panic, sheer terror gives us unknown strengths …' She handed me her notebook, and I read the story from the beginning. Passages from Vi's memoirs, edited and condensed, appear in italics below:

> *Pakea*[4] *was, and still is, probably, one mile wide by one and a half miles long; a sailor's dream of a desert island really. Beautiful sandy beaches with a lagoon out the back, and another island, called* Nawila, *forming the lagoon. At low tide, you can walk across. My grandparents,* Frank *and* Alice Whitford, *settled there in the mid-1890's…*

Vi was a sensitive, quiet child who felt deeply but said little. Above all, she feared the loss of her mother, a single parent around whose tough love and violence Vi thought her world revolved. That Vi began her memoir of the hurricane with the following 'aside' about the Second World War interested me. War and storm, so often metaphors for each other, also introduce alternative sanctuaries. It is as if Vi is saying, 'I wish my mother could have taken the family to the safety of The River during the hurricane, as Grandfather had in World War I.'

> *I've always remembered the feeling when news came that* World War II *had started. [I was eight years old].* Mum *and her ni-Vanuatu foreman, Jumbo, were talking about it and the fact that they were children when* WWI *had ended. I felt it meant the end of the world, a heavy leaden feeling that was hard for me to bear, but I couldn't tell* Mum *about it. Later,* Mum *told me their father had taken them to the other family plantation called* The River, *[on the larger island of*

Vanua Lava] when WWI was on. There might have been a bigger and better house there, I don't know.

As a little girl, Vi was afraid of her Uncle Jim's native wife, Elsie. What frightened Vi initially about Elsie was her foreign-ness, and her tone. Elsie spoke very rapid French and always seemed angry with Uncle Jim.[5] Even before the hurricane cast further doubt on Elsie's judgement, Vi never wanted to be left alone with Elsie. Vi knew that this picture of Elsie could upset Whitford family members, and says so towards the end of the following passage:

Sometime during 1939, Uncle Jim planned on coming to live and work on Pakea. He'd brought [his Melanesian wife] Elsie and left her with us. She was expecting a baby, so rather than move her with a young child, she would be settled with us, and Uncle just had to finish up whatever he was doing elsewhere. The baby was born in September, a girl. Elsie seemed to have taken a long time to pick up. Mind you, she was a "tartar" and extremely volatile. Mum, being fairly no-nonsense where sickness was concerned, must have had her hands full.

Early in the morning of November 25th, 1939, strong winds woke Mum. She checked the barometer and saw it falling. She checked every fifteen minutes; it was falling so fast, she said, and it was only four am. [Indigenous employees] Mabel and Freddy's house was the closest, so Mum went there to wake them, and also to get Jumbo.

By daylight, it was obvious we had a hurricane.6 Jumbo's house was farthest from the beach, so everything that Mum could move was taken there – her mother's sewing machine and our clothes and some personal things. Everything in the sitting room was left, so many things, that her family had collected over the years. The family piano, gramophone and records, all were left. I'd seen Mum dance there in the sitting room to the old gramophone, when she'd had visitors ...

There was an old canoe, without outriggers, on the ground some little distance from the sitting room. I had six small turtles in it, I used to feed them with a little help from [the Melanesian nursemaid] old Bella. When we were moving things, I cried because Mum wouldn't let me carry the turtles.

The hurricane came from the south. By mid-morning, we were all up at Jumbo's house. Elsie was holding her baby. Mum had me on a chair and we were at a window. All the labourers were inside, praying. The labourers' houses were in various stages of being de-roofed; they were only thatch and bamboo walls, coral floors. Our houses were being blown away, sheets of galvanised iron were flying everywhere. Huge seas were pouring into Dudley Channel. Hitting the high cliffs of Vanua Lava opposite us, the waves swept sideways back onto Pakea. Coconut trees were snapped in half like matchsticks. At Jumbo's, the sea was up to the mid-calf on the adults. Jumbo's house was swaying dangerously. Mum told the people to get out before it collapsed on them. Mum told Elsie it was each mother to her child. Jumbo offered Elsie a [length of cotton cloth called a] lava-lava to tie the baby to her. But, she refused saying she couldn't possibly use the 'calico' of a native on her 'master's' baby! Sorry, my poor Whitford cousins, but this was told too often by too many and I was there! Mum was angry with her, and never forgot how Elsie carried on at the time.

As the house was falling, Mum clutching me, Hilda the cook, old Bella, Elsie and the baby all moved through the water watching out for sheets of iron going over our heads. Hilda's husband, Harry, had been sent to the other side of Pakea to check on Mum's launch, the 'Pumpkin.' Seems she rocked and rolled over the waves…

Vi still had vivid memories of what happened next. She could remember the feeling of being knocked down by a falling tree, of being under water. This is where she wept as she read aloud, and could not go on.

A couple of yards from Jumbo's house, Mum, Hilda, and I were hit across the back by a falling coconut tree. We were submerged in salt water, by this time swirling around the adults' hips. Anger, panic, sheer terror gives us unknown strengths. Mum managed to get out and up. Then she and Hilda pulled me out. Mum said we three had sore backs, besides bruises, for some time. During all this, Mum was hanging onto a travel rug she got with Bushell's Tea coupons.

Elsie was screaming she'd lost her baby. She'd prayed loudly earlier, in French, and now she was hysterical. The sea was pouring in quite fast, the wind and rain horrendous. Pakea, being so sandy, we were whipped and blinded by the sand. With hindsight, Mum should have headed towards the back, north easterly, away from the south end. Anyway, we swam, ran and looked for a patch above water. There was Bella, her daughter Mabel, Mabel's husband Freddy, and their daughter Violet, Elsie, Hilda, Mum and I. We all scattered wherever a spot looked likely. We eventually came to a piece of land where a large tree had blown over. The roots formed a canopy. We crawled in there and touched a warm bundle that yelped. It was Harry, who had gone to check on 'The Pumpkin'. He told Mum there was no sign of the boat, she would have been blown out towards the gap between Pakea and Nawila and to her end.

They got some branches, large wild taro leaves, whatever could be salvaged and threaded through the trees roots to make a shelter. We huddled together, Elsie crying and praying in French all the time. Mum put leaves down for me to sleep on, covered me with the wet rug from Bushell's Tea labels. I itched and scratched and complained that something was biting me. Mum, never the 'softie' said, 'Be quiet and go to sleep'. The hurricane had gone on for nearly twenty-four hours. We were all facing something none of the grown-ups had ever experienced, and I'm told to go to sleep as though tucked up in my bed at home!

Next morning, the sun shone bright, clear and scorchingly hot. It is always like that after any hurricane. When we were crawling out of our tree-root shelter, it was found that my bed had been made with nanggalat [stinging nettle] leaves. I felt less guilty about complaining. Mum told this part of the story many times. Though the death of the baby and loss of our homes must have been hard for her, she didn't dwell on them.

Mum's first orders were to find the baby's body before any dogs that survived found it as they would be hungry. In places, water was still deep. Everyone started coming out. When roll call was taken, Elsie's baby was the only fatality. Her body was found not far from where she'd been dropped. Mum got sheets from her suitcase, wet but usable, left at Jumbo's, a matter of clearing through debris and flotsam to find things. The little one was thrown up where the water had receded, bruised, swollen and so, so sad. She was cleaned as best that could be done. Mum got her sodden prayer book out, a grave was dug and where the remains of the family cemetery were, a coffin was made out of scraps of wood. Jumbo was a Godsend to Mum. He got things done and used whatever was at hand. We had a funeral service under the blazing sun before noon that day. Mum read the service. We buried the baby, and all this before we could see if there was any food. No one had eaten since the night of the 24th. This was the 26th.

We drank coconut juice and ate the nuts. It's safe for a couple of days; the two wells were filled with salt water and rubbish. All the tanks had blown over and smashed. There was a hollow coconut tree there that always had water, we used to drink from it, just rainwater running down the trunk. Can you believe how desperate the adults were? Mum was afraid of dysentery with too much coconut juice, and the nuts that were bruised soon turned sour. Some tinned food was salvaged. Rice, flour and everything else had been saturated with salt water. We looked to find where the houses had been. We found the top of the piano under the sand. I dug out the black keys to play with. All we had left was what had been taken to Jumbo's – as the roof collapsed, things didn't get too scattered.

Matches were useless. Wood was too wet. Salt water takes a long time to dry. Things were pretty bad. The first night, some sheets of iron were salvaged and a little humpy was made for Mum, Elsie and I, near the remains of Jumbo's house. Everyone else camped around near us, clothes and such were draped out to dry. We couldn't wash anything. There was no fire or water for three days [until a wild cotton wood dried out enough to make fire by friction.] Freddy's father lived on the mainland [the large island of Vanua Lava]. The spot was called the Point. It directly faced Pakea. It took the old man three days to find something to lash together, bring water and some wild yam and taro over. He told Mum when they looked over the first morning, they felt there couldn't possibly be anyone left alive. Coconut trees were left like broken stumps, every sign of houses gone, the coastline completely changed. Huge trees washed along the beachfront. They [on Vanua Lava] could escape onto higher ground even though their homes on the shores had gone. The waves hit there then bounced back over us.7 Calicos were waved when they saw us coming out. Mum had gone three days without a hot drink, or even a drink of water.

Mum rallied everyone around. Quarters were built out of anything they could find. Coconut leaf branches made walls and roofs. On a higher spot near Jumbo's, Mum had him start a two-room house. She said to me 'One was our bedroom, and a small sitting room'. When we left, it would do for Uncle Jim. We helped by hammering nails to straighten them. We dug in the sand for anything we could find.

On Christmas Day, 1939, we had another hurricane. It didn't have the big seas. We all collected around our partially built house, and waited it out. We'd been semi-camping. Our bathroom was an enamel dish on a box. Elsie was no help to Mum. I can only guess at Mum's heartache, would she have felt despair. Uncle Jim was at Wairua with Aunty May during November – December 1939. When word got to them somehow about the devastation in the Banks, and a child dying at Pakea, he said it 'would be Violet, she is that age to run around and not listen to her mother'. He also said as for his baby, Elsie would tie her to herself. Sadly, that wasn't the case.

When the 'Mirani', Burns Philp's inter-island ship, arrived in January, 1940, the government gave hand-outs to the natives. Mum had to scrounge up money to buy tea, sugar and what else we needed. It was something we talked of many times. Mum had still been paying the labourers a small wage as it was. There were no coconuts for copra, no other means of making money. The copra made before the hurricane had gone. How those men on the ship could do what they did to Mum, I can't begin to think. I really trust they got their just deserts.

Vi's account is a narrative of self, place and culture. It is about herself and her mother, inseparable as a child's memories are from the stories her mother tells her. Like much of the rest of her memoir, the hurricane story points to her mother's strength and courage but also to her lack of affectionate behaviour towards her daughter. The story describes the storm's devastation of the place – the topography, the houses, landmarks, food sources and possessions. Yet rebuilding begins before the story ends. Vi's mother's leadership provides the direction, and Jumbo, the ni-Vanuatu foreman, makes it happen. The cultural dimensions of this story are evident from beginning to end. What was there to destroy? The piano, the gramophone, are emblems of European culture, but there is little else in terms of material culture that distinguishes the Whitford's household from the islanders'. The building in which Vi and her mother lived, for example, was no fancier, and not as strong as Jumbo's house.

Who was there – a European-style mistress, her child, her ni-Vanuatu employees, the ni-Vanuatu mother of her brother's baby – and who was not – for example, other indigenous people, white men – is also culturally revealing. Women like Vi's mother often ran plantations on their own or did so while the men of the family were away for extended periods of time. Vi's mother's partnership with Jumbo is among the best she ever had with a man, but it appears never to have been intimate. Other white people in the islands, such as colonial bureaucrats and missionaries, viewed settlers as mixed race (British) or metis (French), for most settler families had or have some Melanesian members, such as Elsie. Many such families chose to live as culturally European while some, particularly the offspring of Melanesian mothers, chose indigenous lifestyles. We can infer from Elsie's use of French, her relationship with Jim Whitford, and her residence at Pakea, that she had made the choice to live a European lifestyle. But this is a choice that she has to make continuously, as is evident when she refused Jumbo's offer of the piece of cloth that might have saved her baby's life. She does not present herself as above accepting the offer, rather she presents her child – the Master's baby – as too white, in effect, for a black man's calico. This story resonates with Homi Bhabha's notion of being 'white but not quite'; in her concern to defend the borders of her status, and her baby's, Elsie does something that only the most racist European woman would have done.

Even in European-style households like the Whitfords', settlers learned to live self-reliantly off the land. Settler men could have female native partners but a double standard made it particularly unacceptable for settler women to have native lovers. The consequent shortage of 'suitable' men meant that women who thought of themselves as 'European' often lived as single household heads. They had been raised to be self-reliant and many, like Vi's mother, found a scope for their leadership abilities that would not have been available to conventional, European wives.

Like Vi's memoir, the stories about a hurricane on Ambae Island in 1985 can be read as part of ongoing conversations about the experienced self and as what Lila Abu-Lughod (1991) calls the little, local stories through which culture works (see also Ochs and Capps 1996).

Ambae Island, 18 January 1985

In 1985, Bill Rodman and I made our fourth field trip to Ambae, a volcanic island shaped like a capsized canoe. It is a lush, fertile island about 40 km long, with coconut palms, gardens, and scattered settlements in the coastal areas and foothills, then dense rainforest rising to crater lakes at the summit (1,500 m) of the semi-active volcano in the centre of the island. It was October when we returned to the village of Wailengi in the foothills on the southeast side of the island, where we had lived in 1978 and 1982. But the events of 18 January were fresh in local residents' minds. Hurricane Nigel had destroyed 90 per cent of the housing on the island. The national newspaper had reported that the storm 'took everyone by surprise, wrecking ships, uprooting houses and trees, devastating gardens and plantations, ripping off fences, and scorching the greenery of the land, leaving behind a bewildered and shaken people and a trail of total destruction'.[8] In the village of Wailengi, the damage was less severe, but the cyclone was still the worst in living memory: twenty-seven of the thirty-four buildings in the village suffered major damage or were totally demolished. The giant banyan tree that had dominated the central clearing in the village had fallen in the storm. Bill wrote in his journal:

> The fallen banyan by moonlight was one of the eeriest sights I've ever seen: it is so big it looks unearthly, unnatural, part of a fantasy landscape ... [The fallen banyan] looks wildly out of place on the village plaza. The presence of the banyan is a constant reminder of the hurricane, as if anyone needed reminding.

Everyone wanted to tell us what had happened to them in Hurricane Nigel. Narration seemed to be part of the healing process; framing and articulating individual experiences celebrated the survival of the community. Few Wailengi residents were at home when the storm struck; some had gone on foot to attend a wedding about eight kilometres distant, others were about a kilometre away preparing for a mortuary feast, still others were tending to their coconuts or their gardens.

The leader of the village, a man in his seventies named Mathias Tariundu, had spent the night at the wedding. In the morning, he stepped outside the house where he had been sleeping:

> As I stepped outside, I saw three young boys racing down the hill toward the settlement. When they reached me, I said, 'Hey, have you kids gone crazy? You're running so fast you look as if you were being carried along before a flood.'

Mathias's metaphor of the flood set the scene for his recollection in much the same way as Vi did with her evocation of the First World War, though she had a specific historic event as her reference point and perhaps a wistfulness about a safer house. What the boys replied, however, marks another difference with Pakea in 1939, when there was no national radio service:

> Mathias: 'We heard it on the radio,' they yelled, 'There's a hurricane coming!' Within minutes, the wind rose, and what a strong wind it was! It broke off the branches of the trees and sent them flying through the air. We all took shelter in the na gamal [a traditional men's house, several times larger than most dwelling houses, with a wind-resistant roof reaching to the ground]. I looked through the cracks in the wall and I couldn't see anything but whiteness. It was as if the air was filled with sea foam.

Most accounts of the hurricane included a description of the sea foam effect. All of them noted how fortunate it was that the storm hit in daylight. Unlike the Pakea hurricane, this one was of brief duration.

> Mathias: If the storm had hit us at night, I'm sure we would all have been killed. But it struck during the day and it only lasted about two and a half hours, then the wind died down. We stepped outside and we couldn't believe what we saw. The na gamal was the only building left standing.
>
> But the hurricane wasn't over yet. The wind rose again, and we ducked back inside the na gamal. The wind was so strong that it threatened to rip the roof right off! All the men in the building were hanging onto the ridge pole. A gust of wind would get under the roof, and the men hanging onto one end of the pole would be lifted off the ground. Then the wind would shift and the men at the other end of the pole would have their feet lifted off the ground. It looked as if the na gamal would be destroyed just like the other buildings. Women began to cry. Even some men began to cry. I told everyone, 'Don't worry! Don't cry! If God had intended for us to die, we would all be dead by now. Our lives are in God's hands.'

Many people's narratives mentioned the noise of the wind as sounding like a bulldozer, a metaphor that fits the local name for such a wind (siritano) that 'sweeps the earth bare'. The wind was so loud that no one heard the huge banyan tree fall. One little boy expressed his shock after the storm. His father told us how the child worried, 'Papa, Papa, the sea is getting closer and closer!' Looking downhill from where the banyan had stood, the sea was now clearly visible, though it was (still) several kilometres away.

We also heard about the hurricane from a couple who had gone to help prepare for the mortuary feast in a nearby hamlet.

> Stanley: We wanted to go home but we were afraid because the coconut palms were swaying so wildly, bending down to the ground and then snapping back up toward the sky. We told each other, 'Oh! This hurricane is the worst we have ever seen!' All in all we hid in four different houses in the hamlet. First we hid in Edmund's house. Then it blew down. So we went to John Steel's house and it blew down. So we ran to Reuben's house. When it blew down, too, we went to the na gamal because it was the only place left to hide.

During the lull at the eye of the storm, they were able to walk back to Wailengi. They were shocked that all the leaves had been blown off the trees. As the winds resumed, they took refuge in their corrugated iron house, thinking it would be safe.

> Martha: The place was completely white. The rain was like the froth from a wave when it breaks on the shore. It was like sea foam and you couldn't see anything else.
>
> Stanley: I was worried about the church [an older cement and corrugated iron structure] So when the wind died down a bit I went to see how it was doing. I was surprised to find the church full of people. As the storm began to pass and the wind dropped, we put a ladder up and Edmund climbed onto the roof to nail down sheets of iron that were about to break loose. When there was a gust of wind, we would shout 'Hide!' and he would lie flat against the roof. Then when the gust passed, he would get up and hammer in a few more nails.
>
> Martha: The wind gradually dropped and it was extremely quiet because there were no leaves left on the trees to make any noise. The next day was sunny and very, very hot. We decided the best thing for us women to do was to wash all the clothes and mats.
>
> Stanley: The first thing the men did was to build a shade out of coconut fronds to protect the women washing clothes from the very strong sun.

This gendered work was the beginning of a well-organized community response to the disaster. In contrast to Pakea, where there was no fresh water after the storm, Wailengi's rainwater tanks had survived and rock pools on old lava flows brimmed with water. As in the 1939 hurricane that struck Pakea, one life was lost. An older man had died in a nearby hamlet when his house fell in upon him, so the first thing to do was to attend to his funeral.

After that, and after the women had done the washing, the entire community gathered to plan the reconstruction of the village. No one expected government assistance in responding to this crisis, though some rice and tarpaulins did arrive in subsequent weeks. Mathias and the local Anglican priest, a native of the village, led the rebuilding effort but they did so in a way that democratized the process. All adult males were expected to work until everyone's house was rebuilt. The reconstruction proceeded systematically through the village, first repairing the least damaged houses, then building house frames and roofs where dwellings had been levelled. Rebuilding was a salvage process. What was left of roofing iron, woven bamboo walls, and house posts were reused. This process resulted in most houses being smaller and shorter than before the storm. Each family was left to salvage or make the woven bamboo walls needed to finish their homes. The reconstruction process followed the cultural, gendered division of labour: the men did the major carpentry, while the women prepared food to feed the workers. Men and women worked together to complete repairs to their own houses. Within three weeks, the village was rebuilt.

Conclusion

The fear, loss of life, loss of housing and possessions, disorientation, and transformation of a familiar landscape into an alien one were similar in both hurricanes. The process of place making is highlighted both in the breach – as the experienced world is turned into sea foam or submerged – and in its recreation. Part of that recreation occurs in the production of narrative. The world is spoken into being. The multiple narratives of Wailengi emphasize the community on which individual narratives and experiences centred, a community that worked together to rebuild the village. The stories have shared narrative features, notably the rhythm of rising and falling – the wind lifts men holding the ridge pole off their feet and lowers them; Edmund hides and rises – so evocative of the storm winds. The stories also point to the importance of cultural activities in daily life – weddings, funerals, gardening. Because the hurricane on Pakea struck before dawn, everyone was home in bed; had it hit later in the day, the activities in which people were engaged would have been similarly revealing of local cultural patterns: making copra for Vi's Mum, cooking to feed her and the labourers, and so on. Both kinds of activities express the values of a community and encourage its reproduction, but also point to key cultural differences in life on a settler's station and in a Melanesian village.

Environmental factors were important. Ambae and Pakea are hit frequently with hurricanes but rarely by ones as strong as these. A low island like Pakea will always suffer more in terms of salt water inundation, lack of fresh water, and lack of food crops after such a storm. In contrast to Pakea, there was food in the gardens to feed the people of Wailengi, albeit food that would soon spoil from storm damage. There were also enough coconut husks and dry wood to start a fire.

Vi's memoir emphasizes the individual, focusing on her own experience as a child and that of her mother. We have no accounts from other Pakea residents, though the voices of Elsie, Harry and Jumbo are part of her story. The social hierarchy on Pakea was such that Vi's mother appeared to make all the decisions, while relying on Jumbo to carry them out. In Wailengi, the priest and the chief organized the reconstruction, just as Vi's mother did on Pakea, but the Wailengi narratives present the recovery from the storm as a collective recreation of community. The interdependence amongst people on Pakea is as great in its own way as in Wailengi, and in both situations people had to be self-reliant in order to refashion their homes and their identities as settlers and employees, Europeans and Melanesians.

Notes

1. Missionaries in the nineteenth and early twentieth centuries wrote about meanings of place in Melanesia (Codrington 1972; Leenhardt 1984); see also Clifford on Leenhardt (Clifford 1982).

2. Key contributors to the debate about cultural production of personhood in Pacific contexts include Abu-Lughod and Lutz (1990), Kahn (1996), Lutz (1988), Mageo (1998, 2002, 2003), M. Rosaldo (1980, 1984), R. Rosaldo (1989), Shore (1996), White and Kirkpatrick (1985). Nancy Chodorow, writing from a psychoanalytic perspective, also views the self as both interiorized and exteriorized, as both cultural and personal. She critiques the anthropology of self and feeling (especially Lutz 1988; M. Rosaldo 1984; Abu-Lughod and Lutz 1990) for 'making a double and in some ways contradictory argument. They wish on the one hand to establish it is important to study emotions, persons, and selves, but on the other, that emotions and selves are exclusively culturally constructed …' (Chodorow 1999: 167). See Morris (1994) for an overview of socially constructed approaches to meanings of the self, and Csordas (1994) for a more phenomenological approach to the self.
3. Tropical storms are called hurricanes in the western North Atlantic, eastern North Pacific, and the western South Pacific (Vanuatu's location). In the Bay of Bengal, south Indian Ocean, and northwest of Australia, they are called cyclones, and in the western North Pacific, they are typhoons (Kemp 1976: 893).
4. Vi's handwritten memoirs which her daughter, Anne Mooney, typed.
5. When Vanuatu was the Anglo-French Condominium of the New Hebrides, 'in mixed Anglo/Franco company you would have a broken English by the French, a broken French by the British and so on. So, I'd never actually heard two people conversing in French' (Bowhay memoirs, 1998: 28).
6. Unlike Hurricane Nigel, which hit Ambae in 1985, the storm that Vi experienced had no name. Hurricanes were not named until 1950.
7. The hurricane affected most of the islands in the area. In an article about settlers Tom and Sela Stephens' golden wedding anniversary in 1948, *Pacific Islands Monthly* reported that in the 1939 hurricane,

   … great waves swept across Motalava where Mr. Stephens happened to be. The trading station was wiped out, and the trader himself was carried away in the wild rush of waters. He was rescued from death by the natives. All their food and gardens had been destroyed, and he and the natives were very close to starvation when they were found by the Islands steamer Mirani. (*Pacific Islands Monthly*, June, 1948: 39)
8. *Vanuatu Weekly*, 26 January 1985: 1.

Chapter 10

The 'Anthropology of Landscape' as a Research Method

Susanne Kuehling

Introduction

The Anthropology of Landscape is a relatively recent direction in contemporary anthropology. The term became popular through the volumes edited by Hirsch and O'Hanlon (1995) and Bender and Winer (2001). 'Landscape', in this context, refers to the interpretation of space. In other disciplines, like Art History and Geography, 'landscape' is used in a less interpretative way. For that reason, I shall here use the term 'place' instead of 'landscape', in line with Myers (1986), Rodman (1992), Feld and Basso (1996), Fox (1997), Bolton (1999) and Rumsey and Weiner (2001). 'Place', then, is a construct of collective history, memory, personal and shared emotions and experiences, created by social interaction and relationships. It connects persons with the past and individuals with their social group (here, I follow the definition given by Miriam Kahn [1996]).

Research on the social relevance of place has convincingly shown that humans' interaction with their physical environment is a universal feature that anchors individuals in time and space. Places are symbolically charged, on a communal level as well as personally. While the physical environment, of course, consists of land and sea, animals and minerals, weather, resources and seasonal changes (see also Ingold 2000), I will here focus on the relation between persons and place, the meaningful interpretation of, and interaction with, space.

Since place is an entity, the social relevance of which is only partly reflected by individuals, research on this topic requires a wide range of methods and needs to be embedded in long-term fieldwork. By sharing people's lives, learning the appropriate behaviour, becoming competent in their vernacular and experiencing the general pattern of 'culture', anthropologists develop personal relationships that enable them to dive quite deeply into a society. In general, an overall sensitivity is needed to accommodate the emotional side of the relationships between persons and 'their' places of significance.

Previous anthropological work has shown that our Western perception of place does not always allow us to accommodate other peoples' views. The visual bias of our own cultural background sometimes inhibits us from 'seeing' aspects of place that are of great significance to those who inhibit it (see Gell 1995b). Rocks can be 'seen' as ancestors who even move about, as Kahn (1996) has reported from Wamira (Papua New Guinea). The moral dimension of 'good' and 'bad' places refers to agricultural and social levels of understanding that take a foreigner a long time to understand and would not be disclosed easily by local informants who have never reflected on such notions. Similarly, degrees of sociality and solitude are often metaphorically linked to particular places and names are as emplaced as metaphors and myths. Sounds can be the key to an understanding of notions of place, as Feld has convincingly argued (1982, 1996).

Place, then, incorporates sensations of all senses. The poetic dimension of place has been pointed out in a volume edited by Jim Fox (1997) that shows how Austronesian origin myths are emplaced. Alan Rumsey argues that history is most precise when spatially organized (2001: 12). As 'lived experience' (Rodman 1992: 641; Kahn 1996: 173), place is both multivocal and multilocal, because individuals may experience the same space quite differently so that often there are at least 'two places in one' (Kuehling 2002). Notions of place are also relevant as expressions of politics, of power relations, as the work of Barbara Bender has shown (1993, 1995, 1996, 1999, 2001).

Anthropological fieldwork allows for a range of qualitative methods to study the perception and symbolic meaning of place to its inhabitants. These methods are: participant observation, interviews, sketch maps, life histories, kinship analysis, myths, and language studies. Quantitative methods could be used in addition, if the fieldwork situation allows for such forms of enquiry that are closer to 'tests' and 'experiments' than to interactive and personal research techniques. Human Geography is a discipline that would have to offer a variety of tests, for example, based on the paradigm of Time Geography (see below). Psychological tests could certainly be included in a project, too. Except for the most basic of these testing methods, an anthropologist may yield more reliable data if these methods were conducted by appropriate specialists from these neighbouring disciplines.

This paper will mainly address qualitative research methods and illustrate them with examples from Papua New Guinea, where I conducted fieldwork in 1992–94 and 1997 (see Kuehling 2005). Although the main topic of my fieldwork was gift exchange, the social relevance of place became important in various research contexts.

Participant Observation

Living amongst villagers, joining them on their daily excursions and observing spatial behaviour is an appropriate method to gain first impressions of the significance of place. On Dobu, for example, individuals have limited access to most areas of their small island, as the ethical rule of *miabaula* ('staying at home, doing one's work and minding one's own business') inhibits people from roaming about as they like. Land belongs to matrilineages and in precolonial times anyone who was found trespassing could have been killed by landowners or their spouses. The impact of this ethical rule is still obvious when people walk about: whenever they meet someone they immediately explain why they are there, for example, 'We are going to the garden.' As a participant observer, one learns about the restrictions of movement as well as the exceptions, such as night prowling of teenage boys or special rules for feasting times. One also finds out how to circumvent the rule, for example, by creating exchange links that give a good reason for visits. Different rules apply for different individuals, according to gender, status and age. Children, by the way, are a good source of information, as they are always well informed about where fruits await their early-morning picking (and picnicking) excursions and know tales that their parents may not be bothered to tell the researcher.

Interviews

Many interviews happen quite naturally while walking about, waiting for someone to come or for something to happen (a typical fieldwork situation that can madden the impatient researcher but actually gives a good opportunity for talks). Interviews may be used to gain local knowledge on place that cannot be observed, for example because it is 'supernatural'. I learned about a secret tunnel on Dobu, for example, that is used by witches and sorcerers to cross the island secretly, when I accompanied a senior man – and famous sorcerer – to another village. He pointed to a small hole on the beach and told me that witches and sorcerers can get through it, also telling me the place where the exit of this tunnel was believed to be. It is also useful to ask for route descriptions (see Lewis 1976; Klein 1982; Wassmann 1997), because the landmarks mentioned in these descriptions may indicate important places that otherwise would perhaps escape one's attention. Comparing one's own landmarks with those used by the local people, discussing such differences and collecting data on seasonal changes that have an impact on landmarks is another method to gain an understanding of their sense of place, both on an individual level and as a cultural pattern. Generic terms for specific parts of landscape, such as 'settlement', 'cemetery area', 'last years' garden' etc. give further indications of interactional patterns and local categorizations regarding place.

A different approach that I believe is especially illuminating is to ask a friend to show 'places that matter' to him or her. Visiting these places and learning about their particular relevance yields data on several levels that will be further mentioned below. The combination of actual 'being there' with specific interviewing seems to be most useful, I believe, as the place itself triggers particular information that might otherwise not be regarded as relevant, either by the informant or by the researcher. Obviously, such excursions to 'places that matter' need to be conducted with a variety of informants, as notions of gender, age, status, kinship, level of education and personal background are likely to have an impact on the importance of certain places to individuals.

Sketch Maps

As Wassmann (1990) and others have shown, sketch maps are a useful tool to achieve an understanding of the worldview of the local people. Both the physical environment and the degree of Western-type education seem to have an impact on their perception of the world. Valleys may produce the sense of a closed system, rivers may trigger the impression of a lineal world, coastlines and islands rather encourage the view of an endless 'sea of islands' (Hau'ofa 1994). Sketch maps can be used as a background for further enquiries, perhaps group discussions and – if collected from several individuals with varying social backgrounds – give evidence of changes in the worldview. Sketch maps should be combined with interviews in order to avoid (culturally informed) misunderstandings of what they are actually supposed to show. They have to be analysed in relation to the general setup of everyday life, climate, gender and purpose. Inuit sketch maps, for example, focus on hunting grounds in a 'white world' and cannot be analysed as a key to geographic wisdom in our sense (Pelly 1991: 61; Bravo 1996: 21).

Life Histories

As we know, memory is emplaced (see, e.g. Küchler 1993; Strathern and Stewart 2000a). When close friends/informants open the treasure box of their personal life history, particular places are always mentioned. While, to us, a statement like: 'Then I lived at my father's place' might mean rather little, further questioning reveals that a significant set of information is enclosed in this statement. It may refer to specific rules of behaviour, for example spatial restrictions of movement, as in matrilineal and predominantly uxorilocal Dobu, where the graveyard of one's father's hamlet is not accessible for his children, where his relatives have to be treated with special respect and where a person will not inherit any land unless he (seldom she) is adopted into his lineage as a substitute. Life histories also refer to local concepts of 'home' as the place that one belongs to. Contrasting 'homely' with 'foreign places', life histories can help to understand spatial dimensions in relation to notions of familiarity, of emplaced kinship and of emplaced power relations (see Rensel and Rodman 1997). Migration, travel, marital residence and other spatial shifts in life, when recalled in life histories, can yield data along these lines. Young (1983b) has shown how a woman from Kalauna (Goodenough Island,

Papua New Guinea) has constructed her life in terms of places and the menfolk she served in them. Life histories should be combined with interviews to clarify particular points. They should be recorded and transcribed by the ethnographer to avoid intimidation of the informant who shared private thoughts and experiences (see Jaarsma 2002). If such privacy is not maintained, future informants might withdraw their consent to narrate their life history or give a short and less fruitful account.

Kinship Analysis

As place is intimately linked to notions of kinship (either real or constructed), the analysis of relatedness is crucial for an understanding of the significance of place (see also Kahn 1990; Lepowsky 1994: 300; Gell 1995b; Hirsch 1995). On Dobu, for example, land is in many ways linked to a clan system. While hamlet space should only be shared by matrilineages whose clan birds have the same nesting pattern, bodies should only be buried on land of the same clan. Matrilineages manage their land, distributing gardening and fishing rights amongst their members and thereby determining the places that an individual may visit or use. Kinship is far more than ascent and descent, as adoption allows for flexibility in the system. Namesakes and 'substitutes' can be given access to places of other lineages if both sides agree to it. People avoid certain places as part of mourning. Such spatial restrictions are either voluntary or prescribed, depending on the degree of closeness which is expressed in terms of kinship.

Kinship is an analytical tool of anthropologists as well as a rhetorical tool for locals to determine an individual's access to land. Both levels are significant for the study of emplacement: the former can explain particular connections of people with places while the latter gives evidence of symbolic and moral meanings of land. Nevertheless, taking genealogies can be problematic if land is contested – the anthropologist's notes could become legal documents and may support one party in land disputes. I know of colleagues whose fieldwork became very difficult for this reason when they began to note down genealogies, as lineages were accusing each other to tell lies about their descent – at least one genealogy that I took on Dobu was definitely faked (with ten generations in 100 years).

The 'authority of the written word' may have an impact on village conflicts about land use, when such genealogies appear in books and are then turned into forceful tools of power when used by villagers with Western education to turn down arguments of others who rely on oral history. Data on kinship are therefore a delicate issue in many societies and have to be treated accordingly. However, understanding the kinship system and patterns of creation of kin relations through a variety of ways, such as adoption, is crucial for the study of place.

Myths

Recording, transcribing and translating oral history is also an important key to reach an understanding of place. Myths of origin give evidence of sacred sites, just-so stories can be analysed in terms of symbolic and moral connections of place and person. As myths often refer to local cosmology, they help to reach an understanding of spatial dimensions that cannot be researched by participant observation. While I found it difficult to discuss deeper meanings of oral history with local informants, it was easy to use transcribed texts in interviews to raise particular questions about the places and cosmological concepts mentioned in them. Notions like 'bush land', 'beach', 'mountain' and 'sea' were eagerly qualified by informants and helped to raise my attention to different qualities of places and their impact on actual use. In myths, spirits and other 'supernatural' beings are emplaced actors whose behaviour points out to social values that provide further clues to the analysis of place (see Young 1983a).

Language Studies

While legends can be translated with the support of local interpreters, everyday talk, proverbs, grammatical constructions, semantic fields, and genres like swearwords, jokes and hidden talk require a thorough knowledge of the vernacular. Such data can be extremely useful for the anthropological study of place, as they open up a field of subconscious concepts. For example, the deictic terms of 'towards the mountain' and 'towards the sea' have been identified as examples for a geocentric worldview (see also Senft and Widlok in this volume). On Dobu, they are used in a variety of contexts, as motion verbs, directional adjectives, indicators of cosmology (*bwaulaga*: to call up to a goddess); they are strongly related to *kula* exchange and, as suffixes, qualify one's position. While they are essentially geocentric, they also point to egocentric dimensions: 'put the cup down "seawards" from my position (here)' as opposed to 'put it down "seawards" from your position (there)' is distinctive in Dobu language through the particles *-ma* (here) and *-wa* (there).

Local concepts of sickness are often related to place and can be detected through language, as in the case of *gwasa*, a social malaise (or emotion?) that afflicts villagers when someone has left their place. Trespassing is believed to be sanctioned by the affliction of sicknesses called *tabu* on Dobu, causing individuals to reflect on their past movements when falling sick and approaching those landowners, whose place they might have violated with their presence. Movement is hence a matter of consciousness in this society and it does matter which path an individual chooses to use. The wide semantic field of 'path' (*eda*), accordingly, is of great significance, which can only be fully appreciated if noted in the variety of contexts in which it is used (such as in overseas exchange [*kula*], as a metaphor for life, as optional strategy, as actual pathways over land, as connecting particular places). Place names often have a meaning that may not be obvious – or significant – to the local people but that illuminate semantic fields to the anthropologist, as in the name 'Gulebubu' (Rock-ancestor) for the top of the dormant volcano that forms Dobu Island. Terms for places like 'hamlet', 'bush', 'garden', 'old garden', 'deep sea', 'beach' etc. are also used metaphorically and can often be detected only when listening to natural utterances.

Learning the vernacular is also significant for the range of possible informants, as old people and those who are not interested in 'Westerners' do not appear on the anthropologist's horizon unless he or she is a competent speaker and thus able to communicate on an appropriate level. In my experience, such people were most interesting to talk with as they could provide me with insights that others did not have because of their age and lower status in matters 'traditional'. For the study of place, magical spells that I learned from those people added a new dimension of understanding (e.g. a chain of islands perceived as teeth, plants as embodying a spirit), some terms and metaphors they used can be classified as 'elaborate code' and opened up new fields of enquiry, as in names of mortuary gifts that relate to place. I am convinced that a study that attempts to dive deeply into another society's conception of place yields far better results when the researcher is familiar with the vernacular. There are, however, limitations in lexical studies, as semantic boundaries are sometimes not clear. Idiosyncratic interpretations of single informants may not cover the whole range of meanings; comparison of collected material may be 'next to impossible' (Kirkpatrick and White 1985: 17). Specific meanings need to be studied in their context of social relations, actual situations and their connected cultural concepts (see also Stewart and Strathern 2000; Strathern 2004).

Time Geography

I will propose here the use of a modified version of Hägerstrand's model of Time Geography. This method aims at collecting exact data on spatial movement and has to my knowledge not yet been used as a part of anthropological fieldwork. Developed by Hägerstrand in the 1970s (e.g. 1970, 1975, 1982; see also Giddens 1984: 111f.), this method of Social Geography uses data on spatial movements and correlated specific activities in order to come to conclusions on local (or regional) use of space. The method involves following a person for days (and nights?) and noting both the places and times as well as particular data on behaviour. For example, as Beatrice Moral has pointed out, the parallel movement of siblings in Micronesia could well be studied by using this method (Moral and Escobar 2004).

I plan to evaluate its usefulness in a future field study on a small atoll by following informants and marking both time and space (the latter with GPS

– Global Positioning System – or by marking it on a map). As a further qualification of this data material, I will also note particular behaviour. The selection of these qualifying aspects depends on the particular study and on the field situation itself. I could imagine noting body posture, speech and silence, company, clothing and consumption. This expansion of Hägerstrand's method into social activities has not been tested, as far as I know, in Anthropology. Interdisciplinary cooperation with Human Geographers may be needed in order to make the most of this collection of data (see Meusburger in this volume).

The resulting data can be used to show significant movements in space. They can be displayed in diagrams and analysed with computer programs, giving detailed information on actual behaviour that otherwise would have to be presented in text form (which would be boring to read). To collect such data appears to be useful, because it allows for differentiated statements on patterns of spatial use in terms of gender, age, status and degree of Western education as well as on differences in daytime and night time, natural and 'cultural' seasons and in relation to special events.

The parameters of Time Geography will have to be adapted for use in small-scale societies, since it has been developed to be used in urban settings. It also needs to be part of intense fieldwork in order to yield reliable data because informants are required to ignore the researcher while these data are collected.

Conclusions

Anthropological methods for the study of place open up a field of knowledge that is linked to all major areas of society. As such, they are not any different from other anthropological methods aimed at the understanding of a social system and need to be embedded in intensive fieldwork that includes 'classical' areas of enquiry like those mentioned. In addition, quantitative data on the use of space can add to our understanding and provide a means of presentation of data that may be useful. The fieldworking anthropologist may not be appropriately trained to conduct tests derived from other disciplines or might have taken up (or been given) a role that clashes with the role of a conductor of tests. It seems, therefore, preferable to invite colleagues (Psychologists, Linguists, Geographers) to spend a short time with the anthropologists and to make use of his/her insider's perspective for the choice of appropriate samples of informants for their specific approaches. While such interdisciplinary cooperation needs to be planned well ahead, and future co-workers should get to know each other in advance in order to avoid personal difficulties during fieldwork, this approach should yield an interesting set of data, both overlapping and supplementary.

In our globalized world, it is appropriate to compare data from rural areas with those gathered in urban settlements among migrants. Such multisited anthropology (see Marcus 2002) can prevent wrong perceptions about the isolation of contemporary societies and assist a more appropriate perspective of adaptation and syncretism. As humans seem to be universally likely to engage in emotional and practical, mythological and individual relationships with places, such an approach can lead to an understanding of the conditions that encourage ego- or geocentric orientation systems, motivate indigenous people to make commercial use of their land, cause migrants to keep strong ties to their home communities and so on.

Part III
Memory

Chapter 11

Smell, Person, Space and Memory

Bettina Beer

Introduction

Anthropology still lacks a synthetic analysis of the senses and culturally bound perceptions, although perceptions are the foundation for the discussion of distinctions and definitions of strangeness and the foreign and intercultural communication. The outer appearance of a person might be essential for their social status, but in many societies smells also are important for the categorization of people and assumptions about their morality (Beer 2000b, 2002; Oosterhout 2001). For example age, sex and ethnic background are associated with different odours or abilities to smell (Reichel-Dolmatoff 1971,1978, mainly 1985a, 1985b, and 1989). In these societies notions of the person might include the idea that body and soul have characteristic distinguishing smells (Leenhardt 1984; Howes 1988).

Theories developed in cognitive science offer a framework within which research on the senses carried out within different disciplines (including approaches from natural science, e.g. Reyna 2002) could be integrated. Until recently, psychologists have investigated perception and the senses mainly in Western industrialized societies (Sekuler and Blake 1990; Engen 1991; Clark 1993; Gelman and Au 1996; Goldstein 1999), but at the end of the 1980s, the Concordia Sensoria Research Group initiated comparative anthropological research on the senses, their cultural meanings and significance (Howes 1991; see also Classen, Howes and Synott 1994; Classen 1990, 1991, 1993, 1998; Fisher 1997; Howes 1987, 1988, 2000; Synnott 1991). This group of Canadian anthropologists initially analysed existing ethnographies that contained information on the senses, but did not conduct empirical research (for a critique see Poewe 1996). More recently, however, David Howes has published empirical results in *Sensual Relations* (2003), and Katherine Lynn Geurts in her ethnography *Culture and the Senses: Bodily Ways of Knowing in an African Community* (2002).

Research aimed at the empirical investigation of the neglected 'lower senses' (taste and touch) is still scarce (on taste: Kuipers 1991; whereas, for smell see: Gell 1977; Almagor 1978, 1987, 1990; van Beek 1992; Bubandt 1998; Rasmussen 1999; Keifenheim 2000; Oosterhout 2001). Over the last decade, empirical research has still been focused on sight and on acoustic world-*views*, as is to be expected in light of the traditional European hierarchy of the senses.[1] Yet this cultural hierarchy of the senses is not universal; and the ranking of the senses differs between cultures. All senses are crucial to human social life, although some are culturally more or less emphasized in certain domains or are more or less important for actors in given situations. The very question 'which sense is generally the most important?' is misleading.

Smells can be isolated from other sensations only for analytical purposes. Given the emotional significance of smells, their contribution to memory, to the synaesthetic perception and embodied experience of space, this sense needs to be emphasized in future investigations.

Smell, Space and Place

Though the title of this article was originally intended as 'smellscapes: space and memory', I do *not* want to add another '-scape' to 'land-', 'sea-', and 'soundscapes'. The empirical and theoretical insights gained from the debate about landscape as 'a cultural process' are not yet very helpful for the investigation of relations between space and smells (e.g. Hirsch and O'Hanlon 1995; Feld and Basso 1996). Rather, I will simply ask in which ways person, place, smell and emotional memories are interrelated and how this could be investigated in anthropological fieldwork.

Anthropological research on the perception of space has emphasized visual (Keifenheim 2000), acoustic (Feld 1990, 1991, 1996), and recently kinaesthetic (Geurts 2002) perception. Smells and their significance for memory of the perception of space and place have so far been neglected (e.g. Levinson 2002; exceptions are: Cohen 1988; Pandya 1990). Even when the importance of the senses to moods and emotions that affect the perception of space, environment and ritual is discussed, smells are often omitted as if they were not important at all (Nihill 2000; Harrison 2001).

Yet, in natural settings there is no smell-neutral space. Space is always filled with smells produced by components of the environment, influenced by climate and human actions. Human beings produce smells all the time, sometimes consciously and sometimes unconsciously. The olfactory dimensions of space and their perception are a methodological challenge because in most cultures the sense of smell is only loosely linked to and expressed in language (Richardson & Zucco 1989: 358; Hatt 1990: 107; Vroon et al. 1996: 110; Köster 2002: 31). Reactions to smells are often emotional and direct. The perception of smells becomes more conscious if they are: very unfamiliar; unexpected; their context deviates from everyday experience; they are culturally marked; or, questions about smells are raised.

In the remainder of this paper I will summarize some results from psychological research on the connections between smell and memory, and then give two examples of the relations between person, space, place, smell and emotional memories from my own data. Finally, I will describe some implications for fieldwork and methods in the cognitive anthropology of the senses.

Psychological Research on Smell and Memory

The general literature on memory focuses on the visual and auditory system, while smells are seldom mentioned (Anderson 1995; Kluwe, Lüer and Rösler 2003). Psychologists have investigated the connection between smell and memory in four main ways (Engen 1991; Richardson and Zucco 1989; Issanchou et al. 2002). The first group of experiments are recognition tests: (1) Smells are presented and subjects are asked to identify them and give the proper name of the source or smell. Often a subject suddenly has a certain emotion, and can recognize the smell, but not recall the name of it. This phenomenon is known as tip-of-the-nose effect. (2) In other experiments (recall tests) smells are presented and subjects are asked to recall them later after the presentation of other visual or aural stimuli, or after an extended period of time without other stimuli being presented.

Recall tests without presentation of any real smells are the second group of experiments. (3) Subjects are asked (without specific stimuli) to recount memories of smells so that they can be verbalized during free recall. Subjects are also asked to describe accompanying associations (Schleidt, Neumann and Morishita 1988). (4) In other experiments subjects are asked to imagine a smell without visualizing the source, but it is not possible to verify what mental images a person has and whether the subject had in fact retrieved the expected images. It is not yet clear whether the nonvisual imagination of smells is possible at all or how it might work. It is not clear either whether people are able to dream of odours.

Results of Psychological Experiments

Some results on the links between smell and memory are important for anthropological research on the relation between smell, place and memory. Smell identification tests show a high percentage of short-term forgetting, but little subsequent forgetting. The long-term recognition of some remembered smells is usually very good. This is mainly explained by the fact that experiences of smells are represented as unitary, distinctive events, as episodes involving food, significant other people, and places (Engen 1987: 501ff.). Researchers found a consistency of performance across retention intervals of up to one year. Richardson and Zucco conclude that odour memory 'is normally very resistant to retroactive interference, that is, forgetting produced by subsequent learning experiences' (Richardson and Zucco 1989: 356).

Verbal encoding and rehearsal play little part in the storage of olfactory experience. Consistently, the right cerebral hemisphere seems to be more dominant than the left in identification and retention of odours (ibid.: 358).

It is seldom possible to resist negative reactions caused by smells; often they are direct and other types of knowledge have no impact on them. Engen, for example, states that, 'Even when one knows that it was not the smell of bourbon that made one sick but rather an overindulgence in it, the odour remains aversive as though it had been the cause. And such odour memories persist' (Engen 1991: 8). Correspondingly, if associated with positive experience when first detected, smells will be classified as positive later. Moreover, positive and negative reactions to odours, or to associations and memories regarding smells, are much quicker than one might expect (ibid.: 12).

Besides the importance of first experiences with smells, familiarity is important to their evaluation. In experiments a higher percentage of familiar odours are evaluated positively than unfamiliar odours (Engen 1991: 13ff.; Ayabe-Kanamura et al. 1998), which are evaluated negatively even when subjects are sure that they are harmless. Engen reports that 'even in a safe test situation where subjects had been told that no harmful odors would be presented, there was a strong tendency for unfamiliar odors to be viewed with disfavor' (Engen 1991: 14).

People often perceive smells without paying attention to them or without their reaching the threshold of consciousness: 'Awareness of odors is the exception rather than the rule. ... But even when olfactory stimuli can no longer be consciously perceived or are no longer attended to, they continue to exert influences on behavior and mood' (Köster 2002: 31). The capacity of odours to influence behaviour and mood below the threshold of consciousness may, nevertheless, be consciously used. The use of incense in church rituals evokes emotions and memories of former rituals, and the use of pleasant smells in shopping malls stimulates consumption. The perception of situations, places and persons is strongly influenced by odours, but the evaluation of and reactions to smells are also highly context-dependent and situational, which, of course, raises questions about the possibility of generalizing laboratory research results to everyday life.

It remains an issue whether inborn reactions to smell are found among human beings. Herz (2002: 160–62) writes that no inborn reactions exist, and Köster agrees, 'the absence of inborn mechanisms in olfaction is an evolutionary advantage in an omnivore' (2002: 30). Experiments with newborn infants have shown that they make a 'sick face' in the presence of putrid fishy smells (Steiner 1974). Schleidt interprets these results, and the fact that rotten food is labelled as unpleasant in very different cultures, as support for the thesis that certain dislikes are inborn, but perinatal shaping of selective odour responsiveness would also explain these reactions (Schaal, Soussignan and Martier 2002). However, to get into a discussion of whether reactions are learned or inborn would lead us away from our current concern. Suffice it to say that reactions to most smells are learned in early childhood and are quite resistant to later change, which suggests learning and culture have the main impact on the perception and evaluation of smells. For social and cultural anthropologists psychological findings indicate that the absence, presence and use of smells have strong culturally bound effects on emotions, self-perception, memory and identity, as well as on the perception of others and the environment.

Smells and their names are stored in explicit memory (conscious recollection) or implicit memory (resulting from unperceived influences), and in semantic, episodic and procedural memory (see below and contributions in Rouby et al. 2002). One of the most remarkable effects of smells is connected to the episodic memory. Some autobiographically important odours are stored in the memory together with whole episodes, which makes up part of the effect known as 'Proustian Phenomena' (Chu and Downes 2000), which will be described in more detail later on in this article.

To summarize: the associations and memories of smells learnt in childhood and the early stages of the life cycle are quite resistant to change. That means that most evaluations of smells are learned during early stages of socialization and enculturation. The perception of smell is most of the time unconscious. Reactions to olfactory stimuli are quick, emotional, at first nonverbal and often unavoidable. This shows how important environmental odours are, combined with cultural conventions about the use of smells in everyday life or in certain ritual, festive, social and emotional situations, for the perception of space and place. It also reveals methodological problems for research on the perception of smells that are not easy to solve.

Effects of Smell on Space Perception and Emotional Memories

Usually, olfactory perception is not involved in spatial orientation (Köster 2002: 30). However, some ethnographies show that foragers who hunt in forests, where sight is not very helpful, use smells in addition to sound in searching for game (Reichel-Dolmatoff 1985a, 1985b, 1989; Pandya 1990; Keifenheim 2000). In these cases, the sense of smell is not used to perceive space as such; it is used to detect if there is game and helps to focus *all senses* to track it down. Here, smell is less a matter of spatial orientation than spatial perception. A colleague told me that he overheard a conversation between two old men during his fieldwork in Fiji: one said to the other that he would always find his way back to the village at night because of the strong smell of a plant growing near there (Michael Dieckhardt, personal communication 2003).

Space and Emotional Memories

The importance of olfaction for the perception of space and place lies in its power to evoke emotionally charged memories:

> Because of the episodic nature of olfactory memory and its longevity, odors play an important role in making us feel at home in the world. Without any explicit awareness on our part, we link emotionally meaningful situations to them and carry them with us in an implicit memory that renders them up (usually involuntarily) when those odors are encountered again. (Köster 2002: 33)

The way smells (like other sensory data) are remembered differs from verbal memory significantly. They are usually stored in the episodic and not in the semantic memory. The semantic memory does not store perceptions of sensations, but their abstractions, the knowledge of the world. It represents perceptions and information as concepts and language. The procedural or episodic memory stores past events as such, the perception of sensations themselves (Larsson 2002; Reyna 2002: 64ff.). Stephen Reyna proposes a close connection between semantic memory and culture (Reyna 2002: chapter 5), which is a rather controversial theoretical position. The strong relationship between smells, implicit knowledge and the problem of their semantic representation in signs, symbols and language might be one reason for their neglect in anthropology.

Most people know from their own experience of the close relationship between smells, autobiographical recollections and episodic memory, or the so-called 'Proustian Phenomena'. Although autobiographically important, emotional recollections are individual; and some are highly idiosyncratic, as Marcel Proust showed in *A la Recherche du Temps Perdu* (1913). Some are based on collective practices and events (such as rituals, initiation, school, visits to hospitals), which have a shared cultural meaning.

Returning to one's old school, for example, and smelling the odour of floor-polish, unventilated classrooms, or the smell of sweat in the gym, activates emotional recollections – even after a very long time. Although the connections between smell, emotion and remembering are grounded in neurophysiological processes, what is connected by them varies in different cultures, and cultural domains. The example of 'school smells' is taken from European culture. Which smells connect places with emotional recollections in other cultures? And what memories do they evoke?

Examples from Fieldwork

In my research in the Philippines (see Figure 11.1) and in Papua New Guinea[2] I focused firstly on the categorization of smells (using the techniques of free lists and pile sorting, Figure 11.2), then on free associations related to real odour stimuli, metaphors, jokes, and stories people tell about experiences with smells in everyday life. The method of free lists involves persons naming all smells they know. Then words for smells that are recalled more than three times are written on cards. Informants are asked to sort these cards, to say which smell is typical for a category, what the smells have in common and/or comment on their decisions while sorting.

I want to give two examples of the slightly different significance that smell-evoked emotional memories of past experiences can have in another culture. My first example of smell as a marker of the border between the profane and the sacred and as a medium of communication between humans and spirits refers to the Visaya region in the Philippines. In Visayan culture odours are of great importance. In everyday situations, Visayans discuss smells very often and know more specific terms for smells than

German or English speakers do (e.g. *anghit* – smell of perspiration or goat; *angso* – urine or goat's smell; *langsa* – fishy smells; *langog* – putrid smells; *langto* – smell of blood, flesh, dirt; *bakag* – dead bodies, faeces and putrid smells; see Table 11.1 below). Smells mark boundaries not only between members of ethnic groups ('white people' are said to have a strong and very unpleasant body odour), between the sexes (fragrance is associated with femaleness and sexuality), but also between living persons and the dead, between everyday and religious contexts. Smells mark these boundaries, the transgression of them, and the transition from one condition to another.

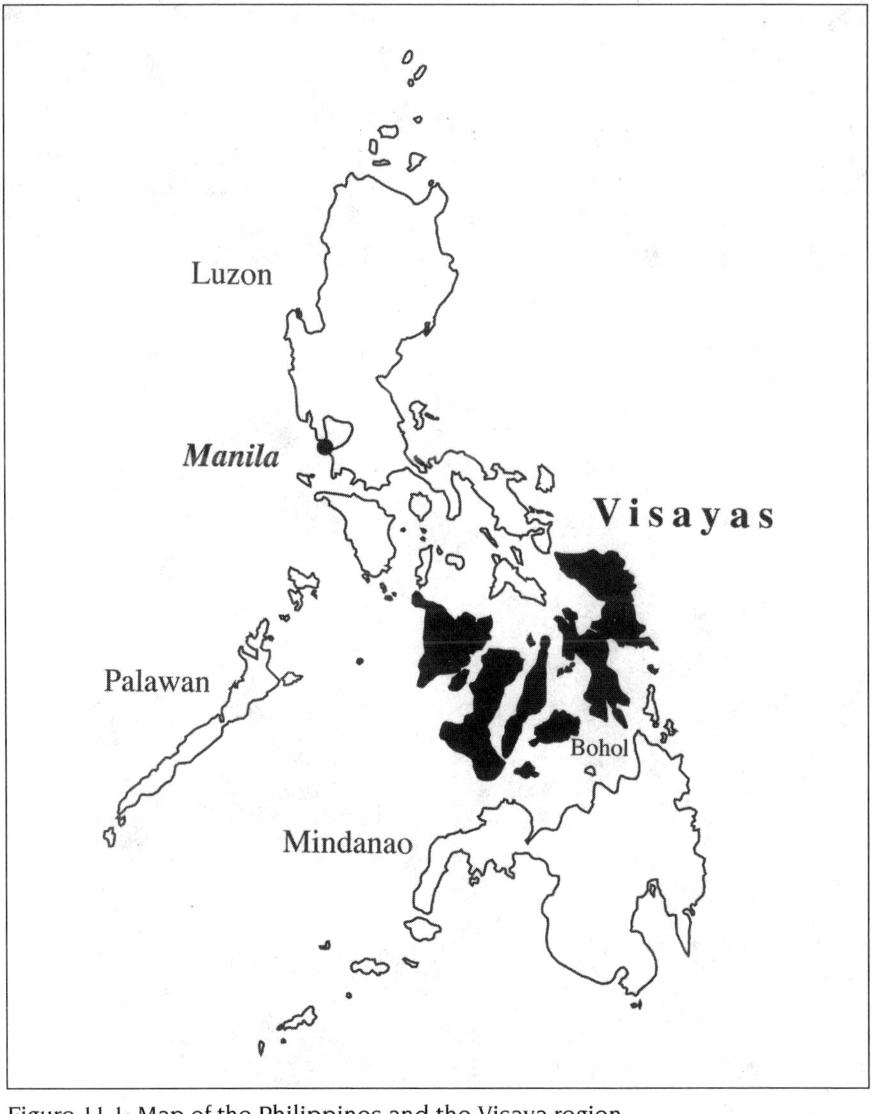

Figure 11.1: Map of the Philippines and the Visaya region.

Figure 11.2: An informant doing a pile sort task.

Figure 11.3: Incense and herbal medicine burned in a half coconut shell.

In Catholic Philippines the burning of incense is a perceptible indicator that transforms the church into an especially holy space. Also, however, burning incense together with certain herbal medicines is integral to traditional methods of curing. This smoking (*palinάan*) drives out the harmful spirits responsible for many sicknesses. Against certain diseases (cough, fever), particularly of infants, incense, or native resins, and dried herbs are burned in a half coconut shell (Figure 11.3). It is also used to purify people after a visit to the cemetery. Herbs and dried leaves are burnt and the person has to step over the fire's smoke to ensure a safe return from the spirit world.

On certain occasions, food is offered to the spirits of the dead, who are believed to consume and enjoy the food through its smell. Communication between the visible world and the invisible world inhabited by harmful and benevolent spirits is made possible with the help of smells, which encompass and connect both worlds.

The figures of saints, which are carried through the village particularly during the Easter processions, are decorated with flowers and their clothes are 'perfumed', usually with the latest fashionable brands (Figures 11.4 and 11.5). Also, the figure of Christ is perfumed when it is laid out in the church at the Easter weekend. On the Saturday after Good Friday, when believers kneel down in order to kiss his feet, they perceive his 'holy smell'.

Figure 11.4: Fragrant flowers are prepared for the decoration of saints during Easter Week.

Figure 11.5: Saint Veronica with a perfumed robe and decorated with fragrant flowers for a procession.

Smells are usually used in situations in which intensive contact with the other world exists, and these are sometimes highly emotional situations, which most Visayans have experienced from early childhood. Smells are more consciously perceived when they deviate from the everyday olfactory environment. The recollection of smell-connected experiences and their autobiographical significance has great power to change mood and evoke highly emotional responses.

My second example shows the close relationship between smell and emotional memories, which is familiar from numerous descriptions in European literature. Above all, childhood memories and very personal emotional experiences are evoked by smells. In the European context, nostalgia, emotion, memory and smell are firmly connected. Thus, it surprised me that in the Philippines my question about special experiences connected with smells elicited no childhood memories of the sort Germans report. Focused inquiries did not incite people to provide these nostalgic childhood narratives. Instead, Filipinos came often up with stories about encounters with spirits like the following:

> It was at a celebration of the anniversary of the death of my grandmother. Many relatives were with us to attend the celebration and I helped in the kitchen. I was alone in the kitchen and suddenly I smelled certain flowers, which my grandmother always liked very much. But there weren't any of them. I could not find flowers. I knew that the spirit of my grandma was with me. I said to it: 'Grandma, I am finished with cooking and want to take a bath, and you should go back again there, where you live now.' (Interview with T.R., 2002)

If Filipinos notice the smell of certain flowers that the deceased liked during their lifetime, or they smell a burning candle without a visible source, they see it as a proof of the presence of spirits. Spirits of ancestors are invisible, but now and then they can be perceived through smells. They live in a parallel world of ancestor spirits. The strong bond between emotional memories and smells has the same basis in affective cognition as in Europe, but Philippine stories are more strongly connected to religious contexts and/or encounters with spirits. Even more importantly, these memories are often interpreted as evidence for a still-existing parallel world in the present.

These differences in the meaning of memories probably have to do with concepts of person, the individual and identity. In the Philippines one's autobiography consists less of egocentric individual childhood experiences, with which one justifies one's personality as an adult and expresses one's individuality and identity. Narrations of autobiographic smell experiences in most cases refer to other humans, or different kinds of spirits, and to affective relations with them and to their persistence in the present.

Implications for Fieldwork and Methods in a Cognitive-oriented Anthropology of the Senses

Psychological research and my own fieldwork experiences give hints about how methods in a cognitive-oriented anthropology of the senses should be developed.

1. The senses vary in importance in different societies and certain contexts. The connection between language and sensual perception differs between cultures as well. Therefore it is necessary to elicit the basic sense vocabulary as a first step. Different languages might have more or fewer specific terms for sensations, which for example do not refer to the source of a smell but to the characteristics of the smell itself (as an example of non-comparative, specific smell terms see the list in Table 11.1). In everyday life discourses on sensations should be investigated.

Table 11.1: Noncomparative, specific smell terms in German, Visaya and Wampar.

German	Wampar (Papua New Guinea)	Visaya/Boholano (Philippines)
Geruch (smell) *Gestank* (stench) *Aroma* (aroma) *Duft* (fragrance) *Mief* (fug, stink)	*Muteran* (stench) *Rene waso* (smell, fragrance) *Ntsedz* (smell of roasted food, mainly meat) *Pupu* (urine) *Eboboron* (to smell fishy) *Ufin* (body smells) *Epaparats* (to smell like faeces)	*Baho* (smell) *Bahô* (stench) *Kahumot* (fragrance) *Kaamyon* (aroma, fragrance) *Alimyon* (fragrance, aroma) *Anghit* (smell of perspiration [only from axillary glands]) *Angso* (urine) *Langsa* (fishy smell) *Bakag* (smell of dead bodies [especially dead toads], faeces, putrid smells) *Langog* (putrid smell) *Langto* (blood, flesh, dirt)

2. To come to a fuller understanding of the significance of smells it is necessary to extend research beyond the psychologists' focus on 'veridical labels' (see Dubois and Rouby 2002). Dubois and Rouby summarize:

> When scoring the verbal performances of subjects in identification tasks and in diagnostic tests, words such as 'orange', 'apple', and 'vanilla' are considered the 'the correct odor names' or the 'veridical labels' (e.g. Engen 1987, Doty 1992). However, little consideration has been given to what, precisely, a 'word' is and what it means to researchers and subjects. (Dubois and Rouby 2002: 47)

The complicated, ambiguous and still obscure links between smells and words will not yield to such simple treatments when it comes to smells.

To grasp smells' affective connotations and their connections to emotions, values and morality, it is necessary to focus on metaphors, jokes, figures of speech, idioms, and stories as well (D'Andrade and Strauss 1992; D'Andrade 1995; Strauss and Quinn 1997). Von Karstedt suggests the application of 'cognitive schemata' as a methodical foundation for an anthropological approach to the study of the senses (von Karstedt 2003). 'Scripts', 'frames', 'idealized cognitive models', or 'schemata' refer to mental structures – that organize and store abstract information. They are neither passive archives, nor static entities because new information is added and structures are constantly updated; reactions are based on them. Knowledge based on perception and sensation is stored in individual and cultural schemata. Von Karstedt sees cognitive schemata as a starting point in investigating knowledge in particular cultural domains (such as 'smell' or 'taste'), their relations to one another and wider cultural contexts. They are methodological units, and fundaments for theories rather than theories themselves (ibid.: 308).

3. Language *and* real smells should be used in investigations, for olfactory sensations are not directly stored in semantic memory. Therefore, they are not immediately accessible to informants. Nonverbal reactions like gestures, facial expressions and behavioural reactions must be taken into account (Figure 11.6).
4. Many odours are volatile and short-lived, though in most environments more permanent smells exist as well (e.g. the smell of the forest or the sea). In mapping space and describing important cultural sites, smells and their dynamics should be taken into account (on the problem of culture-specific cartographies see Pandya 1990). Trees or plants with fragrant flowers could, for example, mark culturally meaningful places. Places where food is prepared, animals butchered, the dead buried, and ritually important spaces are probably also marked by smells. Odours add new dimensions to places in connecting time, emotion, moods and space.
5. Participant *and* systematic observations of everyday behaviour are necessary to grasp the significance of odours, their avoidance and use in everyday life. In particular, smell (and taste) sensations and their evaluations are very dependent on situation and context. To understand them in the laboratory or in experimental situations means only understanding how persons perceive them in these highly artificial settings (Köster 2002: 39).

   Comparative empirical research in the Philippines and in Papua New Guinea has shown that smells mark aspects of personhood, and spaces and places (as, for example, inhabited by ancestors), as spiritually important, dangerous, gendered, social, or erotic. Here it is argued that during fieldwork, participant observation is basic for discovering subtle aspects of the olfactory experience of space. The 'olfactorially significant' situations and spaces thus identified should be investigated further through more systematic observations and interviews.

Notes

1. The critique of the Western privileging of vision has itself been challenged (see Jay 1993).
2. My data are based on fieldwork in a rural area at the island of Bohol in the Visayan region of the Philippines (1993, 1994, 1996, 1997, 1999, 2003) and on fieldwork in Gabsongkeg village in Morobe Province, Papua New Guinea (1997, 1999/00, 2003/04).

Figure 11.6: An informant sniffs at a smell sample.

Chapter 12

Memory Measurement

Edgar Erdfelder and Martin Brandt

Memory has been an important topic in philosophy, rhetoric and the arts since ancient times (Finger 1994: chapter 23). However, it took until the end of the nineteenth century for the science of human memory to evolve from its philosophical roots. Influenced by the pioneering work of Gustav Theodor Fechner (1801–87), founder of psychophysics and ingenious inventor of sensation measurement methodology, Hermann Ebbinghaus (1850–1909) advocated the use of experimental control techniques and quantitative measurement methods to study the laws of human memory empirically. Unlike his predecessors, Ebbinghaus believed that scientific measurement methods could be applied successfully to higher mental processes and not just to elementary processes such as sensation or perception. His methods and empirical results, for the most part derived from experiments on nonsense syllables and lengthy poems in which Ebbinghaus served as his own and single subject, were published in the now famous book *Über das Gedächtnis* (Ebbinghaus 1885).

A wealth of different memory measurement methods has been proposed in the past 120 years, including refinements of the methods originally developed by Ebbinghaus (1885). Applications are not limited to nonsense syllables and poems any more. Contemporary memory research refers to stimulus materials of various types, for example, numbers, letters, words, names, sentences, texts, stories, sounds, smells, colours, scenes, pictures, movies, locations and actions. The goal of the present chapter is to provide both a system and a synopsis of modern memory measurement methods,[1] along with a discussion of the strengths and the weaknesses of these methods. We present examples and suggest approaches to overcome the weaknesses of the methods. In doing so, we specifically focus on possible applications in cross-cultural research. It is our belief that modern memory measurement methodology provides an indispensable toolbox for ethnologists, anthropologists and cross-cultural psychologists interested in uncovering the cognitive processes underlying cultural differences (cf. Beer, this volume). Conversely, we also believe that the significance and relevance of cultural differences has been considerably underestimated in past research on memory and cognition. Until recently, cognitive psychology has been dominated by the assumption that perceptual representations and elementary cognitive processes are universal and do not depend on culture (e.g. Rosch Heider 1972; Rosch 1973). However, as will be shown below, more recent research using refined measurement methods converges on results that tend to contradict universalistic positions. Instead they emphasize the importance of cultural differences. Joint efforts of ethnologists, anthropologists, geographers and psychologists are required to reveal the sources of such cultural differences in cognition.

We suggest a system of memory measures that is based on the distinction between direct and indirect measures (Johnson and Hasher 1987). Following Richardson-Klavehn and Bjork (1988: 477), 'we define as *direct* those tasks in which the instructions at the time of memory test make reference to a target event (or target events) in the personal history of the subject … . The subject is deemed successful in such tasks when she/he gives behavioural evidence of knowledge concerning that event.' In other words, in direct memory tasks subjects are required to remember some information experienced in the past. Typical examples include recall and recognition tasks referring to a list of items (e.g. vocabulary) they have previously learned. By definition, direct tasks always involve the intentional use of memory at the time of test. However, direct memory tasks must not be confused with intentional learning tasks. The distinction between *intentional* and *incidental learning* refers to the time of learning. In intentional learning situations, subjects are informed about the forthcoming memory test, in incidental learning situations they are not. In contrast, the distinction between *direct* and *indirect memory tasks* refers to the time of the test. As exemplified by a surprise class test in school, it is definitely possible that information is learned incidentally in the first place and intentionally retrieved later on in a direct memory test. Thus, direct tests may refer to both information learned intentionally and information learned incidentally.

Again following Richardson-Klavehn and Bjork (1988: 478), 'we define as indirect those tests requiring the subject to engage in some cognitive or motor activity, when the instructions refer only to the task at hand, and do not make reference to prior events'. Hence, in indirect tasks subjects are not asked to remember information acquired in the past but to engage in some other activity that indirectly reflects memory traces of past experiences. Consider Ebbinghaus' (1885) 'Erlernungsmethode' (method of complete mastery) and his 'Ersparnismethode' (method of savings) as examples. The method of complete mastery involves repeated intentional learning trials of a list of items with serial recall of the complete list on each trial. By definition, the number X of learning trials required until perfect mastery of the list is a direct measure of memory. It reflects both the difficulty of the to-be-learned list and the (lack of) memory capacity of the subject. Ebbinghaus' second method, the method of savings, requires repeated applications of the method of complete mastery at some time points t_1 and t_2 with $t_2 > t_1$. Let $X(t_1)$ and $X(t_2)$ denote the number of learning trials required for complete mastery of the same list of items at t_1 and t_2, respectively. A typical observation is $X(t_2) < X(t_1)$, that is, fewer learning trials are required if a person learns a list of items for the second time. Because no reference was made to the first learning episode in the second episode, the savings measure $X(t_1) - X(t_2)$ is an indirect measure of memory at t_2 for the first learning episode at t_1. Typically, the measure $\%S = ((X(t_1) - X(t_2)) / X(t_1)) \cdot 100$ is reported in empirical studies and interpreted as 'percentage of savings'.[2] As one might expect, %S decreases when the time difference $t_2 - t_1$ increases. Ebbinghaus' (1885) famous logarithmic law of forgetting over time was derived from this result.

Some authors refer to direct memory tasks as 'explicit tasks' and to indirect tasks as 'implicit tasks'. We do not follow this common praxis. Rather, we prefer to use the terms 'direct' and 'indirect' as defined above. In our understanding, the notions 'explicit' and 'implicit' indicate memories with and without the subjective experience of conscious recollections, respectively (Bredenkamp and Erdfelder 1996; Richardson-Klavehn and Bjork 1988). Explicit and implicit memories are perhaps associated with different memory systems in the human brain and must not be confused with tasks designed to measure them. As will be outlined below (see section 2.2), performance in direct memory tasks may be influenced by both explicit and implicit memories. The same holds for indirect measures.

The first part of this chapter is devoted to measures derived from direct memory tasks, the second part to those derived from indirect tasks. In each part, we will illustrate and discuss the most influential and important memory tasks of each type. Examples from cross-cultural research are limited to direct measures of memory, reflecting the fact that research on cultural differences in indirect measures is still in its infancy. Nevertheless, we consider it important that ethnologists and anthropologists are aware of these methods as options for assessing unconscious forms of memory. We therefore include them in our review.

For most of the memory tasks, we first describe ad hoc measures based on the raw scores typically obtained for that task and then proceed to more sophisticated model-based measures, which aim at separating several cognitive influences on the raw scores (cf. Bredenkamp and Erdfelder 1996 and Puff 1982 for other reviews of the literature). Model-based measures of memory provide remedies to several problems of ad hoc measures and thus form an important part of the scientific progress that has been made since 1885.

1 Direct Memory Measures

Roughly speaking, all direct memory measures are derived from two types of memory tasks: recognition and recall tasks. In recognition tasks, persons are asked to recognize a test item that may or may not have been experienced in the to-be-remembered target episode. In recall tasks, no test items are presented. Rather, persons are asked to recall information from a target episode, either with (cued recall) or without (free recall) the aid of retrieval cues. We will discuss recognition tasks first and then proceed to recall tasks.

1.1 *Recognition Tasks*

Most recognition tasks are yes-no or forced-choice tasks. In yes-no recognition tasks, participants receive a randomly arranged list of test items sequentially, consisting of some items that were experienced in the target episode before ('target items' or 'old items') and some other items that were not part of the target episode ('distractor items', 'new items', 'foils', or 'lures'). For each of the test items, participants are asked to indicate whether the test item was ('yes') or was not ('no') part of the to-be-remembered target episode. Forced-choice recognition tasks differ from yes-no tasks in that target and distractor items are presented together in sets. For each of these sets, participants are required to indicate the target item. More precisely, in m-alternative forced-choice tests (m-AFC tests, m = 2, 3, ...), one target item is presented along with (m-1) distractors, and the participant's task is to select the target. Consider Experiment 3 of Rosch Heider (1972) as an example. Influenced by the work of Berlin and Kay (1969), Rosch Heider was interested in testing the hypothesis that colour categories – or, more precisely, focal colours such as a prototypical red, green, yellow and blue – are universal. In other words, the cognitive representations of focal colours are not mediated or modified by language, culture or ecological context. To test this hypothesis she investigated colour memory in twenty native English speakers and twenty-one Dani of West Papua. According to Rosch Heider (1972: 16), the Dani language does not provide for more than two colour names, namely, 'mili' (roughly 'dark') and 'mola' (roughly 'light'). Rosch Heider (1972) reasoned that if focal colours are universal, memory differences for focal (e.g., green or blue), internominal (e.g., green-blue) and boundary colours (e.g., greenish green-blue) should be similar in US undergraduate women and Dani women. If, in contrast, the cognitive representations of colours depend on language, culture or ecological context, memory data for US and Dani women should differ markedly.

Rosch Heider (1972: 15–18, Experiment 3) presented single Munsell colour chips from each of the three colour categories (focal, internominal and boundary) for 5 seconds, always followed by an unfilled delay of 30 seconds. Next, a randomly ordered array of 160 Munsell chips was presented simultaneously, consisting of the target chip plus 159 distractor chips. The task was to indicate the target chip seen 30 seconds before. This procedure was repeated for eight target chips from each of the three categories. Technically, this task corresponds to a 160-AFC recognition test of visual short-term memory.

Rosch Heider (1972) analysed the number of correct recognition judgements for each of the three colour types separately for US and Dani participants. Although the US participants outperformed the Dani in each of the three conditions, the performance rank order across conditions was similar for Dani and US participants. Focal colours were recognized best in both groups, followed by the other two colour types that did not differ significantly. Hence, Rosch Heider (1972) concluded that focal colours are universal and identically represented in people from different cultures, even if some of these cultures lack unique linguistic labels for focal colours.

Can we be sure that this conclusion is correct? First, there is the problem of *lucky guesses*. In 2-AFC recognition tests, the a priori probability of a correct response by pure guessing is as high as 1/2; in m-AFC tests with $m > 2$ it is still $1/m$. Moreover, if the distractor items are chosen inappropriately, it can occur that the participant rules out some or all of the response options except one. Note that the probability of a correct response can approximate 1 in such cases even if a participant has no memory for the target items at all. Although these are serious problems for small m (see Bredenkamp and Erdfelder 1996 for some precautions in such cases), they cannot explain the results of Rosch Heider (1972) based on m-AFC tests with $m = 160$.

However, there is a second serious problem we need to reflect on, the problem of possible *response biases*. Differences in proportions of correct recognition judgements can arise even if recognition memory does not differ at all. Consider the example of a yes-no recognition test applied in two experimental conditions that differ only in one respect. In Condition 1, participants are correctly informed that 80% of the test items are old; in Condition 2 they are correctly informed that 20% of the test items are old. Knowing that 'old' responses will be correct in 80% of the cases, the first group will most likely adopt a more liberal response criterion and thus call more items 'old' than Group 2. As a consequence, the so-called 'hit rate' (HR), that is, the proportion of target items correctly called 'old', will be larger in Group 1 compared to Group 2. This happens as a consequence of different response tendencies in both groups even though recognition memory does not differ at all. Note that the same problem may apply to Experiment 3 of Rosch Heider (1972): since focal colours are more salient in the environment (e.g., because of their ecological frequency or their high level of saturation) and thus more familiar than nonfocal colours preexperimentally, there may be a strong response bias to select focal colours in the recognition test even if there would be no memory for the target colour chip at all. As a consequence, we cannot be sure that the recognition advantage of focal colours is due to better memory for focal colours.

How to cope with this problem? Most memory psychologists routinely use one of two model-based approaches to disentangle memory and response bias contributions to recognition judgements: Signal Detection Theory (SDT) or the Two-High Threshold Model (2HTM). Because both approaches are very important for appropriate analyses of recognition memory data, we will briefly summarize the assumptions and the methods of data analysis associated with these models. Readers not interested in these details can skip the following paragraphs and turn to the applications described below.

The standard version of SDT is based on the assumption that the subjective familiarities of old and new test items are normally distributed with a common standard deviation $\sigma = 1$. Whereas the distribution of new items represents preexperimental familiarities (see Figure 12.1b), the distribution of old items mirrors the preexperimental familiarities *plus* the familiarity increase d' caused by previous processing of the items in the learning episode (see Figure 12.1a). Thus, if the mean familiarity of new items is chosen as the zero point of the familiarity scale, the mean familiarity of old items is $0 + d'$. Given these assumptions, parameter d' can serve as a measure of the memory strength of old items at the time of test.

Figure 12.1 illustrates the typical dilemma of a participant in a yes-no recognition task. For a given test item he or she is asked to decide whether the item is old or new. The only evidence available to the participant is the subjective familiarity of the test item. However, the item familiarity often

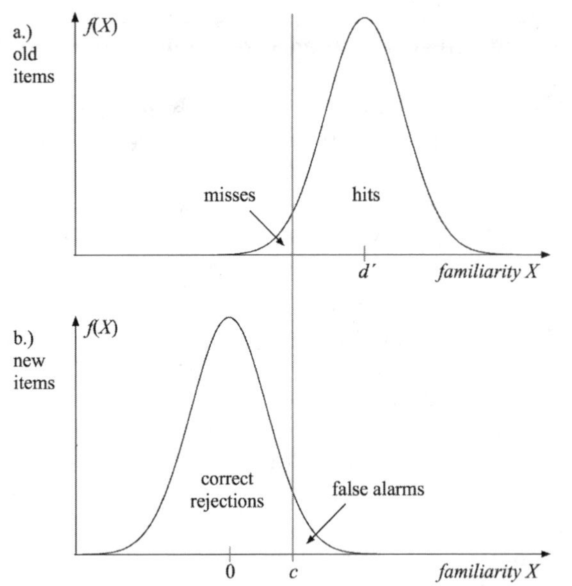

Figure 12.1: Assumptions underlying the standard version of Signal Detection Theory (SDT) applied to recognition judgements. X denotes the familiarity of test items, $f(X)$ the frequency (probability density) of X. The parameters d' and c denote the mean familiarity of old items and the response criterion for responding 'yes', respectively. Hits = correct 'yes' responses, False alarms = incorrect 'yes' responses, Misses = incorrect 'no' responses, Correct rejections = correct 'no' responses.

does not allow for a definite decision because the familiarity distributions overlap to a considerable degree. SDT assumes that participants solve this dilemma by setting a response criterion c on the familiarity dimension. If the subjective familiarity X exceeds the response criterion c for a given test item, the participant responds 'yes', thereby producing a certain proportion of correct 'old' decisions for target items ('hits', H) but also some incorrect 'old' decisions for distractors ('false alarms', FA). Otherwise he or she responds 'no', resulting in correct decisions for distractors ('correct rejections', CR) but incorrect decisions for targets ('misses', M). Whereas d' indicates the memory strength of target items, c measures response bias. Large values of c reflect a strict response criterion, small values of c a liberal criterion. Liberal criteria correspond to many 'yes' responses, both correct and incorrect.

Given these assumptions, the probabilities of hits, $p(\text{H})$, and false alarms, $p(\text{FA})$, are easily expressed as functions of the model parameters c and d':

$$(1) \quad \begin{aligned} p(\text{H}) &= 1 - \Phi(c - d') \\ p(\text{FA}) &= 1 - \Phi(c). \end{aligned}$$

In equation (1), (.) represents the cumulative standard normal distribution. Solving for d' and c, we obtain:

$$(2) \quad \begin{aligned} c &= z_{(1-p(\text{FA}))} \\ d' &= z_{(p(\text{H}))} - z_{(p(\text{FA}))}. \end{aligned}$$

Thus, by replacing $p(\text{H})$ with the empirically observed hit rate HR and $p(\text{FA})$ with the observed false alarm rate FAR, we can estimate the parameters d' and c as z–scores or z–score differences corresponding to these proportions. z–scores are defined by the inverse of the standard normal distribution function, that is, $p = \Phi(z) \Leftrightarrow z = \Phi^{-1}(p)$, and can be found in almost any introductory statistics book (e.g., Hays 1972: Appendix C).[3]

In a conceptual replication of the classical experiment by Rosch Heider (1972), Roberson, Davies and Davidoff (2000: 380, Experiment 2a, Table 5) used SDT to test for possible response bias effects. For another hunter-gatherer culture, the Berinmo from Papua, New Guinea, these authors successfully replicated Rosch Heider's result that the proportion of correct recognition judgements was highest for focal colours. However, when false alarms (i.e., incorrect decisions for focal colours) were taken into account and analysed by means of SDT, the recognition advantage of focal colours vanished in Berinmo participants. More precisely, response bias differed between conditions, producing more false alarms for focal than for nonfocal colours. In contrast, d' did not vary significantly for the Berinmos. For English-speaking participants, the results were reversed. Hence, whereas English-speaking participants show superior short-term recognition memory (d') for focal colours, Berinmo speaking participants, who lack appropriate linguistic labels for discriminating between these colours, do not show this memory advantage. Despite the fact that the raw data (i.e., the proportions of correct recognitions) are highly similar for the Dani and the Berinmo participants, the conclusions to be drawn from the SDT results of Roberson et al. (2000) are obviously exactly opposite to those drawn by Rosch Heider (1972).

We cannot rule out, of course, that these SDT results are only valid for the Berinmo and cannot be generalized to the Dani. In fact, there are some differences in colour vocabulary between both cultures that may account for the results (cf. Roberson et al. 2000: 370). Importantly, however, the evidence presented by Roberson and collaborators clearly shows that any analysis of recognition memory data is incomplete unless response bias has been taken into account explicitly.

An alternative method of separating effects of memory and response bias on recognition memory judgements is based on the Two-High Threshold Model (2HTM). Although SDT is used more often than the 2HTM in memory research, comparative model evaluations have shown that the latter model is as good as the former (Snodgrass and Corwin 1988). The 2HTM is an attractive alternative to SDT because (a) the 2HTM measures are more easily computed and (b) the 2HTM easily generalizes to more elaborated recognition paradigms (see, e.g., Bayen, Murnane and Erdfelder 1996). For yes-no recognition tasks, there are four basic 2HTM assumptions. First, the 2HTM assumes that the familiarity dimension can be subdivided into three intervals: an interval of low familiarity below the first threshold ('distractor detection threshold'), an interval of intermediate familiarity between the first and the second threshold, and an interval of high familiarity exceeding the second threshold ('recognition threshold'). Second, it is assumed that response probabilities depend on the familiarity interval only, not on the familiarity value X within intervals. Third, for target items, familiarity exceeds the recognition threshold with probability p_r. If this happens, a 'yes' response is given in the recognition task. If the target item familiarity does not exceed the recognition threshold with probability $(1-p_r)$, familiarity is assumed to be intermediate. In this state of uncertainty, the participant is assumed to guess 'old' with probability b_r and 'new' with the complementary probability $(1-b_r)$. Thus, hits can arise either by recognition or by guessing so that

(3) $p(\mathrm{H}) = p_r + (1 - p_r) \cdot b_r$.

Fourth, distractor items are assumed to attain familiarity values below the first threshold with probability p_r, resulting in correct rejections ('distractor detection'). If this does not happen with probability $(1-p_r)$, the familiarity is again assumed to be intermediate, resulting in 'old' and 'new' guesses with probabilities b_r and $(1-b_r)$, respectively. Consequently, false alarms can arise only by guessing:

(4) $p(\mathrm{FA}) = (1 - p_r) \cdot b_r$.

The two thresholds presumed by the 2HTM are 'high' in the sense that target items can never attain familiarity values below the first threshold whereas distractor items never exceed the second threshold. In other words, each item class is associated with familiarities in only two of the three familiarity intervals. If all these assumptions hold, p_r and b_r can be used as measures of recognition memory and response bias, respectively. By solving equations (3) and (4) for p_r and b_r,

$$(5)\quad \begin{aligned} P_r &= p(\mathrm{H}) - p(\mathrm{FA}) \\ b_r &= p(\mathrm{FA})/(1 - (p(\mathrm{H}) - p(\mathrm{FA}))), \end{aligned}$$

and replacing $p(\mathrm{H})$ and $p(\mathrm{FA})$ with the empirically observed hit rate HR and false alarm rate FAR, respectively, we can easily estimate p_r and b_r.[4] Reanalysing the data of Roberson et al. (2000) with the 2HTM would lead to the same conclusions that were also derived from the SDT analyses.

Importantly, both the SDT and the 2HTM are based on assumptions that may turn out to be false in some applications. In fact, generalized SDT models with unequal familiarity standard deviations for targets and distractors (Macmillan and Creelman 1991) and models with recollection parameters in addition to familiarity parameters (Yonelinas 1994) have been proposed in the literature and were found to be useful extensions of SDT in many applications. The 2HTM is also based on some restrictive and perhaps inappropriate assumptions. For example, at least for some applications, it might be reasonable to provide for the possibility that the probability p_r of target item recognition differs from the probability that distractor item familiarity falls below the distractor detection threshold.

How to test between these generalized models and the original versions of SDT and 2HTM? Receiver-Operating Characteristics (ROCs) are convenient tools to accomplish this goal. ROCs arise from pairs of hit and false alarm rates obtained when recognition memory is held constant but response bias varies. By plotting the hit rate HR on the ordinate against the false alarm rate FAR on the abscissa, typical ROC curves arise that allow one to test between different models (Macmillan and Creelman 1991). A convenient method for obtaining such ROC curves is to use confidence ratings in addition to yes-no recognition judgements. For example, a 6–point recognition confidence rating scale ranging from 1 = 'definitely new' to 6 = 'definitely old' generates five ROC points, one point for each of the five response criteria associated with the five category boundaries (for details and examples, see Macmillan and Creelman, 1991; Bredenkamp and Erdfelder 1996).

Next to *m*-AFC tasks, yes-no tasks, and confidence rating scales, source monitoring tasks are very useful extensions of recognition memory methodology. Source monitoring tasks resemble recognition tasks in that participants are asked to classify each test item as either 'old' or 'new'. However, for each of the items classified as 'old', participants are additionally required to indicate the source associated with this item in the past. As in *m*-AFC tasks, *m* different sources can be considered ($m \geq 2$). Consider experiments on bilingual memory as an example. Saegert, Hamayan and Ahmar (1975) presented bilinguals with a mixed list of French and English words in the study phase of their experiment. One group of participants saw each of the words in isolation (word group), the other group saw each word as part of a sentence (sentence group). Later on, participants received a source monitoring test including old French and English test words intermixed with new test words. All test words were presented in English. For each of the words they classified as 'old', participants were additionally asked to indicate its source during study, that is, whether this word was originally presented in French or in English.

Saegert and collaborators aimed at testing single-code theories of bilingual memory against dual-code theories. According to single-code theories, lexical information is stored in a single, abstract representation, irrespective of the language in which it was originally presented. Thus, memory for the language of information acquisition is possible only if 'language tags' are stored in episodic memory in addition to the lexical information. Dual-code theories, in contrast, assume that separate mental lexica are used for each of the languages. Saegert et al. (1975) argued that if lexical information is successfully retrieved, memory for the language should generally be good according to dual-code theories because participants simply need to identify the mental lexicon they used to retrieve the information. According to single-code theories, in contrast, memory for the language should be poor, especially if semantic processing of the information prevents persons from storing 'language tags' in episodic memory.

Saegert et al. (1975) measured recognition memory as the proportion of items correctly recognized. As expected, they found poorer recognition memory in the sentence group compared to the word group. Source memory (i.e., language memory) was measured as the proportion of correct language judgements, conditional upon correct recognition judgements. Matching the results found for recognition memory, language memory also proved to be much poorer in the sentence condition. Thus, in line with the reasoning outlined above, Saegert et al. (1975) concluded that their results are supportive of single-code theories and inconsistent with dual-code theories.

Is this conclusion justified? Remember that analyses of recognition memory data should always take response bias into account. Of course, this also holds for source-monitoring paradigms. Source-monitoring measurement models separating effects of (1) item recognition memory, (2) source memory, (3) item response bias and (4) source guessing bias, all of them generalizations of the 2HTM discussed above, were suggested and evaluated by Batchelder and Riefer (1990) and Bayen et al. (1996). Batchelder and Riefer (1990) reanalysed the data of Saegert et al. (1975) with their high-threshold source-monitoring model. The results matched those of Saegert and collaborators with respect to item recognition memory. For source memory, however, Batchelder and Riefer (1990) obtained a completely different pattern of results: for recognized words, source memory for the language was equally high for both experimental groups, consistent with what dual-code theories would predict. Thus, as in the case of Roberson et al. (2000), the interpretation of the results again reverses if memory measures based on sophisticated model-based approaches replace ad hoc measures derived from raw scores. If correctly analysed, the data actually support dual-code theories of bilingual memory although at the first sight the raw data appear to be supportive of single-code theories.

1.2 *Recall Tasks*

In contrast to recognition tasks, recall tasks are characterized by memory recall instructions in the absence of the to-be-recalled information. Recall tasks are either free recall or cued recall tasks. In cued recall tasks, certain cues are provided that might help in retrieving the target information. Such cues are lacking in free recall tasks. We will discuss free recall tasks first, followed by a discussion of cued recall and the relation between recall and recognition performance.

1.2.1 Free Recall

In free recall tasks, participants are asked to recall all pieces of information from a target episode, irrespective of order. A typical example would be free recall of a list of words learned previously. Recall of childhood memories from the first four years of life would be another example. A recall measure typically employed is the number X of items recalled from the target episode. If the total number N of items in the target episode is known, percentage recall measures such as $(X/N) \cdot 100$ are also quite common.

Next to X or $(X/N) \cdot 100$, the number of intrusions I should also be considered and reported in publications. I resembles the false alarm rate FAR in recognition tests in that it indicates the number of items recalled that were *not* part of the target episode. Just like high values of FAR, high values of I indicate a strong guessing bias. If X and I correlate across conditions, one should be very careful in interpreting differences in X as differences in recall memory. In such situations it is quite likely that X is inflated by guessing, so that differences in X may actually reflect differences in guessing rather than differences in recall memory. In general, it may be a good idea to use $X - I$ rather than X as a recall memory measure, similar to the recognition memory measure p_r of the 2HTM (see equation 5).

Special problems arise if the target episode is unknown to the researcher so that neither N nor I are defined. As an example, consider the studies by MacDonald, Kimberly and Hayne (2000), who asked Chinese and New Zealand participants from three cultural backgrounds (European, Maori and Asian) to recall and date their earliest personal memories. They found that Maori adults recalled significantly earlier memories than European adults. Asian participants reported the latest memories, an effect that was especially pronounced in women. Guessing is hard to assess in studies on autobiographical memory because the target episodes are unknown, except to participants themselves.

How to cope with this problem? We would recommend a control-group design to control for possible guessing influences. In such a design, a randomized control group is required for each of the participant groups of interest. Participants of the control groups are not asked to recall own memories but to guess the earliest childhood memories of another unknown person from the same culture. If the reported dates of the guessed childhood memories in the control groups correlate with the reported dates of the recalled childhood memories in the experimental groups, we would be very cautious in interpreting the effects. If recall differences between cultures also show up in the control groups, stereotype and schema influences on autobiographical memory reports are quite likely. Since MacDonald and collaborators did not run control groups, we cannot rule out that their results do not reflect differences in actual childhood memories but differences in cultural stereotypes instead.

If the to-be-recalled target episode includes separate categories of information, for example, words of different semantic categories, the number *C* of categories recalled at least once and the number *IPC* of items recalled per category are also of interest. If *C* is low but *IPC* is high this points to possible problems in retrieving information, whereas storage of information within categories appears to have been successful (Tulving and Psotka 1971). If, conversely, *IPC* is low and *C* is high this points to possible problems in item storage, whereas category retrieval appears to be less of a problem. Note, however, that more elaborated, model-based approaches exist for separating storage and retrieval influences on free recall data (e.g., Batchelder and Riefer 1986). For the sake of brevity, we cannot discuss these or other approaches to the analysis of free recall data here (see Puff 1982 and Bredenkamp and Erdfelder 1996 for more extended reviews of free recall methodology).

1.2.2 Cued Recall

In cued recall, participants receive cues to aid recall from the target episode. Retrieval cues can basically be of any type. Thus, semantic cues ('Which type of beverage was mentioned in the list?'), rhymes ('Which target word rhymed to 'hear'?), word stems ('Which target item started with 'be..'?), perceptual attributes ('Which target item tastes bitter?') and semantic associates ('froth'), among others, can serve as retrieval cues for the target item 'beer'. Moreover, sounds and smells form an important class of retrieval cues in everyday life (see Beer, this volume). Cues can be both explicit and implicit. For example, an explicit cue for the to-be-recalled sentence 'The sun shines on the snowman' would be the word 'sun', whereas the consequence 'Therefore, he melts' would be an implicit cue. Retrieval cues need not necessarily refer to verbal information only. For example, several of the Nijmegen space games described by Senft (this volume) can be conceived as cued recall tasks, provided that (short) retention intervals elapse between presentation of the target information and the cue-aided response. Thus, a unifying definition of retrieval cues is neither possible nor desirable. In fact, different types of retrieval cues are often used in the same experiment to assess their differential capacity in aiding successful recall.

Typical measures of cued recall are frequencies (*R*) and percentages (%*R*) of correct cued recall judgements. Of course, guessing is an even more serious problem than in free recall because retrieval cues typically narrow down the possible answers to small sets. Thus, even if participants cannot retrieve the target information, there is a quite high chance of lucky guesses. Therefore, control conditions should be run in which certain pieces of information (e.g., 'beer' in the example above) are *not* part of the target episode. These control conditions can be used to assess the percentages of intrusions (%*I*). Again, it is better to refer to %*R*-%*I* as a measure of cued recall accuracy than to %R alone.

Serial recall (SR) and paired-associate learning (PAL) are special cases of cued recall (Bredenkamp and Erdfelder 1996: 13). In SR, participants are asked to recall the target information in exactly the same order in which it was originally presented. SR instructions can be understood as sequences of cues ('Please report the first item, then the second item, the third item etc.') and thus form a special case of cued recall. In PAL, participants study A-B pairs of items and later receive the first item A as a retrieval cue for the second item B. Consider another classical study by Rosch (1973) as an example. To further test the theory of universal focal colours, she taught sixty-eight male Dani participants, unaware of colour names other than 'mili' and 'mola' (see above), to respond to Munsell colour chips with the name of a Dani *sib*. A Dani *sib* is something like a Dani clan. Each sib name was associated with three different colour chips as stimuli. The colour triples were of three types: (1) focal colours as central members of the category (i.e., natural colour categories according to Berlin and Kay 1969), (2) internominal colours as central members and (3) boundary colours as peripheral members of the category. As expected, Dani participants learned the colour-sib pairs of the first category faster and with fewer errors than the other two category types. Errors for focal colours were generally lowest, matching the results previously found for short-term recognition of colours (Rosch Heider 1972).

It should be acknowledged that Roberson et al. (2000, Experiments 3a and 3b) attempted to replicate these results for the Berinmo. The authors made sure that response stimuli (words in Experiment 3a and pictures in Experiment 3b) were randomly assigned to colour stimuli across participants, thus ruling out the possibility that some stimulus-response pairs are easier to learn than others. As in the case of their visual short-term memory study, the replication of Rosch's results failed. The single exception was the focal red colour chip (5R 4/14 in Munsell notation) that was consistently learned faster than other focal and nonfocal colours. As outlined by Roberson et al. (2000: 385), it is difficult to decide whether this is due to the salience of this colour or due to the fact that this focal red is also at the centre of a linguistic colour category of the Berinmos.

1.2.3 Relation between Recognition and Recall

Free recall, cued recall and recognition tasks are often interpreted as varying in retrieval difficulty, with recognition tasks involving memory storage only and no memory retrieval at all. According to this interpretation, free recall and cued recall tasks are more difficult than recognition tasks because they require memory retrieval of the critical information in addition to memory storage. In fact, the generate-recognize theory of recall (Baddeley 1997) posits that recall consists of a generate stage in which candidate items are mentally generated, followed by a recognition stage in which candidate items are classified as 'old' or 'new'. Only items that were generated in the first stage and classified as 'old' in the second stage are recalled. According to this theory, recall implies recognition, that is, it is impossible to recall an item that cannot be recognized.

It is important to know that this view of the relation between recall and recognition is definitely false. More precisely, recall does not imply recognition and retrieval problems may arise even in recognition tasks. Consider the famous recognition-failure paradigm studied by Tulving and Thomson (1973). Weakly associated A-B pairs of words are presented for study. Later on, the isolated B words are presented in a yes-no recognition test along with several distractors, followed by a cued recall test of the B words given the A cues. In this paradigm, it often happens that participants fail to recognize a B word in the recognition test but successfully recall it later on given the retrieval cue A. Thus, for some words at least, recall can be successful when recognition fails. Tulving and Thomson (1973) explained this surprising result by the principle of encoding specificity. Encoding specificity means that human memory does not store items or other pieces of information in isolation. Rather, memory stores items *plus* relevant aspects of the encoding context of the item. As a consequence, retrieval is easy when the retrieval context matches the encoding context; it is difficult when retrieval and encoding context mismatch. Applied to recall and recognition, a recall task can be made easy by making the retrieval context similar to the encoding context. Conversely, recognition can be made difficult by selecting a retrieval context that significantly differs from the encoding context in important respects. This is the essence of the encoding specificity demonstrations of Tulving and Thomson (1973).

Riefer and Batchelder (1995) suggested a model of storage and retrieval processes in human memory that is based on the principle of retrieval independence in recognition and cued recall tasks. This model successfully accounted for many results obtained in the recognition failure paradigm and thus shows that recall and recognition should be seen as independent rather than dependent retrieval processes.

Numerous results can be explained by the encoding specificity principle. For example, information acquired under water is recalled more easily under water, whereas information acquired on land is recalled more easily on land (Godden and Baddeley, 1975). Similarly, autobiographical memories acquired by bilinguals in language A are better recalled later on in the same language compared to their second language B. The reverse result holds for autobiographical memories acquired in language B (Schrauf 2000; Schrauf and Rubin 1998, 2000). Thus, evidence from autobiographical recall memory converges with evidence from source-monitoring studies in that both support dual-code theories of bilingual memories.

2 Indirect Memory Measures

Indirect memory tasks differ from direct tasks in that subjects are not instructed to remember a target episode in the personal past. Rather, they are asked to perform some cognitive or motor task that can be solved without consciously retrieving information from the past. Nevertheless, performance in this task can be influenced by a specific past episode even if subjects are unaware of their memory being tested for this episode. Like direct memory tests, indirect memory tests have a long history (cf. Schacter 1987), for a review of the historical roots).

The typical procedure of an indirect memory task is as follows. In the first phase, subjects in the experimental group process the target material, typically (but not necessarily) under incidental learning instructions. Subjects in the control condition perform the same task, but with different material that is irrelevant for the forthcoming second phase. In the second phase, subjects of both groups work on a second task that may or may not differ from the task of Phase 1. However, the material processed in the second phase must be identical or related to the material processed by the experimental group in the first phase. Importantly, the Phase 2 task must not presuppose the processing of the material in the first phase, and the Phase 2 instructions must not emphasize retrieval of the first phase. In such experiments, subjects in the experimental group typically perform better or faster compared to the control group. Differences in performance between both groups are often called 'priming effects'. Priming scores serve as standard memory measures in indirect tasks.

If control groups are lacking, indirect measures cannot be interpreted unequivocally. For example, Ames and Stacy (1998) studied the relationship of implicit memory and drug abuse in different ethnic groups. They used a cue-association task as an indirect memory measure. Their participants had to respond to a semantically ambiguous cue with the first word that came to their mind. The number of drug-related responses was used as an indirect memory measure. Ames and Stacy interpreted their results as showing that the strength of indirect memory for drug-related topics predicts drug abuse better than gender or the ethnic group of the subjects. However, this conclusion might be questioned because both their memory measure and the self-reported drug abuse most likely are affected by the willingness to talk about drug-related topics. Note that this problem is similar to the response bias problem in direct memory measures. To avoid this problem, we strongly recommend control-group designs in indirect memory research. In Phase 1, for example, the participants could be asked to solve a set of drug-related anagrams (e.g., 'NIORHE' that can be solved as 'HEROIN') in the experimental condition and a comparable set of drug-unrelated anagrams in the control condition. Ames and Stacy´s cue-association task could then be used in Phase 2. Given this procedure, the priming score (i.e., the experimental-control difference in the number of drug-related associations in Phase 2) would be uncontaminated by the willingness to talk about drugs and thus provide a better indirect measure of memory for drug-related information.

2.1. *Types of Indirect Tests*

In this section we briefly describe some well-known indirect memory tasks and the corresponding memory measures. We do not aim at an exhaustive enumeration of indirect tasks. Rather, we want to illustrate the variety of tasks used to study memory phenomena indirectly. Given the variety of indirect memory tasks, it is useful to classify them along some dimensions. One possibility is to refer to the processing demands of the tasks (e.g., Tulving and Schacter 1990). According to this view, indirect tasks can be classified into conceptual tests and data-driven tests. *Conceptual tests* require semantic processing of the target materials, similar to what is also often involved in direct memory tests. *Data-driven tests*, on the other hand, focus on the perceptual processing of the stimulus. Another classification, the dichotomy of indirect memory tests and indirect learning tests, is suggested by looking at the knowledge structures assessed by the tests. In *indirect memory tests*, preexisting knowledge structures are changed temporarily. In *indirect learning tests*, new knowledge structures or motor programmes are acquired. Examples will be presented in the subsequent sections.

Different types of indirect tests may be important for cross-cultural research because they provide ideal tools for answering questions like 'Are cross-cultural differences in cognition limited to conceptual processes? Or do they also show up in data-driven tests and indirect learning tasks that are more closely linked to perceptual processes or motor programmes?' To our knowledge, convincing answers to these questions are still lacking.

2.1.1. Conceptual indirect memory tests

In *word completion tasks*, subjects first perform a task on lists of words (e.g., pleasantness ratings). In the second phase, an incomplete letter string is provided and participants are asked to complete this string to form any word. Suppose that in Phase 1 the word HOUSE was presented to participants of the experimental group but not to participants of the control group. In Phase 2, the letter string HO_ _ _ (*word stem completion*) or _ O _ _E (*word fragment completion*) is presented. Note that there are several solutions to this task (e.g., HOUSE and HORSE) and that the instructions do not emphasize the retrieval of words that were shown in the first phase. A priming score is computed by subtracting the percentage of cues completed to target words (e.g., HOUSE)in the control group from the percentage of cues completed to target words in the experimental group.

In *lexical decision tasks*, a letter string is presented in Phase 2 and the subject's task is to decide as quickly as possible whether the letter string is a regular word or not. To make this task nontrivial, only half of the presented letter strings conform to words. The latency of this decision is typically reduced if the same word was previously presented to the subject in the experimental group. A priming score is computed by subtracting the response time of the experimental group from the response time of the control group. If the same words are used in both phases, priming is called *repetition priming*. However, priming can also be observed if the words are only semantically, graphemically or phonemically related. Such priming effects are referred to as *associative priming* effects.

In *category-exemplar production tasks*, subjects are required to produce an arbitrary exemplar of a given semantic category (say, birds). Former processing of a specific exemplar in the experimental group (say, robin) enhances the probability that this exemplar is produced in the subsequent category-exemplar production task compared to a control group.

2.1.2. Data-driven memory tests

In *perceptual identification tasks*, a stimulus (word or picture) is presented in Phase 2 in a perceptually degraded form. The subject's task is to identify the stimulus as quickly as possible. Usually, the degree of degradation is reduced continuously until the stimulus is identified. In the experimental group, the stimuli were previously processed in a nondegraded form in an initial study phase. Again, priming scores are computed as differences between response times of the experimental group and the control group.

In *evaluation tasks*, subjects rate the pleasantness of stimuli in Phase 2. Prior exposure to the same stimulus material typically increases the pleasantness ratings. Priming scores can be computed as differences between pleasantness ratings in the experimental group (with prior exposure of the stimulus) and the control group (without prior exposure of the stimulus).

2.1.3. Implicit learning tests

In *procedural learning tasks*, subjects are required to learn a new motor task. In the *pursuit motor task*, for example, the subject tracks a light spot with a pen on a complex path. In each trial, the distance between pen position and light spot can be used as a performance measure. Performance in this task improves across trials. Another often-used task is *mirror drawing*. In this task subjects are asked to draw a line within a complex form (e.g., a star). This task is rather complex. Subjects can neither see the form nor their hand directly, only via a mirror. In this task, errors can be counted each time the line leaves the given form. As in the pursuit motor task, the subject's performance improves over several trials, thus showing a memory effect. Interestingly, amnesics (e.g., the patient HM, Milner 1965) also show such improvements in perceptual-motor tasks although they do not remember having done this task previously.

In *classical conditioning*, an unconditioned stimulus (US) that reliably triggers an unconditioned response (UR) is coupled with a neutral stimulus that does not evoke the UR (or responses similar to the UR) before the experiment. After some trials, the neutral stimulus triggers a response similar to the UR in the absence of the US. Thus, the neutral stimulus has become a conditioned stimulus (CS) that triggers a conditioned response (CR). For example, an air blast on the cornea (US) causes the closure of the eyelid (UR). If the blast is preceded by a tone several times, the tone alone will become a CS and trigger the closure of the eyelid (CR). In this paradigm, memory for the CS–US association shows up as a behavioural change in response to the CS.

In some implicit learning experiments subjects perform rather complex tasks. For example, subjects might process letter strings generated by an artificial grammar in Phase 1 (Reber 1967). This grammar is based on a set of rules for valid letter combinations (e.g., a 'B' is always followed by a 'B' or a 'T'). In a second phase, subjects are required to classify new letter strings as 'valid' or 'invalid'. Despite the fact that subjects are typically unaware of the precise nature of the rules that define the artificial grammar, they can often classify valid and invalid letter strings at an above-chance level. Note that this task resembles yes-no recognition tasks. Hence, response bias should be taken into account. As in direct memory tasks, signal detection methods or threshold models are appropriate tools.

Another complex learning task is the *serial reaction task*. Subjects are required to press a sequence of target buttons in response to certain stimuli (e.g., lights or sounds) as fast as possible. This procedure is repeated several times without a break. The participant is only informed about the response buttons to be pushed given different stimuli; he or she is not informed that the order of the stimuli (and the associated buttons) is not randomly determined but follows a certain order. Despite their being unaware of the order of the required responses, participants will generally become faster in their reactions across trials. To test whether subjects have implicitly learned something about the order of the stimuli, a critical trial is implemented after several trials. In this trial, the stimulus order is changed. Reaction times in critical trials typically increase again, showing that the subjects must have learned something about the order of the stimuli. In contrast to many other indirect measures, memory for stimulus order shows up as a *decrease* in performance in this paradigm.

2.2. Experimental Dissociations

What are the benefits of indirect measures? Why not rely on direct measures alone? Indirect measures are crucial to a central problem in memory research – how many different memory systems are there (e.g., Tulving 1985; Buchner and Brandt 2003)? When a new memory system is proposed, the typical assumption is that it is functionally different from other memory systems. In general, different memory systems are thought to be associated with different sets of memory tasks. For example, direct tasks are assumed to be associated with an explicit memory system and indirect tasks with an implicit memory system (e.g., Tulving and Schacter 1990). Alternatively, direct and indirect tasks are sometimes thought to be associated with different memory processes (e.g., Roediger 1990). One of the most fascinating findings in this area is that amnesiacs (patients with memory disorders due to neurological impairments, typically in the medial temporal lobes or in the diencephalon) show dramatically impaired performance in direct memory tasks whereas performance in indirect tasks is more or less unaffected (e.g., Warrington and Weiskrantz, 1970).

Similar types of dissociation have been established in research with normal subjects, too. For example, Tulving, Schacter and Stark (1982) found that longer retention intervals impair performance in direct but not indirect tests. Such empirical dissociations seem to corroborate the idea that direct and indirect memory tasks tap functionally different memory systems. Jacoby (1991), however, convincingly argued against this interpretation. He argued that each cognitive task is typically affected by more than a single cognitive system or process. For example, consider the word-stem completion task outlined above. Although the instructions do not prompt the subject to recollect words of the study phase (i.e., Phase 1) they still might do so. But if a subject consciously tries to remember words of the first phase, explicit memory is clearly involved! Similarly, subjects can intentionally try to remember letter strings or parts of letter strings from the first phase in artificial grammar learning. Again, this task might be affected by intentional retrieval processes. On the other hand, direct memory tasks might be influenced by the implicit memory system as well. In a yes-no recognition task, for example, a subject might recollect a specific detail (e.g., 'when I saw this item in the study phase I had to think of my shopping list'). Alternatively the subject might rely on an unspecific but strong feeling of familiarity. In both cases the subject would classify the test item as 'old', but the underlying processes would be quite different. These examples demonstrate that ad hoc measures of performance in direct or indirect tasks cannot generally be interpreted as pure measures of a single memory system or process.

2.2.1. Process dissociation models

Jacoby (1991) introduced a process dissociation framework to obtain process-pure measures of cognitive processes. In doing so, Jacoby's focus was on controlled retrieval processes (associated with explicit memory) and on automatic processes (associated with implicit memory). The process dissociation framework consists of specific experimental procedures and an associated measurement model. The key idea is to contrast two conditions. In one condition automatic processes are supposed to ease the task. This is usually the case in indirect memory tasks. Since automatic memory processes and controlled processes both lead to correct responses, this condition is called the facilitation condition. In a second condition, automatic processes are supposed to make the task more difficult, so that poorer performance is expected. Therefore, this condition is called the interference condition. Consider the word-stem completion task as an example. The facilitation condition is identical to the classical word-stem completion procedure. Thus, subjects are instructed to complete the cue to any word that comes to mind. In this condition, subjects might engage in controlled retrieval processes to recollect words from the incidental learning phase of the experiment. Alternatively, the word stem can be completed to the target word automatically in the absence of conscious recollections. Thus, in the facilitation condition, both controlled and automatic processes work in concert because both processes result in a 'correct' completion (i.e., a completion to the 'old' word processed earlier). Subjects in the interference condition, in contrast, receive slightly different instructions. They are told to complete the word stem to any word that comes to mind *except* to words processed in the first phase. Thus, controlled retrieval processes result in a correct completion to a nontarget word. Automatic processes, however, enhance the probability of an incorrect target word completion when conscious recollection fails.

The process dissociation measurement model suggested by Jacoby (1991) assumes independence of automatic and controlled processes. Let r denote the probability that the word is retrieved via controlled recollection processes, and let a represent the probability that the word is retrieved automatically. Given independence, the word stem is completed to an old word in the facilitation condition with probability

(6) $p_f = r + (1-r)\,a.$

In the interference condition, however, the word stem is completed to a target word with probability

(7) $p_i = (1-r)\,a.$

Solving for r and a we obtain:

(8) $r = p_f - p_i$

(9) $a = \dfrac{p_i}{1-r}.$

By replacing the probabilities p_f and p_i with empirically observed proportions in process dissociation experiments, estimates of r and a are easily obtained. For example, Jacoby (1998) studied word stem completion under full and divided attention conditions. In the facilitation condition, a priming score of .25 was obtained under full attention. Under divided attention the priming score was .17 only (Jacoby 1998: Experiment 1). Thus, if we assume that the priming score reflects automatic memory processes selectively, we would conclude that divided attention impairs automatic processes. However, if the data are analysed in the framework of the process dissociation model, a completely different picture emerges. Automatic processes seem to be unaffected by divided attention (a = .54 and a = .55 under full and divided attention, respectively) whereas controlled recollection (as measured by parameter r) is impaired under divided attention (r = .29 and r = .14 under full and divided attention, respectively). Again, model-based memory measures favour conclusions opposite to those derived from ad hoc memory measures alone! This could also happen in cross-cultural applications of these methods. For this reason, cross-cultural researchers should be aware of the process dissociation procedure as an option for collecting and analysing their data on cultural differences in explicit and implicit memory.

The process dissociation model can easily be adapted to other indirect or direct tasks. Several improvements of the basic model have been suggested. Some of them provide solutions to the problem of response bias influences on the parameters in process dissociation models and to the problem of possible correlations between controlled and automatic processes (e.g. Yonelinas 1994; Buchner, Erdfelder and Vaterrodt-Plünnecke 1995; Vaterrodt-Plünnecke, Krüger and Bredenkamp 2002;. Erdfelder and Buchner 2003).

2.2.2. Reliability of memory measures

A simple dissociation is defined as a data pattern with a significant difference in one measure (typically the direct measure) and a nonsignificant difference in the other measure (typically the indirect measure). For example, Warrington and Weiskrantz (1970) showed that amnesics perform significantly worse on direct memory measures but do not differ significantly from normal control persons on indirect memory measures. This result has been interpreted as evidence for an impaired explicit memory system (as measured by direct tests) and an intact implicit memory system (as measured by indirect tests) in amnesics. However, this interpretation is quite problematic. Buchner and Wippich (2000) pointed out that the probability of detecting a population difference with a statistical test – other things being equal – depends on the reliability of the memory measure used. Moreover, they also showed that indirect memory tests tend to be less reliable than direct tests. Thus, even if direct and indirect memory tests tap the same memory systems, simple dissociations would be quite likely. Hence, simple dissociations found in memory research or cross-cultural research would not provide strong evidence for different memory systems. One should keep in mind that this problem is not restricted to the comparison of indirect and direct measures. It applies to any comparison of different memory measures, for example the comparison of recognition and recall measures or different indirect measures. Because group comparisons in clinical research often rely on very small group sizes, reliability differences are a very serious problem.

There is no standard solution to the problem of different reliabilities of memory measures. When designing a study, one should choose tasks that maximize reliability. For indirect memory measures, Buchner and Wippich (2000) argued that memory tasks with specific goals tend to have higher reliabilities compared to tasks with unspecific goals. For example, an evaluative judgement measure in a mere exposure paradigm tends to be less reliable than a lexical decision priming measure, because in the lexical decision task (but not in an evaluative judgement task) the correct answer is clearly defined. Another reasonable way of addressing the reliability problem is to increase group size in order to maximize the power of the statistical test. In any case, it is useful to report the psychometric properties of the measure used. Split-half reliability coefficients are often easily computed from the data.

The reliability of the memory tests is not critical if a double dissociation is observed. A double dissociation refers to the situation in which one group is significantly better in task A whereas the other group is significantly better in task B. Unfortunately, this data pattern is found quite seldom in memory research.

3 Conclusions

We have summarized, illustrated and discussed several direct and indirect memory tests along with their strengths and weaknesses. Several examples have shown that memory measures *may* be powerful tools for cross-cultural research on cognition and communication, provided that different cognitive influences on memory judgements are separated by means of reliable and valid measurement tools. However, if easily obtained raw scores in memory tasks are uncritically used as ad hoc measures of memory, misleading and perhaps even wrong conclusions cannot be ruled out. This is especially true in research areas where cultural stereotypes or other schemata have strong impacts on human judgements. Guessing and response bias influences on memory judgements are especially likely if retrieval processes are guided by stereotypes (e.g., Erdfelder and Bredenkamp 1998; Bayen et al. 2000). In such cases use of memory measurement models capturing the effects of different types of response biases is mandatory.

As we have seen repeatedly, the interpretation of empirical data can reverse entirely if we replace poor methods of data analysis by better methods. Interestingly, concerning the examples we have considered in the present chapter, better methods tend to produce results that contradict universalistic positions such as those put forward by Rosch (1973). Rather, they support relativistic theories of cognition in different cultures. Thus, language and culture appear to have stronger impacts on perceptual representations and cognitive processes than previously thought by many cognitive psychologists. It remains to be seen whether this conclusion stands up to future research.

Edgar Erdfelder and Martin Brandt, Lehrstuhl für Psychologie III, Universität Mannheim, Germany. The work on this chapter has been supported by grants from the Deutsche Forschungsgemeinschaft (SFB 504, Project A12) and the Bundesministerium für Bildung und Forschung (DIP-D 4.2). The authors are grateful to Rüdiger Pohl, Carolina Küpper-Tetzel and Erhan Genc for helpful suggestions and to Ben MacKenzie for the preparation of Figure 12.1. Correspondence concerning this article should be addressed to Edgar Erdfelder, Lehrstuhl für Psychologie III, Universität Mannheim, Schloss, D-68131 Mannheim, Germany (email: erdfelder@psychologie.uni-mannheim.de).

Notes

1. Importantly, we refer to memories of human individuals only, for example, to short-term and long-term episodic memories and to the general knowledge or procedural knowledge of individuals. Memories shared by generations, social groups or cultures (Assmann 2002) and memory mechanisms in animals (Mazur 2002, chapter 11) are not covered in this chapter.
2. A disadvantage of this measure is that it cannot attain the value %S = 100% even if a subject needs no relearning at t_2. Assume that $C = 2$ error-free trials are defined as the criterion of complete mastery. If a subject needs $X(t_1) = 20$ and $X(t_2) = 2$ trials, the last two of them error-free, the measure would be %S = ((20 – 2) / 20) · 100 = 90% rather than 100% although no error occurred at t_2. For this reason, Hilgard (1952) suggested using the refined measure $\%S' = (((X(t_1) - C) - (X(t_2) - C)) / (X(t_1) - C)) \cdot 100$, with C being the criterion of error-free trials. Applied to our example, we obtain the more appropriate result %S' = ((18 – 0) / 18) · 100 = 100%.
3. Assume, for example, that HR = .70 and FAR = .20. By replacing p(FA) with .20 and p(H) with .70 in equation (2), we estimate c as .84 and d' as the difference .525 – (–.84) = 1.365 (cf. Hays 1972: Appendix C). In other words, the mean target item familiarity exceeds the mean preexperimental familiarity by 1.365 standard deviation units.
4. For the example considered previously, HR = .70 and FAR = .20, we immediately obtain .70 – .20 = .50 and .20 / (1-.50) = .40 as estimates of p_r and b_r, respectively.

Chapter 13

The Nijmegen Space Games: Studying the Interrelationship between Language, Culture and Cognition

Gunter Senft

Alles ist Spiel.
J. Huizinga *Homo Ludens*, p. 230

1 Aims of the Nijmegen 'Space Project'[1]

The Cognitive Anthropology Research Group (since 1998 the 'Department of Language and Cognition of the MPI for Psycholinguistics'), which started its research in 1991, tries to bring the Cognitive Sciences – especially Psychology and Linguistics – into dialogue with Cultural Anthropology – especially Cognitive and Linguistic Anthropology.

The central aim of the group is to conduct further research into the relationships between culture and cognition by conducting fieldwork on leading issues of common interest to anthropology, psychology and linguistics (see also Levinson 1992, 1996a&b; Senft 1994a, 2001). We try to investigate interdependencies between language, culture and cognition empirically via the following stratagem (Brown and Levinson 1993: 1):

> (a) first, pick a conceptual domain; (b) second, find two or more languages which contrast in the semantic treatment of that domain (i.e., where very different semantic parameters are employed); (c) third, develop non-linguistic tasks which will behaviourally reveal the conceptual parameters utilised to solve them; (d) compare the linguistic and non-linguistic representation systems as revealed by (b) and (c), and assess whether there is any correlation between linguistic and non-linguistic codings in the same domain.

The (first) conceptual domain we have 'picked' is the domain of 'space'. Thus, our initial major goal of research was to investigate the conceptualization of space and spatial reference in a cross-cultural/cross-linguistic perspective (see Pederson et al. 1998).

2 Methods to Elicit Verbal Reference to Space

To do this, we had to develop methods to build a comparative data base through parallel field research in different languages and cultures. This data base should serve us as a kind of 'natural laboratory' for testing and revising theories in psychology and theoretical linguistics. For the purposes pursued here I will describe only some of the many methods we developed for data elicitation (see Senft 1994a). All these methods make use of various sets of interactive 'games' which are used to elicit task-oriented verbal descriptions in native speakers of the language under study. Most of these tasks involve the recognition or the construction of spatial arrays from systematic sets of two- or three-dimensional stimuli (see Senft 2001: 522ff.).

The interactional games for focused linguistic elicitation were especially developed for our research purposes[2] (see also Hill 1993). They all involve a 'director' consultant who is allowed to see a certain stimulus, and a 'matcher' who is not. The players are sitting side by side with a screen separating them so that they cannot see each other's stimuli. The orientation of the players is taken note of, and the field researcher instructs the players what to do in their own language – all instructions are standardized. Moreover, the field researcher encourages the players to interact verbally, especially if they think they have difficulties in understanding each other. On the basis of the verbal descriptions given by the 'director' in the game, the 'matcher' is asked to reproduce three-dimensional models involving familiar objects with intrinsic orientations, like a human statuette in various body poses and mini-landscapes inhabited by model farm animals, as well as unfamiliar and abstract

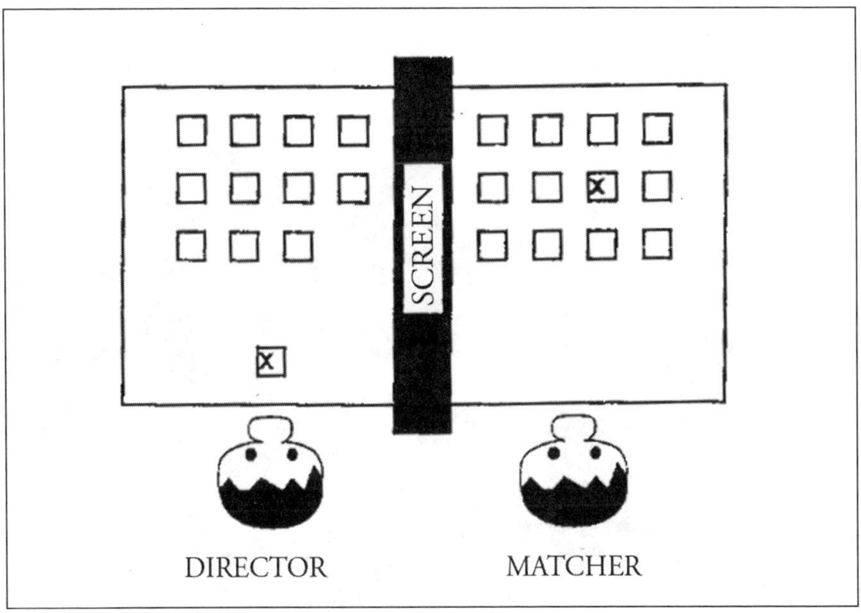

Figure 13.1: Diagram showing set-up of game.

objects. Some games also involve the matching of photographs on the basis of verbal descriptions; these photographs systematically cover certain spatial oppositions. All games are played at least three times with two consultants in two runs each. In the second run the matcher of the first run takes over the role of the director, and the director of the first run becomes the matcher. Figure 13.1 illustrates the basic idea of these games.

In what follows I will describe a selection of these interactional games.

The 'photo-object game' is played with three-dimensional plastic toys and photos depicting a certain spatial configuration of these toys. The director describes the photo, and on the basis of this description the matcher uses the toys to rearrange the spatial configuration. Figure 13.2 presents an example of such an arrangement.

The 'wooden-man game' requires that the director, on the basis of photos or on the basis of a wooden human statuette with flexible angles, describes certain body poses. The matcher has to adjust his or her statuette in such a way that the resulting body pose matches the description. Figure 13.3 presents an example of such a bodypose.

In the 'Tinkertoy games' the matcher – with the help of a building system for children – has to build a number of three-dimensional configurational and nonconfigurational constructions on the basis of the director's description which itself is based either on the same object or on a photo of the object to be constructed (see Senft 1994b). We built eight representational and nonrepresentational constructions. Each has different advantages. Representational constructions give a clear goal in construction and speakers may have already available (body-) part terms for reference. This should make it easier for most people. However, both constructions are fairly complex. Nonrepresentational constructions are more likely to make the informants rely on 'abstract' descriptions which may involve, for example, angles, planes, etc. This task is obviously an

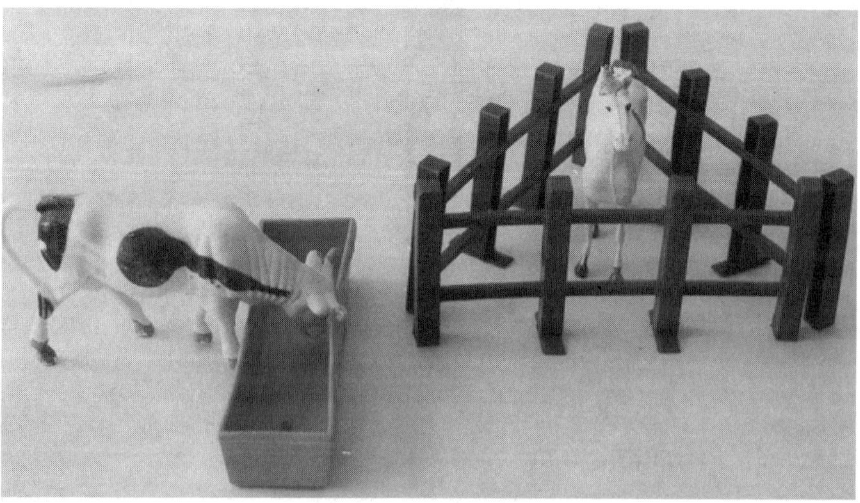

Figure 13.2: Photo-object game.

'unnatural' task and may be quite challenging for many speakers. Figure 13.4 presents an example of a nonrepresentational Tinkertoy construction.

The 'photo-photo game' (also called 'men-and-tree game') consists of four series of 2 x 12 photographs; here the matcher has to select one photo on the basis of the director's description (see Pederson et al. 1998). The photos depict certain localizations and configurations of objects with and without intrinsic features in four directions on the horizontal plane (see Pederson 1993). Moreover, the set contains a number of distractor photos, so that it did not become too obvious to the players to hypothesize about what we were after with the game.

The two sets of 12 photos are each shuffled before the game is played, then they are laid out in front of each player, in a grid. Thus, the photographs in front of the two players are randomly ordered – on each side of the screen. After this the director is asked to pick a photo and to

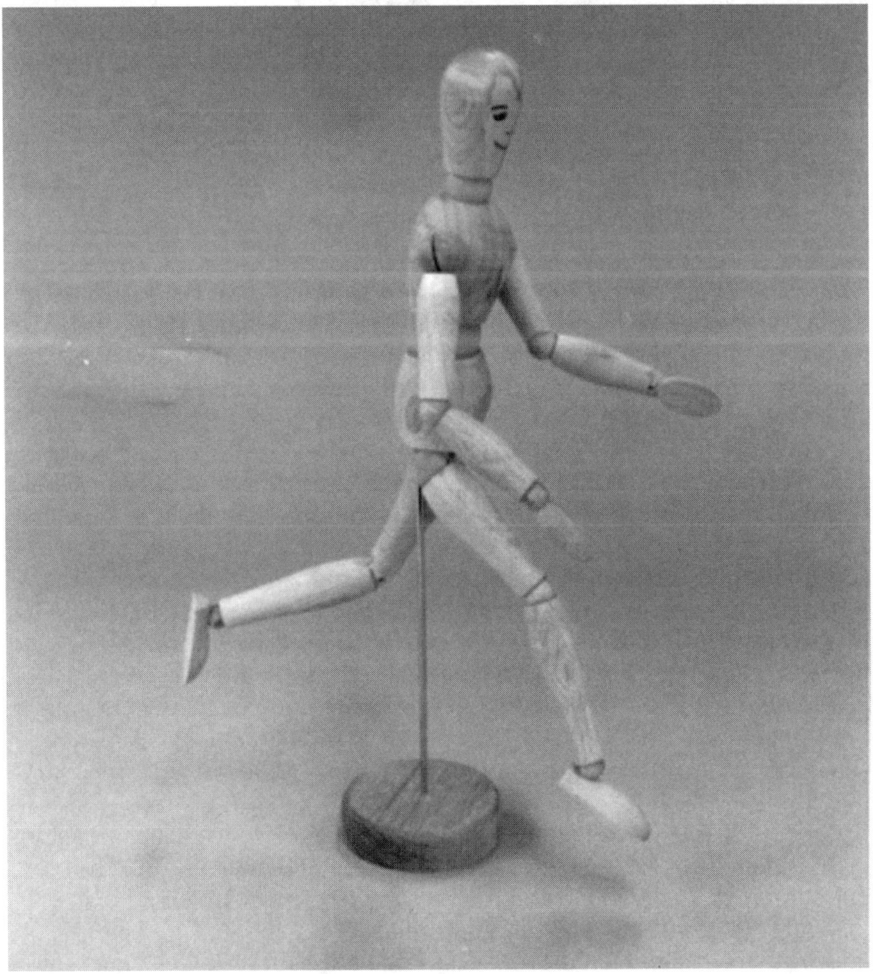

Figure 13.3: Wooden-man game.

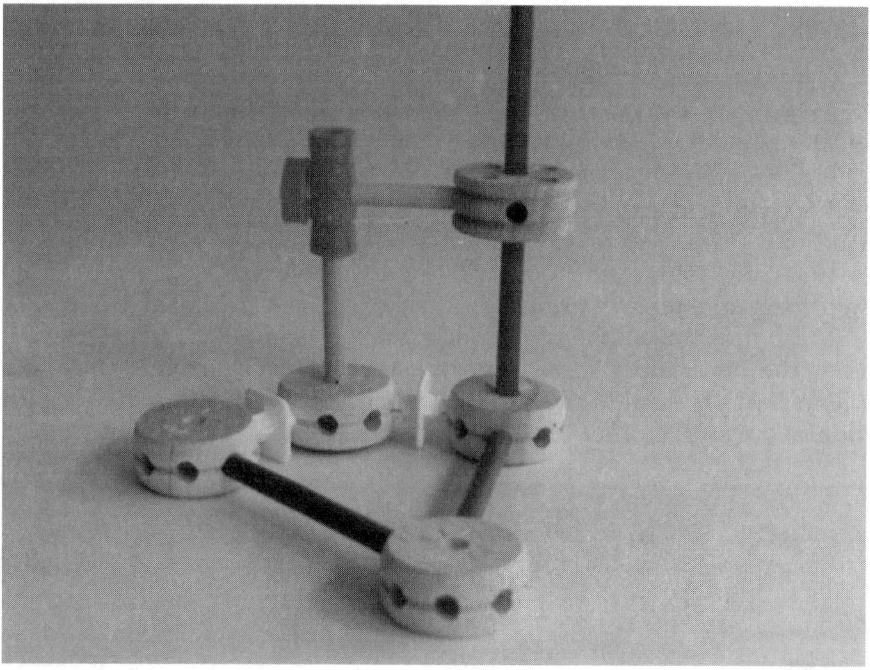

Figure 13.4: Tinkertoy games.

describe it. Both players are encouraged to interact verbally, if they are having problems with the matching task. If the matcher thinks s/he has found the photo described by the director, the players are asked to pick up the photo from their grid and to put it upside down on a pile beside this grid. After the game, i.e. when all 12 photos have been described, the piles are compared with each other and mismatches are discussed with the players. The games should be at least tape-recorded; ideally they should be filmed as well.

The first set of photographs is a training set that should make the consultants familiar with the basic idea behind the matching task.

The second set was developed to elicit spatial reference to relationships in the horizontal plane between two unfeatured objects (balls) and between a featured (man) and a nonfeatured (tree) object.

The third set was developed to elicit spatial reference to relationships in the horizontal plane between two men (featured objects) presented in same orientation, i.e., the two men are always looking in the same direction.

The fourth set was developed to elicit spatial reference to relationships in the horizontal plane between two men presented in different orientation, i.e. the two men are oriented in opposite directions.

Figure 13.5 presents a picture from this last set and figure 13.6 presents drawings of certain 'Man and Tree' configurations from the second set of photographs.

We also developed games to elicit the description of motion events and thus certain notions of motion. They consist of a route direction game and film description tasks.

In the 'route-description game'[3] the director and matcher play with two dolls and two identical 'stylized' landscapes in front of them (see Senft 2000). These landscapes are organized symmetrically around a central axis, with objects on each side of the central axis being of identical shape and colour. This symmetry within the base design should force the consultants to differentiate directions in the cross-axis (the so-called left-right axis). Three pairs of different objects – in the foreground two roof shapes on each side, in the middle ground two simple bridges on each side, and in the back two towers on each side – form the constant objects of the basic scene. Other objects that are symmetrical in one axis – two short fence rails, a truck, a Duplo stair structure, and a rectangular fence structure – can be placed with their axis of symmetry along the central axis. Thus, one way of varying scenes in the landscapes in which routes have to be described is by placing this limited number of symmetrical objects along the axis of symmetry. All these objects were selected to elicit certain notions of motion like 'going around, circling, going through, going along, going over/under, going up/down, passing through/under, moving to the back/front of, moving on the left/right side of, climbing over' etc. (for a more detailed description see Danziger 1993: 15–28). There are four basic scenes to be elicited in the route description elicitation task. Mixed in with these are three distinct conditions:

Figure 13.5: 'Man and tree' photographs.

The first condition is one of pure symmetry in the scenes and applies to the first two routes (see Figures 13.7 and 13.8). The first path starts off to the left and never moves back towards the speaker, but finishes at the other end of the table on the far yellow square. The second path moves off to the right and is a long return path back to the starting point.

Figure 13.6: Drawings of 'Man and tree' photographs.

The second condition applies to the third route. Here the researcher is to use the open spaces in the design to place (e.g. local) objects which destroy symmetry. The only constraint is that these objects do not block the route that is to be described (see Figure 13.9). The point of this condition is to see whether the new objects in the scenes are taken up as landmarks which resolve any of the original problems which may have arisen in the symmetrical condition.

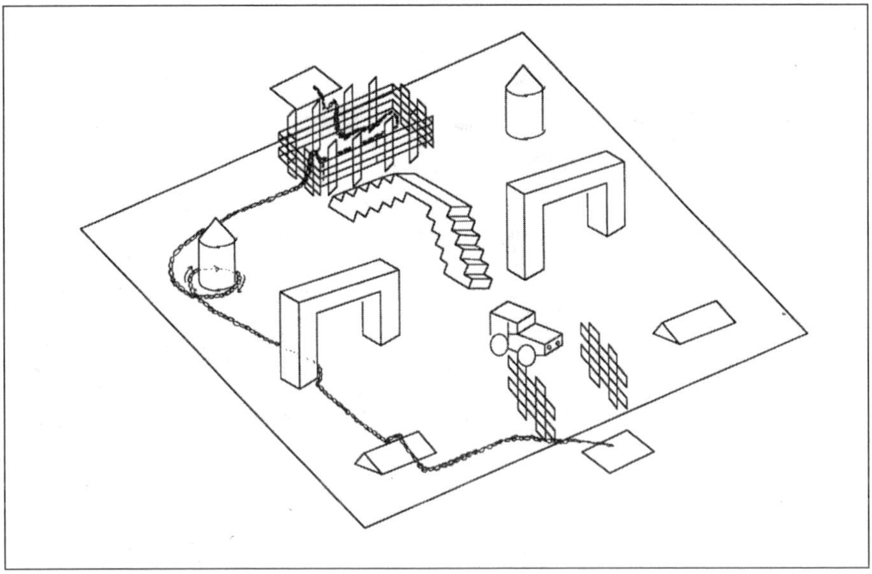

Figure 13.7: Path 1 (set-up for director).

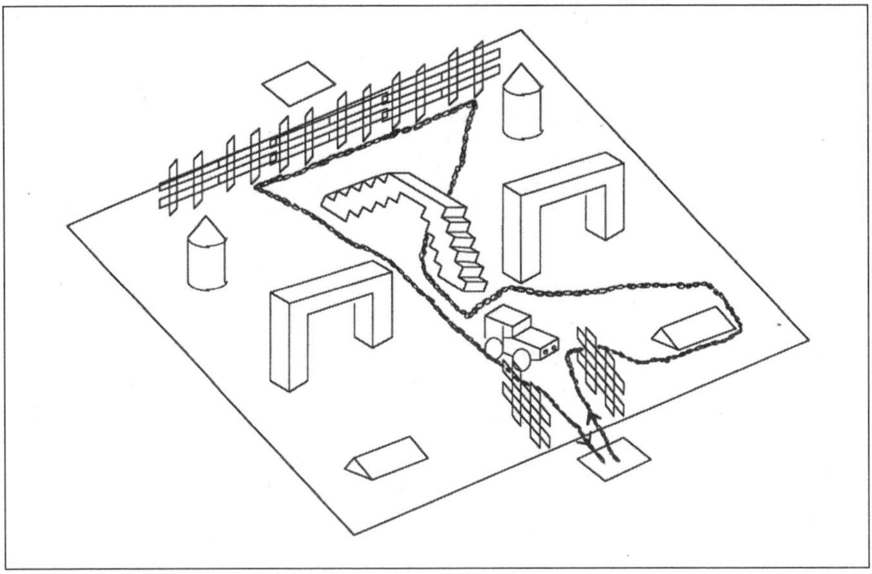

Figure 13.8: Path 2 (set-up for director).

The third condition applies to the fourth route (see Figure 13.10). On either side of the mat, the researcher has to place relatively large objects that are not to be construed as part of the scene itself (like, e.g., a coconut or an apple to the left of each mat and a bush knife or a bowl to the right). The point of this condition is to see whether ad hoc local landmarks – basically outside the scene – will be used to facilitate the route description.

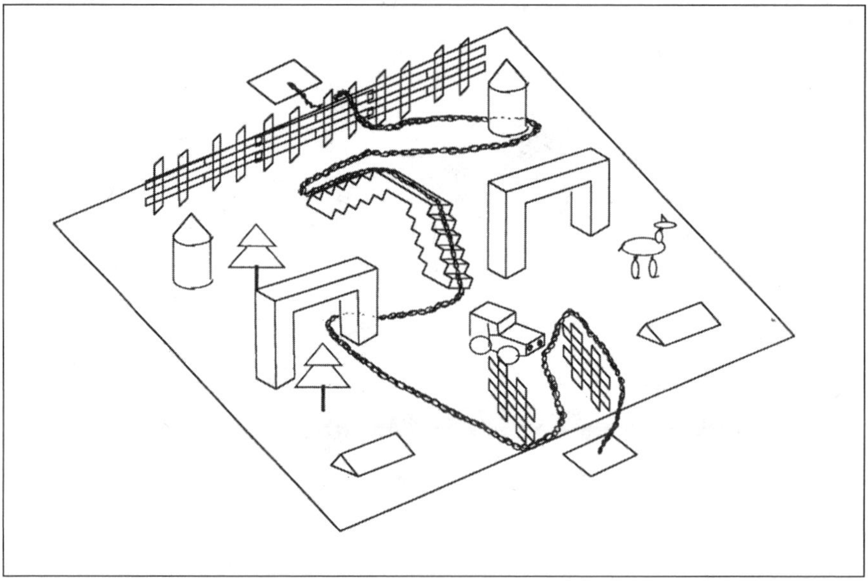

Figure 13.9: Path 3 (set-up for director).

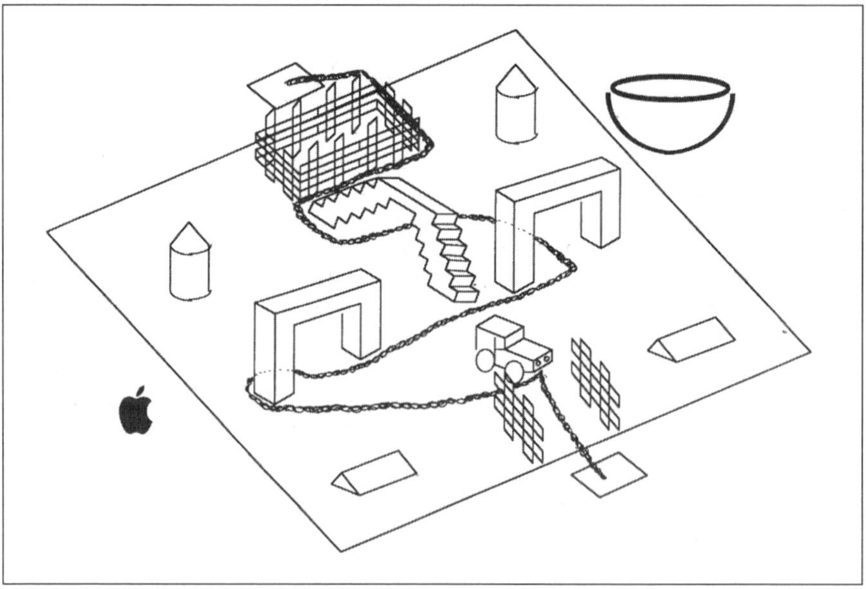

Figure 13.10: Path 4 (set-up for director).

The director is asked to let his/her doll 'walk' and describe a route that is indicated by a small chain on his or her landscape so that the matcher can let a small doll 'walk' this route on his or her landscape. (see Weissenborn 1986; Wilkins and Hill 1995; Senft 2000).

'Staged events' is a data elicitation tool which combines research interests in serial verb constructions, in event typicality, and in event complexity (see van Staden et al.2001). It is designed to collect descriptions of complex events in order to examine how these are segmented into macro events, what kind of information is expressed and how the information is ordered in the descriptions. It contains many different motion events. The tool consists of two tasks:

1. A description and recollection task, designed to elicit:
   - elaborate descriptions of complex events for the description task; and
   - concise equivalents for the recollection task.
2. A reenactment task of some of the scenes on the basis of descriptions given in task 1.

Task 1 consists of two sets of video clips and stills (on DV tape and digitized on a CD). Set 1, a subset of Set 2, consists of 53 clips and 53 stills. Set 2 consists of 86 clips and 86 stills. The video clips depict various scenes with human actors and recognizable objects (for example, an actor fetches an axe and chops wood, an actor bumps into another actor who drops a plate which breaks, an actor plays a guitar over his head, scenes from a football (soccer) game, etc.). The scenes depict many different motion events. The clips are arranged in a specific order. Every seven or eight video clips for the description task are followed by seven or eight corresponding stills for the recollection task. These stills were selected by Alex Dukers from the video clips and depict a crucial moment in the event staged in the clip from which it was chosen.

The researcher elicits these data with two consultants: one acts as the addressee who has not seen the clips and stills before, and one acts as the describer who first describes the clips and then the stills. The researcher makes the addressee ask, 'What happened?' (using a language/culture appropriate phrasing that focuses on the action) and the describer knows that her or his description must be such that the addressee knows what happened. After seven or eight video clips the researcher presents the stills to the describer and asks him or her to describe from memory which scene the picture belongs to, using the appropriate equivalent of the question, 'Which clip was this?'.

The task is run on a laptop computer or on a DV camera. A minimum of six pairs of consultants is recommended. It takes about 40 minutes per consultant to run set 1 and at least 60 minutes per consultant to run set 2. The elicitation session should be video-recorded.

The reenactment task aims to test whether the information contained in the descriptions yielded by the first task is sufficient for a hearer to reenact the scene correctly, but it is also designed to check which parts of a complex scenario are left to inferences based on 'stereotypicality' of events (for instance, if a scene is described as 'a man throws an apple to a woman', does this imply that the apple is caught by the woman?). This second task requires that the researcher selects one representative description from the data collected during the first task for 14 scenes depicted in the video clips there. Moreover, the researcher needs some objects (a shawl or cloth, a fruit, a guitar, a chair, a table, a ball) that are necessary to act out the described scenes. The researcher either plays the tape-recorded description or reads it out himself to a pair of consultants who are asked to reenact what they have just heard in this description. Not all scenes require two actors. Then the actors themselves may decide who is the actor. When two actors are required, they may decide for themselves who acts which part.

For this task a minimum of six pairs of consultants is recommended, and it takes about 30 minutes plus optional discussion time. Again, the elicitation session should be video-recorded.[4] The 'staged events' tool and other elicitation devices developed for the Language and Cognition group's 2001 field season can be found on our website under the following URL: http://www.mpi.nl/world/data/fieldmanuals.

The descriptions of these methods should suffice for the purposes pursued here. In what follows I will briefly describe the methods we developed to investigate nonverbal spatial cognition (see Danziger 1993). However, before I can do this I have to briefly present the insights we needed for developing these methods.

3 Methods to Investigate Nonverbal Spatial Cognition

The analyses of the data we gathered in the languages researched by members of our group with the help of methods like the ones described above revealed fundamental differences in how these languages refer to space (see Senft 2001: 526ff.). For describing these differences we use a typology of spatial systems or frames of spatial reference. This typology defines three such systems. We refer to them as 'relative', 'absolute' and 'intrinsic' (see Senft 1994a: 419; Levinson 1996a: 359, 365–373, 1996b; see also Bühler 1934). They differ with respect to how angles are projected from the 'ground' (or 'relatum') in order to situate the location of the 'figure' (or 'theme') that is referred to (Talmy 1978: 627; see also Klein 1991: 78; Senft 1997: 10).

Relative systems are viewpoint-dependent: localizations in space are derived from, and described on the basis of, the position and orientation of the speaker. In these systems a sentence like 'The tree is to the right of the man' is understood from the speaker's point of view only – i.e., this reference completely neglects the orientation of the man.

Absolute systems operate on absolute concepts of direction (which may be linear or defined by quadrants). They are based on conventionalized directions or other fixed bearings that can be derived from meteorological, astronomical, or landscape features. In these systems (and in our data) we find sentences like e.g., 'The tree is to the west of the man/uphill from the man/seawards to the man.'

Intrinsic systems utilize inherent, intrinsic features of an object to derive a projected region or to anchor the spatial reference to an object in these features. In these systems a sentence like 'The tree is to the man's right' is understood as follows: A man is an object with a front and back, a left and right side assigned to it. Thus, in intrinsic systems this sentence refers to the position of the tree on the basis of the orientation of the man – the tree is at the right side of the man, then – the orientation of the speaker does not play any role whatsoever and is – within this system – completely irrelevant for the understanding of this sentence. However, we want to note here that speakers using intrinsic systems for their spatial references also refer to the same configuration with the sentence we already mentioned above, namely: 'The tree is to the right of the man.' Thus, languages can be ambiguous with respect to whether they use an intrinsic or a relative perspective in their spatial references. Sentences like the last one presented can only be disambiguated in the actual situation and context.

All three systems can be found in a given language, and they can be utilized for spatial reference; however, many of the languages we have been studying so far frequently seem to prefer one frame of reference in a particular context. Because of these observations we came up with the following hypothesis: If speakers of a language preferentially use one reference system in a particular spatial domain, then these speakers will rely on a comparable coding system for memorizing spatial configurations and making inferences with respect to these spatial configurations in nonverbal problem solving.[5]

To falsify or verify this hypothesis we developed a number of experiments to test the interrelationship between space and cognition. First of all we had to explore the cognitive implications of the three systems of verbal spatial reference. Relative (R), absolute (A) and intrinsic (I) systems differ with respect to their dependence (+) or independence (–):

- with respect to the speaker's location and orientation,
  R + A – I –
- with respect to the rotation of the spatial configuration,
  R + A + I – and
- with respect to the rotation of the ground
  R – A – I+.

Based on these differences of the three frames of spatial references we developed five different nonverbal experiment-like tests. These experiments for the investigation of nonverbal spatial cognition explore the nature of the spatial coding for memory and inference, and make it possible to determine whether this nonverbal coding has certain specific properties. These properties can then be compared to the verbal codings elicited by the first kit to see whether there is a correlation between the verbal and the nonverbal systems of spatial coding (see Danziger 1993; Senft 1994a: 420–27; also Senft 2000 and 2001: 527ff.).

I will only very briefly describe these experiments and the considerations they are based on here (for detailed descriptions see Brown and Levinson 1993; Danziger 1993; Senft 1994a). First I want to mention that all the five tasks attempt to investigate the opposition between two different coding systems of space, namely between what we call the 'relative' coding system or frame of reference that uses expressions like 'left/right/front/back' for spatial references, and the 'absolute' system or frame of reference that uses expressions like 'north/south/east/west, uphill/downhill, seawards/ landwards, upriver/downriver', etc., for spatial references. All five tasks within this kit have the same fundamental design. The consultants are shown a stimulus on one table (table 1) and are instructed to memorize what they have seen. After a short delay they are rotated 180 degrees and led across to another table (table 2) at a certain distance which faces in the opposite direction from table 1. The consultants are now asked to reconstruct the same array, or to select the same array from a set provided. The stimulus arrays are so designed that they have either a left/right or a front/back asymmetry when viewed on table 1.

Suppose the consultants see an arrow on table 1 that is pointing from their point of view to the right. After a short pause and after having been turned for 180° they are led to table 2. There they find two arrows; again, from their point of view one arrow is pointing to the right and the other arrow is pointing to the left. The consultants are asked now to choose the arrow that resembles the one they just saw half a minute ago on table 1. Consultants who memorized the orientation of the arrow on table 1 on the basis of a relative system of spatial coding will select at table 2 the arrow that – from their point of view – is also pointing to the right. Here the fact that standing in front of table 2 the consultants have turned 180° is of crucial importance. Consultants, however,who use an absolute system of spatial coding memorize the fact that the arrow on table 1 pointed, e.g., towards north. They will then select the arrow which is also pointing towards this direction at table 2, completely independent of the fact that they have turned 180° (see Figure 13.11).

The first experiment – 'animals in a row' – tests recall memory for spatial configurations (see Figure 13.12). Subjects look at three animals in a row, should concentrate on the relative order of animals, are then twisted 180° and asked to remake the same assemblage. The experimenter, however, is really only interested in the alignment direction.

The second experiment – 'red and blue mazes' (also called 'red and blue task') – tests recall and recognition memory for spatial configurations (see Figure 13.13). This test uses five identical cards with two differently coloured circles of different size. Subjects look at a card and should concentrate on the orientation of the circles; they are then turned 180° and are asked to select the card with the same orientation of circles from among four choices of identical, but differently oriented cards.

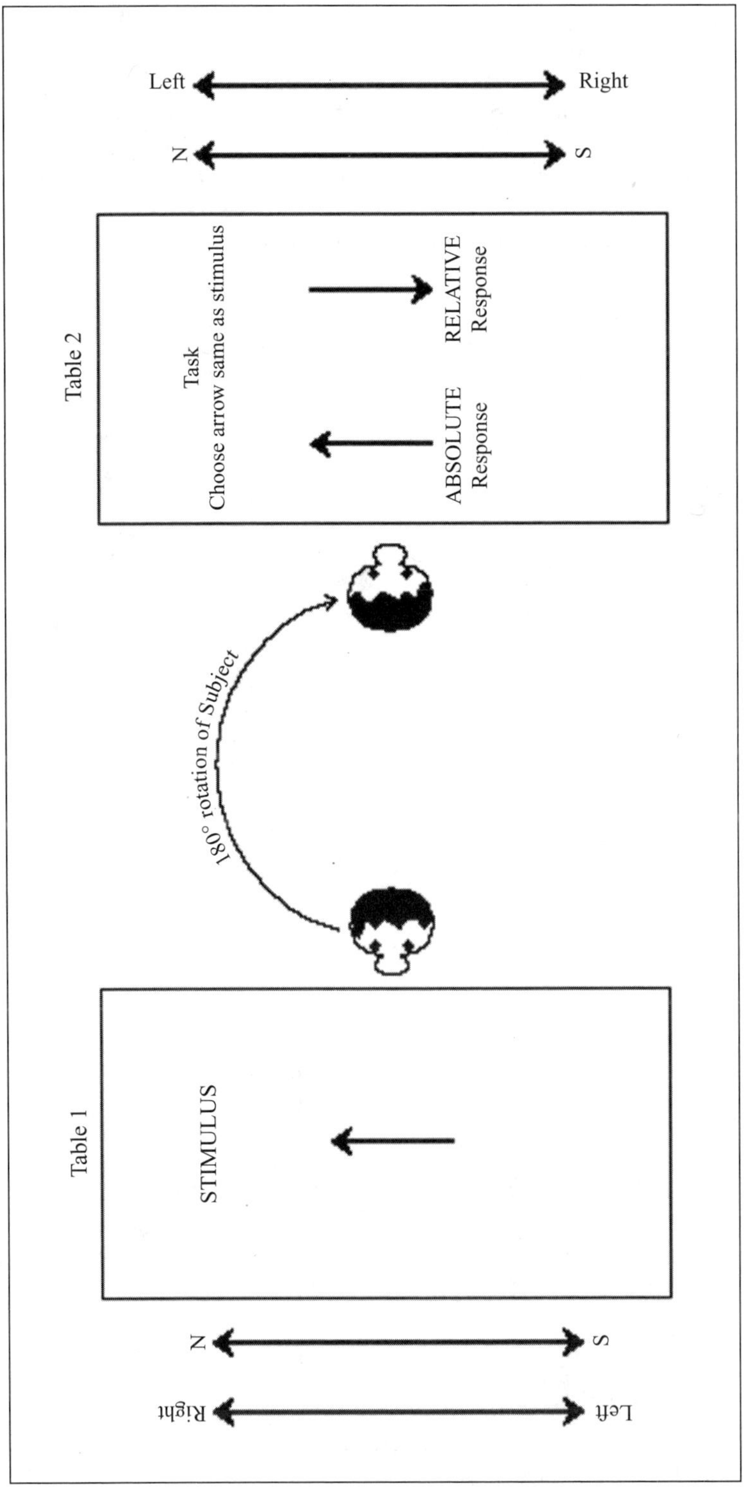

Figure 13.11: Absolute and relative frames of reference.

The third experiment – the 'maze task game' (also called 'Eric's maze task' because it was developed by Eric Pederson in cooperation with Bernadette Schmitt) – tests recall and recognition memory with respect to movement in space and the transformation of movement into the construction of a path (see Figure 13.14). The test consists of a figure resembling a little man and a maze. Subjects look at the little man being walked by the experimenter in a certain complex path and should remember this path; they are then turned 180° and are asked to select the

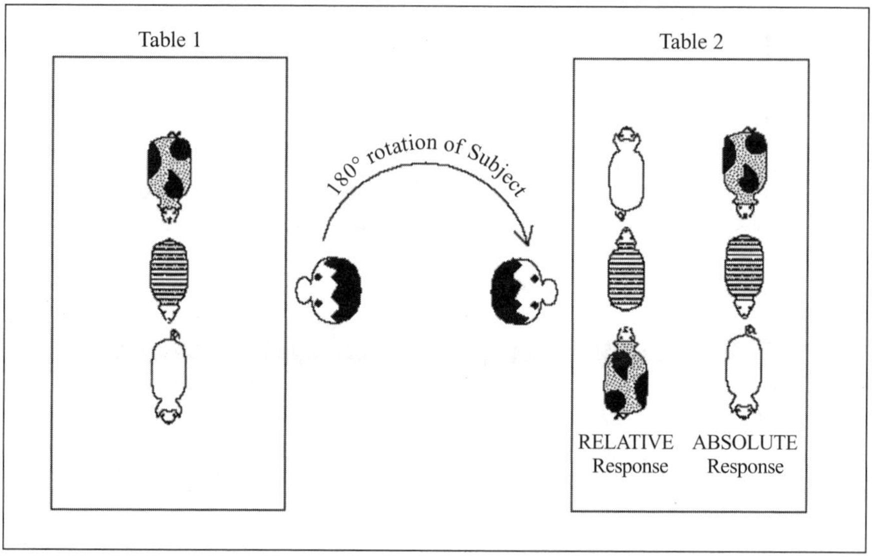

Figure 13.12: 'Animals in a row' game.

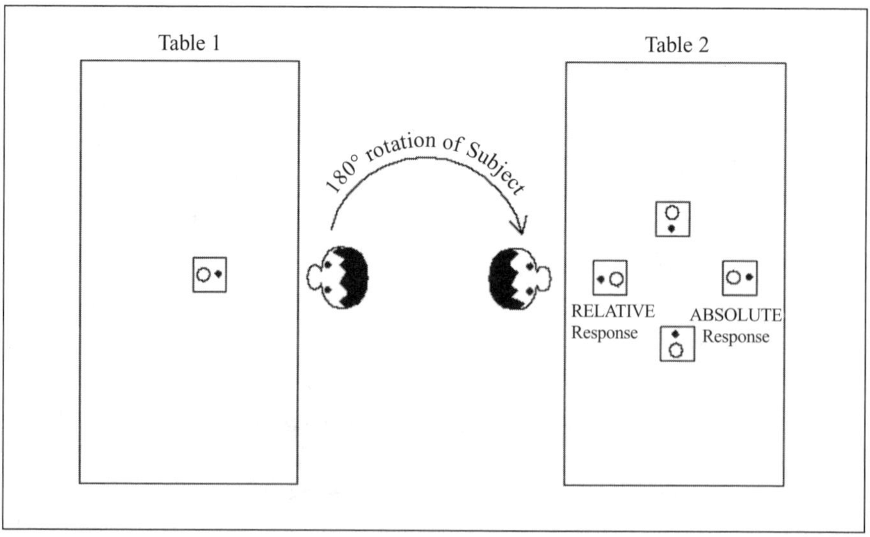

Figure 13.13: 'Red and blue mazes' games.

endpoint on a maze where the little man would end up if he had followed that path and not others on the maze (see Pederson and Schmitt 1993).

The fourth experiment – 'the scout game' (also called 'Steve's maze task' because it was mainly developed by Stephen Levinson) – tests the ability to finish an incomplete path and recognition memory (see Figure 13.15). It consists of five separate maps with three cards – a distractor with a path that does not complete the path seen on the map, a card with a path that is

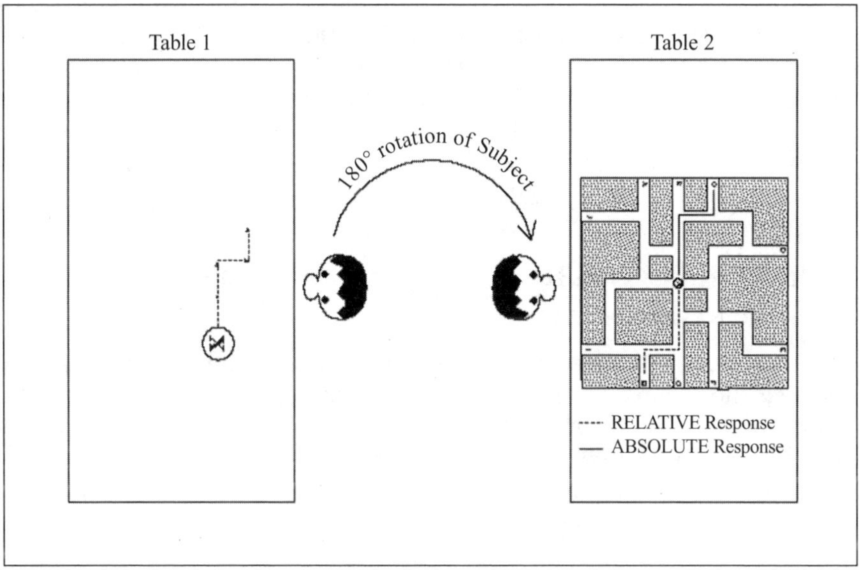

Figure 13.14: 'Eric's maze' task.

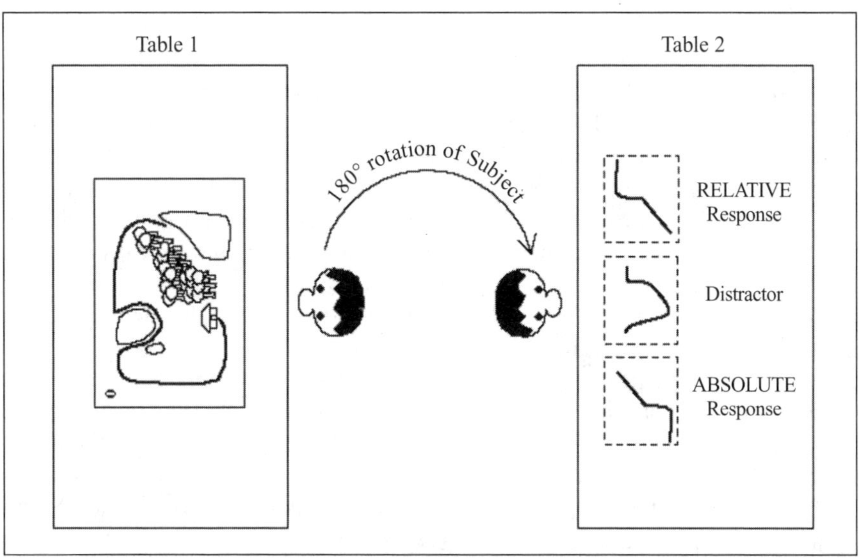

Figure 13.15: The 'scout' game.

chosen in an absolute reaction and a card with a path that is chosen in a relative response. Subjects have to look at the map and memorize it; they are then rotated 180° and are asked to chose one of the three cards that will finish the incomplete path seen on the map.

The fifth experiment – 'the transitivity task' – tests the ability to make transitive inferences. Widlok (this volume) discusses this test in detail, therefore I just mention this task but I will not describe it in more detail here.

With these experiment-like tests we collected a relatively broad data base on nonverbal problem solving in the spatial domain. Having presented all the methods to investigate nonverbal spatial cognition and a representative selection of the methods to elicit verbal reference to space we developed in Nijmegen, I will now discuss these methods with respect to their ecological validity (see also the contributions by Mishra and Dasen, Funke, and Widlok to this volume).

4 'Everything Has Its Price': Costs and Benefits of the Space Games

It turned out that all members of our group had no difficulties whatsoever playing the games we developed for testing the relationship between space and cognition. Actually, our consultants enjoyed playing all the games presented here. Another problem, however, was how to interpret inconsistencies in the overall behaviour of our consultants. I will illustrate this problem on the basis of the data I collected with my consultants on the Trobriand Islands.

In four of the five memory experiments my consultants preferred absolute solutions. This was not surprising, because these games asked (among other things) for the memorizing of the orientation of certain objects/paths – and analyses of the linguistic data have shown that Kilivila speakers prefer an absolute ad hoc landmark frame of reference system for referring to the spatial orientation of objects in a given spatial configuration (Senft 2001: 545). However, in the maze task my consultants clearly preferred a relative solution of the problem. How is this to be explained? As mentioned above, this experiment tests recall and recognition memory with respect to movement in space and the transformation of movement into the construction of a path. The test consists of a figure resembling a little man and a maze. Subjects look at the little man being walked by the experimenter in a certain complex path and should remember this path; they are then turned 180° and are asked to select the endpoint on a maze where the little man would end up if he had followed that path and not others on the maze. As already stated elsewhere (Senft 2001: 550f.), it may well be that the scale of the maze and especially the man-like toy figure with its inherent intrinsic features made the consultants memorize the figure's movements using a 'left/right' system that in their solution to this memory task transformed into a relative system of spatial reference.

However, I concede that this is a rather speculative attempt to explain the data. It refers back to the problem that we have, indeed, many uncontrollable parameters (like, for example, in this task the possible influence of the size of the maze and of the movement of the figure) within the tests and experiments we devised to investigate nonverbal systems of spatial coding and their interrelation or correlation with verbal spatial coding systems. However, if we neglect the results of the 'motion maze task' experiment, we can conclude that the results of the non-verbal experiments show a rather strong correlation with the results of the verbal codification of spatial configurations in Kilivila. This example is typical of some of the problems we had with the analysis and the interpretation of such – relatively minor – inconsistencies in our data. Nevertheless, I think I can state that with respect to the methods for investigating nonverbal spatial cognition we had only marginal costs, but high benefits.

However, this does not hold for all the methods we used to elicit verbal reference to space. One of the games that turned out to be somewhat problematic for my consultants on the Trobriand Islands – at least at the beginning, was the 'man and tree' photo-photo matching task. This task was the first space game I played with my consultants on the Trobriands. This kind of elicitation situation was completely new for them – I had never before used any comparable device for eliciting linguistic data in Kilivila. It may well be that the stunning new situation with these absolutely unknown elicitation materials detracted my consultants' concentration or otherwise influenced them in a negative way with respect to their performance. It is certainly not an everyday activity for Trobriand Islanders, who only rarely deal with photographs (except with those they get either from tourists or from me), to scan 12 pictures, a subset of which depicts rather subtle differences with respect to the spatial configuration and orientation of objects like a plastic tree and a small plastic toy figure. Thus it is quite understandable, I think, to assume that the matchers decided for a picture as soon as they thought their choice would more or less match the directors' description. Moreover, I also assume that the directors did not monitor all the photographs from the very beginning of this game and that they were not aware of the fact that they really had to follow my instructions and be as exact and as exhaustive as possible in their descriptions. However, these considerations refer to only a few of many uncontrollable parameters for the field researcher in such an elicitation situation. This argumentation implies, of course, self-criticism with respect to our methodology for data elicitation. It especially questions the adequacy of the elicitation tool used, i.e. the photographs developed for this game. It may well be that these photographs as elicitation tools are much too abstract (see Nüse 1996: 91f.) and probably too Eurocentric in their sophisticated depiction of subtle spatial relationships, and thus prove inadequate for eliciting verbal data in small-scale face-to-face speech communities like the community represented by the Trobriand Islanders in the village Tauwema. Another problem we had with these pictures was that some consultants, like for example the Australian Aboriginal speakers of the language Mparntwe Arrernte, (a language with a very sophisticated absolute frame of spatial reference), would only start to play the game if the hardly visible shadow on the photographs caused by the flashlight while taking the photos corresponded to the actual position of the sun. David Wilkins, who researched this language, reported that his consultants changed their orientation and the orientation of the playground, until the shadows in the photographs corresponded with the actual lighting conditions in their environment. Some of his consultants also criticized the fact that the pictures only showed 'whitefellas' and not people with black or brown skin.

I am absolutely aware of the fact that these are general problems for linguistic data elicitation (see Senft 1995), but I would like to note here that, especially in research projects with a comparative orientation and comparative aims, it is extremely difficult to create elicitation devices that can be used in many speech communities that are quite different from one another. Researchers who try to do this see themselves in a position somewhere between Scylla and Charybdis.

Right from the beginning of our research we were aware of the fact that these games have certain inbuilt restrictions and constraints with respect to what kind of data on spatial language is elicited with them – however, this was obviously the price we had to pay for eliciting comparative data. We attempted to design these games so that they really elicit as broad a range as possible of the vocabulary for spatial reference that can be found in the speech community under study. Moreover, playing these games requires a certain familiarity of the researchers with their fields and with the languages they research and speak themselves. It is only on the basis of their experience and their competence in the languages under study that researchers can adequately use these games for linguistic elicitation, that they can give the instructions to their consultants and that they can decide whether or not the gathered data represent – at least in part – the everyday usage of these expressions for spatial reference or whether the elicited data have to be regarded as 'artefacts' of the elicitation method. However, counterchecking already existing text corpora with the elicited data, the members of our group are convinced that this last possible interpretation of the data does not hold for the data elicited for our research purposes. Thus, like Odysseus, we feel that in the end we escaped the dangers of both Scylla and Charybdis.

To sum up, we have always been aware of what psychologists usually refer to as the 'ecological validity' problem of gathered data. We are convinced though, that the tools described here, the space games developed for focused data collection, provide the researcher with an adequate and reliable sample of speech data on spatial reference. With the space games we elicited corpora of contextually anchored yet complex interactive texts in different languages and cultures. These texts incorporate many examples of spatial language. This corpus constitutes our cross-culturally and cross-linguistically comparative data base for the research on verbal reference to space. Aiming for comparable data had – and always will have – its costs, but we are convinced that the benefits of the corpus collected with these space games by far outweigh them.

Notes

1. This paper is based on research done by the following former or present members and guests of our group. The languages on which the members work and the respective LANGUAGE FAMILY are given in brackets following the researcher's name: Giovanni Bennardo (Tongan, AUSTRONESIAN), Balthasar Bickel (Belhare, TIBETO-BURMAN), Penelope Brown (Tzeltal, MAYA), Jürgen Bohnemeyer (Yukatek, MAYA), Niclas Burenhult (Jahai, MON-KHMER), Gabriele Cablitz (Marquesan, AUSTRONESIAN), Eve Danziger (Mopan, MAYA), Susan Duncan (Chinese), Michael Dunn (Touo, PAPUAN, Solomon Island Pijin, CREOLE, and Chuckchi, PALEO-SIBERIAN), Nick Enfield (Lao, SINO-TIBETAN). James Essegbey (Ewe, NIGER-KORDOFANIAN), John Haviland (Guugu Yimithirr, PAMA NYUNGAN and Tzotzil, MAYA), Deborah Hill (Longgu, AUSTRONESIAN), Kyoko Inoue (Japanese), Elizabeth Keating (Pohnpeian, AUSTRONESIAN), Anna Margetts (Saliba, AUSTRONESIAN), Sotaro Kita (Japanese), Lourdes de Leon (Tzotzil, MAYA), Paulette Levy (Totonac, TOTONAC), Sabine Neumann (Kgalagadi, BANTU), Eric Pederson (Tamil and Bettu Kurumba, TAMIL), Eva Schultze-Berndt (Jaminjung and Ngaliwurru, NON PAMA NYUNGAN), Gunter Senft (Kilivila, AUSTRONESIAN), Miriam van Staden (Tidore, PAPUAN) Christel Stolz (Yucatec, MAYA), Angela Terrill (Lavukalave and Touo, PAPUAN), Jürg Wassmann (Yupno, PAPUAN), Thomas Widlok (Hai//om, KHOISAN), David Wilkins (Mparntwe Arrernte, PAMA NYUNGAN), and Roberto Zavala (Oluta Popoluca, MIXE-ZOQUEAN). The director of the 'language and cognition' department is Stephen C. Levinson (Guugu Yimithirr, PAMA NYUNGAN, Tzeltal, MAYA and Yélî Dnye, PAPUAN).

   To administer the described and discussed elicitation methods in the field required specialists with sound knowledge in both the language and the culture of the groups studied. This requirement explains the arbitrary selection of languages and cultures researched in the project.

   We would like to thank all the institutions involved in granting us the permission to do research in their countries and we express our deep gratitude to all the native speakers of these languages, to our friends and consultants in our fields, for their friendly and patient cooperation.

2. These games were piloted for, and introduced to our group by Lourdes de Leon (1991) and John Haviland with inspiration from Clark and Wilkes-Gibbs (1986), von Stutterheim and Carroll (1993), and others. The method was further developed and revised by other members of our group, especially by Eve Danziger and Eric Pederson, and also by Penelope Brown, Stephen Levinson and Gunter Senft.
3. Based on an experiment devised by Jürgen Weissenborn (1986) to elicit route descriptions in children within the framework of an interactive game, and based

on a pilot study I did on the Trobriand Islands in 1992. David Wilkins in collaboration with Eve Danziger, Gertie de Groen, Deborah Hill, Steve Levinson, Paulette Levy and Eric Pederson designed the version of the route elicitation task presented here.

4. For detailed information on this elicitation tool see van Staden et al. (2001). Some of the staged event clips were designed as 'staged event' equivalents of another elicitation tool, the so-called 'Ecom clips'. These clips are designed by Jürgen Bohnemeyer and Martijn Caelen. For further details see Bohnemeyer and Caelen (2001: 168). For another film description task with which we elicited expressions describing 'ENTER' and 'EXIT' events see Senft (1999).
5. For a refined version of this hypothesis see Senft (2001: 545f.).

Chapter 14

The Perception of Space from a Psychological Perspective

Joachim Funke

Introduction

Time and space are fundamental categories for our understanding of the world. Despite the fact that the two categories cannot be separated from one other (time occurs always at given places in space, and space is always space at a certain point in time), I will concentrate in this paper on space (even if recent research shows how deeply space and time are intertwined; see Boroditsky and Ramscar 2002).

There are two levels at which space can be seen from a psychological perspective: (a) the surface level and (b) the deep level. The surface level deals with perception of space based on physiology, mainly physiology of the visual system. At this point we should not forget that we can also locate places by means of hearing, feeling or smelling – accordingly, the physiology of the other sensory systems might also be of interest. The surface level deals with the reconstruction of the environment in the head of a sensing person. But making sense out of perception is more than constructing a three-dimensional inner model of reality. The deep level is occupied with the functional aspects and regards space under aspects like finding food, finding mates, or finding secure places, that is, giving perceptual objects a meaning. The deep, functional analysis deals also with the role of space for our memory (space as external memory) as well as with memory for places. Without the deep level an analysis of space would miss the central issues. Therefore, while anthropologists have to know about the surface level, for a fundamental analysis the deep level is of central importance. However, I will now begin with the former.

Space Perception: the Surface Level

In psychology textbooks on perception (like the famous one from Goldstein 2002), space perception is conceived as an analysis of depth cues, which are available to the eye. There are three types of depth cues that can be used by our eyes: (a) oculomotor cues, (b) monocular cues and (c) binocular cues. We will look at examples of all three kinds of cues.

Oculomotor Cues

Oculomotor cues come from the inward motion of our eyes when we look at nearby objects and the normally parallel viewing axes of both eyes converge; this is called 'convergence' (see Figure 14.1). A second cue is called 'accommodation' and goes back to a tightening of eye muscles, which change the shape of the lens and help to focus on objects we want to see clearly. The ciliary muscles contract the crystalline and thus change the focal length (see Figure 14.2). Both cues – convergence and accommodation – are proprioceptive information, which sometimes (in extreme situations) can be felt consciously.

Monocular Cues

Monocular cues are those which work with one eye alone. Two classes are differentiated: pictorial cues, which can be present in pictures, and movement-based cues, which go back to a head movement.

Pictorial cues are present in almost all pictures. They include occlusion (i.e., one object or part of an object is hidden by another), relative height,

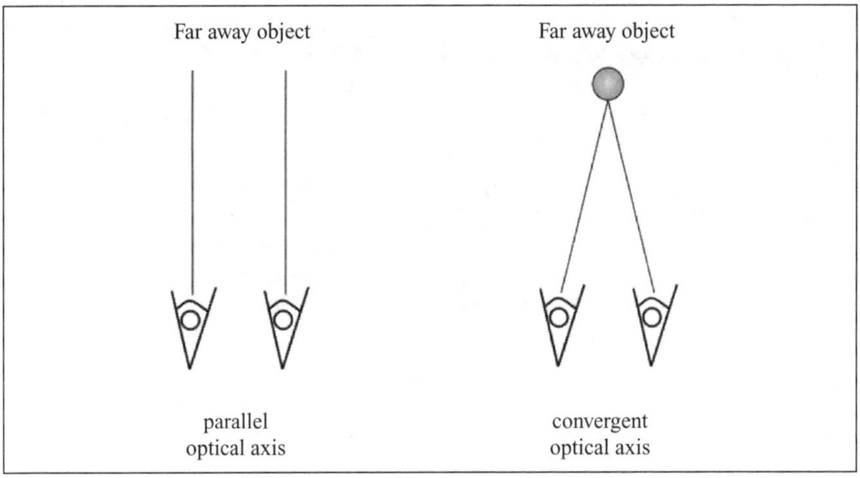

Figure 14.1: Convergence of the eyes: parallel axis for far-away objects (left), convergent axis for near objects (right).

Figure 14.2: Accommodation of the eye. If looking far away (left), the lens is flat; if looking near (right), the focal length of the crystalline changes by means of ciliary muscles.

relative size, cast shadows, linear perspective, atmospheric effects or texture gradient, to mention only the most important ones.

Movement-produced cues are (a) deletion and accretion in overlapping objects which occur during movement, as well as (b) motion parallax which you experience, for example, if looking out of a moving car: near objects glide rapidly past us, more distant objects appear to move more slowly. The speed of the moving objects gives us a signal to their distance; far-away objects move slowly and nearby objects move fast.

Binocular Cues

There is only one depth cue available that requires both eyes: stereopsis. Stereopsis happens because both eyes differ slightly but systematically in what they see respectively. This is due to the fact that the two eyes are around 6 cm away from each other. Fixated points are mapped on corresponding retinal points, whereas objects before or behind the fixation line (which is called a horopter, an imaginary circle that passes through the point of fixation) produce disparity. Figure 14.3 illustrates this effect. This

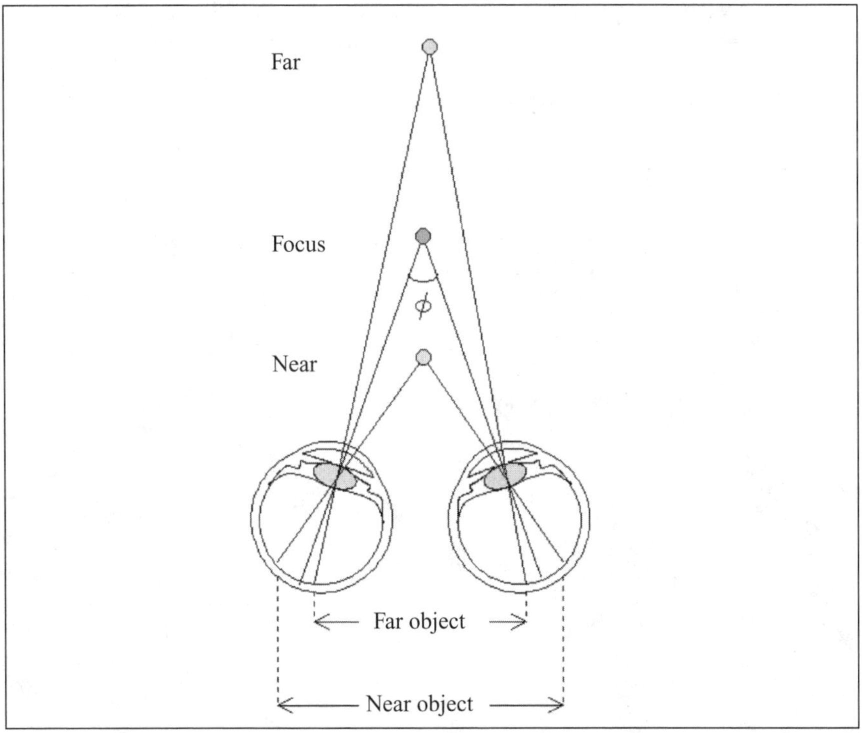

Figure 14.3: A focus point, a near and a far point and their respective placement on the retina. Only the focus point is on corresponding places of the two retinae; the near and far points produce disparity (i.e., noncorresponding places on the retinae), namely the more disparity the further away from the focus point.

disparity can be analyzed in the visual cortex by means of disparity detectors. These detectors allow us to perceive depth in random-dot stereograms in which no other depth cue exists than disparity. You might know random-dot stereograms from those famous books called 'Magic Eye', which contain many of these stereograms; most readers need some experience to see the third dimension; some never reach it because they do not have the ability for stereopsis.

Range of Effectiveness

It is important to know that most depth cues are useful only in a range of maximum 30 metres, some (like accommodation and convergence with maximum 3–6 metres) even less. This means that our visual senses are not made for perception of objects far away, but are specialized for the discrimination of objects in a near radius of 5–30 metres.

Development of Depth Cues

The first experimental studies on this issue were run by Eleanor Gibson in the 1960s with an apparatus called 'Visual Cliff'. This device consists of a glass-covered table and a central platform, from which babies are encouraged to crawl (see Figure 14.4).

Patterns are placed below the glass to create the appearance of a shallow side and a deep side. When trying to get the babies to crawl across the glass, the babies readily crossed the shallow side; but nearly all babies reacted

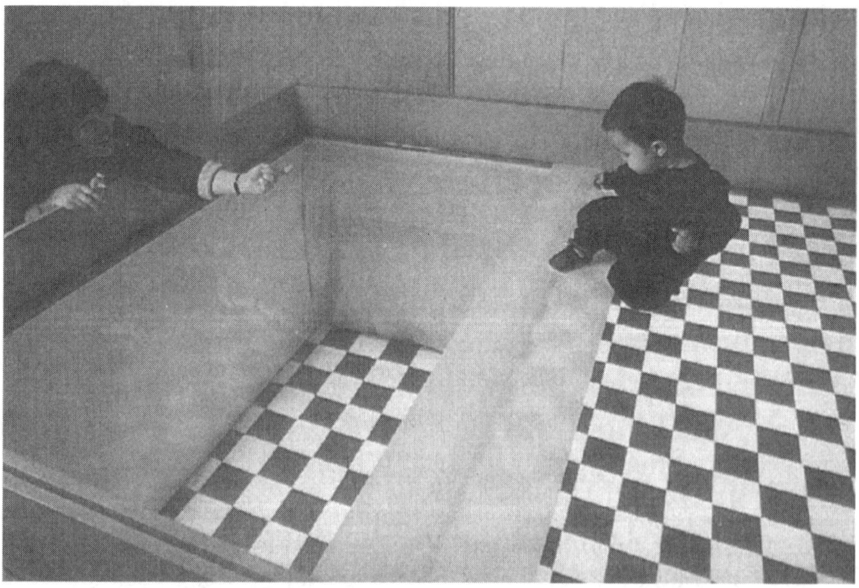

Figure 14.4: A 'Visual Cliff' showing a baby attracted by its mother. A sheet of glass would allow the infant to crawl across the deep section but they stop due to perceived depth.

with fear to the deep side. Gibson concluded that at the time infants crawl, most of them have an understanding of depth perception. Binocular sensitivity emerges in a sensitive time period between two to three months and gradually improves within six months. At seven months, pictorial depth cues lead to specific grasping reactions. Learning of depth cues seems to take place at a very rapid rate during the first three to four weeks of life. It is not completely innate because vision impairments can be compensated for by learning in later years ('vision therapy'). This would not be possible otherwise.

Space Cues from Auditory Information Processing and Other Sources

Up to now, only visual cues for depth perception have been mentioned. But there is also information from both ears, which help to locate objects and say something about the depth structure of one's surrounding based solely on one's hearing. Blind people develop this ability, which is of minor importance if you can see something. But one need not go into clinical examples to get an impression of the power of the ears. Also, space cues can be gained from smell and tactile cues as well (see the chapter by Bettina Beer in this volume).

Gell (1995) reports about the annoying experiences during his fieldwork in Umeda (West Sepik, Papua New Guinea), where he lived for 14 months limited to a sight field not deeper than approximately 10 metres – there was no possibility of seeing the landscape from a panorama perspective. In the dense forest, vision would not help much in hunting but listening did (if you knew what to listen for).

An In-between Summary

What have we learned so far? Perception of space takes place on the surface level depending on our senses. The visual sense mainly, but also hearing (and to a certain degree also movement), contributes to the extraction of certain stimuli from the perceived environment. An anthropologist has to know these facts about the perceptual system – but will that really help to understand space perception? As we will see, a better understanding of space perception will not occur until we enter the deep level.

Space Perception: the Deep Level

To go into the deep level of space perception, one has to go beyond the purely perceptual level. One has to analyse the role of space for a human being. Instead of our previous focus on function, we now have to focus on meaning and significance (which are communicated by language). In a sense, we move on from depth perception to a deeper understanding of spatial cognition (e.g., Montello 2001). The American psychologist Edward Tolman coined the term 'cognitive map' in 1948 to refer to internally represented spatial models of the environment. But as early as the 1920s, the term 'Lebensraum' (in the sense of life-space) was used by the German developmental psychologist William Stern to describe the space in which one lives.

In this early research, some interesting dimensions of life-space have been described which seem worthwhile knowing even today. I follow their description given in Graumann (2002).

Intersubjectivity

Acting in space can become a matter of intersubjectivity. Think, for example, of walking down a street together with your friend. You kick a tin can to him and he responds by kicking the can back. Within seconds, you both agree that the tin can might be used as a soccer ball and the street can be used as your playground. This is possible only because both partners share knowledge about a popular sports game and its requisites. Shared knowledge is equivalent to shared concepts. From there, it is a short way to language, which is the base for communication and therefore the base for establishing intersubjectivity.

Age Specificity

Space perception on the deep level is also dependent on age. Taking an example from Graumann (2002), a junior might climb upon or jump over an object which is seen as an obstacle from a senior person's perspective. Older people (especially those suffering from impaired mobility) explore their environments quite differently from toddlers who enter every place they can. These examples demonstrate how affordances of objects change with age.

Gender Specificity

In the same way that people of different age perceive the space around them differently, so also do the two sexes perceive space in a gender-specific way. If we talk about gendered environment, we make reference to the gender-specific attractiveness of certain places. In a big store, gender-specific preferences can be observed very easily with women strolling through the clothes departments and men exploring the tools section (I apologize for stereotypes!). The attractiveness of certain places like brothels is clearly gender-specific. Therefore, certain places are explored more probably by men than by women; other places the other way round – more by women than by men. In a certain sense, this gender-specificity reflects the differentiation of gender in a given society. Also, religious systems sometimes make prescriptions about places that are reserved for one sex only.

From Space to Place

Some researchers think of space as chaos and of place as order. I would argue that there is a change from anonymous space to personal places by a process called 'Aneignung' (appropriation) by Graumann. Appropriation of space occurs, according to Graumann:

> by marking, naming, defining, categorizing, and evaluating space as appropriate or inappropriate, owned or free, by signs, words, rules, regulations, and laws; but also by regular locomotion resulting in paths and roads; by the cultivation of nature as subsistence or supply of resources; by the domestication of animals; by the conquest of foreign land and the subjugation of other people(s); by building, constructing, and settling; but also by the artistic and scientific representation of space; and finally, by the overcoming of distance by developing means of communication. (2002: 104)

This enumeration of different types of 'Aneignung' demonstrates the many ways for humans to make places out of space. They are all ways to give space a special meaning. Therefore, space perception at the deep level requires an analysis of personal memories for certain places. At this level, one cannot separate the perceptual processes from human memory. In a sense, research on space perception is research on memory for places.

'Space Games': Psychological Considerations

This last part of my paper will give some recommendations to researchers who plan to work with 'Space Games' (see, e.g., Levinson 1992). The Space-Games Kit contains special instruments to elicit task-oriented verbal space descriptions in cross-linguistic fieldwork situations. Following the description given by Senft (2003; see also Senft in this volume), those games involve a 'Director' consultant who sees a certain stimulus, and a 'Matcher' who does not see it. Director and matcher are sitting side by side separated by a screen so that they cannot see each other's stimulus (Fig. 14.5).

In the Tinkertoy Games, for example, the matcher has to build a three-dimensional configuration with the help of the Tinkertoy building blocks, based on the director's description of the original placed before him (either as an object or – even harder – as a photo of the object).

In the Animal Game, as another example, three different animals are placed in a row by the director and have been reproduced by the matcher on a table which is turned by 90° or 180° to the original. Based on the reconstruction of the original, geocentric or egocentric representations can be assumed (see Wassmann and Dasen 1998).

The following recommendations concerning the use of Space Games may seem to be trivial to psychologists, but may be helpful for anthropologists. I have formulated eleven theses for this purpose.

Thesis 1: Memory Is a Representational System

Memory is not a storage or accumulator of events which you put things into to get them out later. Memory is an *active* system, which uses representations: for words, for pictures and for sequences. Representations for words can be found in propositional systems, for example, in associative priming (Tulving and Schacter 1990). Representations for pictures can be

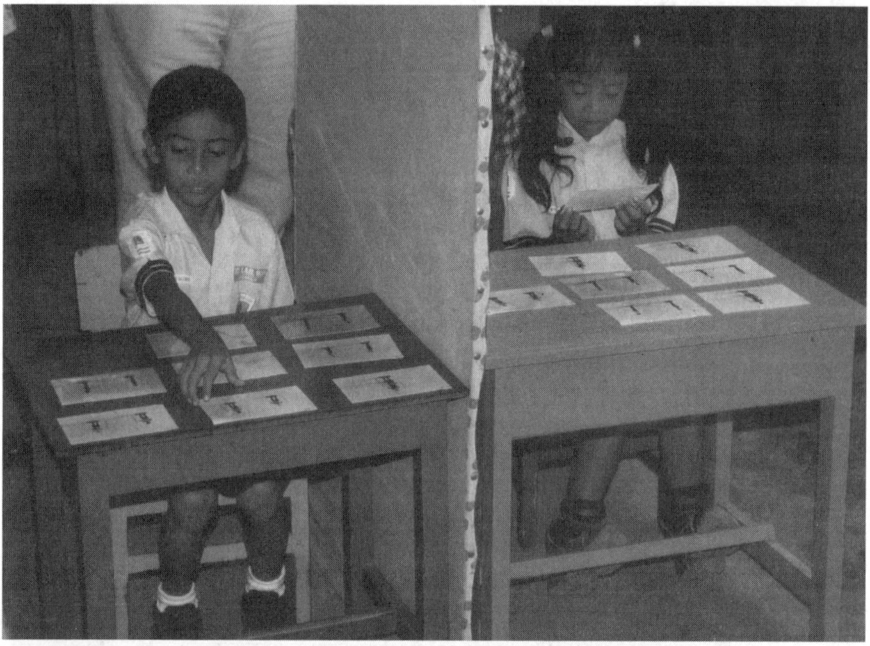

Figure 14.5: Director and matcher working with photographs on both sides of a screen.

found in analogue systems, for example, in mental rotation (Shepard and Metzler 1971). Representations for sequences are necessary to store the succession of events, for example as scripts (Schank and Abelson 1977). Besides these three representational formats, which go back to Anderson (1985) and others, there might be additional representational formats for specific memory entries (e.g., for motor representations). But one should be careful not to assume too many representational formats because it cannot be expected that nature is organized around a representational 'zoo' with many different exemplars.

Representations transform reality. They transform reality during encoding and during recollection under control of attention and proto-/stereotypes. This is the reason why some people say memory is 'schematic'. Different types of representations support different types of processes: if you want to represent the relation between different objects in space, analogue representations might be much more useful than propositional ones. That is the reason why a picture says more than thousand words.

Implications for Space Games: Because Space Games require memory and representations one has to think about the necessary type of representation in a certain experimental condition. It should be clear that the importance of analogue representations might be greater than that of propositional representations if the stimulus material contains analogue dimensions (e.g., different sizes).

Thesis 2: Not All Cognitive Processes Are Conscious

Since Freud, it is not only psychologists who are aware of the importance of unconscious processes, even if we nowadays are sure that his restriction of the unconscious only to sexual impulses is too narrow. The distinction between conscious and unconscious parts of knowledge has been elaborated by Ryle (1949), who said that we have to distinguish between 'knowing that' ('Wissen'), which is declarative knowledge, and 'knowing how' ('Können'), which is procedural knowledge (or even better, 'knowing'). That knowledge does also have a spatial dimension (in terms of concentration of knowledge and professional skills at certain places), is explained in more detail by Meusburger (2000).

Today's research on implicit and explicit learning as well as on implicit and explicit memory (Reber 1967, 1989, 1992; Schacter 1987, 1992, 1999) shows us impressively strong influences of unconscious representations on human behavior (see also Stadler and Frensch 1998).

Implications for Space Games: They measure to an unknown distribution explicit (conscious) and implicit (unconscious) processes at the same time. One has to think about procedures (nonverbal tests), which would differentiate between implicit and explicit parts of the shown behaviour. That automatic processes are unconscious and not available to the subject implies the uselessness of verbal answers – if one asks one *will* get an answer but would one really trust in it?

Thesis 3: Also Nonverbal Tasks Tap Language

There is a difference between verbal and nonverbal tasks in terms of the output system, but not of the internal system. 'Mentalese', a term coined by Steven Pinker (1994) for the language of the brain (which is independent of spoken language), is always needed if someone does something in a conscious, deliberate way (for a fundamental critique of this conception see Fodor 2001).

Implications for Space Games: Researchers should think about automatic and controlled processes. Automatic processes (e.g., perception) can never be verbalized. Control processes *can* be verbalized (e.g., problem solving) but might be disturbed due to verbalization.

Thesis 4: Instructions Need to Be Understood

In every field situation, subjects have to be instructed how to behave. These instructions have to be given for verbal and for nonverbal tasks. One lesson that can be learned from psychology is that subjects can misunderstand or interpret instructions differently from how the experimenter wants them to be understood.

Implications for Space Games: Researchers should ask subjects afterwards how they understood the task and what procedure for solving the task seemed appropriate to them.

Thesis 5: Not Every Answer of a Subject Is True Data

In psychology, there has been for a long time an awareness of the fact that measurements can be decomposed into true values and measurement error. If a subject, for example, shows an IQ of 100, nobody would believe that this is the true value – the true value lies in a range between 95 and 105, depending on the reliability of our measurement instrument. In eyewitness testimony, psychologists are aware of the fact that in recognition tests not every recognized item reflects true recognition. Signal detection theory (SDT; see, e.g., Swets, Dawes and Monahan 2000) offers a methodology to separate the sensitivity of a subject from his or her bias. By means of 'catch trials' it can be assessed how precisely a subject reacts to the stimulus material. Counting the hits, the misses, the false alarms, and the correct rejections makes it possible to simultaneously assess sensitivity and bias of the responder.

Implications for Space Games: Researchers should think about and allow for measurement error and they should think about application of tools like signal detection theory to separate sensitivity from bias. A 'yes, this one' can be everything from a blind guess to secure knowledge.

Thesis 6: 'Labilization' of Subjects Might Occur

Holzkamp (1972) was one of the first who criticized psychological experiments because of their social asymmetry. He argued that the experimenter is always a person who knows about the next steps and who can manipulate the subject for the next 30 minutes, or whatever time the experiment might take. This social asymmetry leads to what he called 'labilization', a situation in which a dependent subject looks for any cue that might help to reduce his or her uncertainty.

Implications for Space Games: Space Games are a highly social situation because the director or manager has the knowledge and the power, gives commands, and has hidden plans. The matcher is a dependent person, is prepared to follow commands, and is uncertain about the plans of the director. This will influence the collection of data, the amount of data as well as the type of data.

Thesis 7: Not Only Data But Also Models Are Needed

Anthropologists like to collect data in the field but sometimes it is necessary to become aware of the fact that data become powerful only in light of theories or hypotheses. Such hypotheses could be related to the assumed representation of the task or to processes, which might be possible with a certain task representation (e.g., mental rotation). Hypotheses might also help to clarify the effect size which determines the amount of subjects needed: if I look for a very small effect size, then I need larger samples whereas if I look for very large effects these can be found with relatively few subjects.

Implications for Space Games: Useful instruments, which might help to think in terms of models, could be task analysis (e.g., is a task positional where space is primary, or is a task relational, where relations are primary?) and cognitive modeling. Cognitive modeling might help to answer the question how would a machine solve a task and what would a machine need to solve a task. Also, lessons from robotics are interesting. The famous 'Robo Cup' (see http://www.robocup.org/) gives an example of how machines which play a kind of soccer communicate their position on a soccer field. According to Christian Freksa (C Freksa 2003), it turns out that geocentric data are easy to communicate but egocentric data are cheap to update. This differential function should be kept in mind.

Thesis 8: Contextuality Is a Major Feature of Human Beings

Humans are highly contextual beings. A simple experiment by Olson (1970) gives a good example of that. Subjects are given a wide round block and are asked to name this object. If it is presented alone, subjects call this object 'the block'. If it is presented in the context of a black block it is called 'the white'. If it is presented in the context of a white square, it is called 'the round'. If it is presented in the context of different other objects, it is labeled the 'white round'. There are many other studies showing the high dependence on context of the cognitive system.

Implications for Space Games: Look out for context effects even if you do not think that context might influence your results. For example, change the animals in the animal-in-a–row task to see if subjects are sensitive to this kind of context.

Thesis 9: Take Care for Perspectivity

Perspectivity is also a general feature of human beings. This feature has been described in more detail in the previous section and does not need elaboration here. As mentioned before, intersubjectivity, age specificity and gender specificity are important aspects. The research done by Wassmann and Dasen (1993) gives good examples of the control of different perspectives.

Implications for Space Games: Researchers should provoke *different* perspectives, for example, compare rural versus village people, young versus old people, male and female people. Also, social status might establish a certain perspective, which might influence data collection.

Thesis 10: Emotionality of Space Has to Be Accepted

Even if it can be assumed that Space Games are primarily a cognitive task we are now sure, for psychological reasons, that even in purely cognitive tasks emotions are present and show their influence. There is a strong connection between cognitive and affective processes.

Implications for Space Games: One should always think of potential influences from the affective side. One could think of using intentionally affectively laden objects to explore the reactions of subjects when confronted with objects in the Space Games that are not neutral.

Thesis 11: Episodic Memory Implies Space and Time

The term 'episodic memory' was coined by Tulving (1972) in contrast to what he called 'semantic memory'. Episodic memory was conceived as that part of memory which makes the recollection of an episode possible in terms of time and place. Semantic memory is the kind of memory that contains knowledge that is not primarily related to the place and time at which it was learned.

Implications for Space Games: Users of Space Games should be aware that it is not only space that plays a role in these games but also time – for example, which of the animals was placed first in the row, which was placed second and which third. This implicit time axis might be made explicit if one changes the task, for example, so that subjects have to remember what was the first animal placed on the desk.

Conclusions

What implications do these psychological insights have for anthropologists? Is there anything to learn from psychologists' research on space perception?

Psychological research about the surface level of perception demonstrates the existence of a large number of pictorial depth cues, most of which seem to be the result of learning processes. The use of pictorial stimuli takes for granted that subjects can use these cues as we do. If we hear that the use of perspective in painting was discovered as late as the fifteenth century in our Western culture, we have to acknowledge the fact that this type of depth cue is not self-evident and requires an environment full of these cues!

Anthropologists sometimes ask if space representation is geocentric or egocentric. This may be the wrong question. A flexible system like the human brain can switch between different modes of representation depending on their usefulness for a given task. Therefore, it is much better to ask under which conditions native subjects prefer which sort of representation and to find out which type of representation is dominant in a certain culture (e.g., Wassmann and Dasen 1998). One also has to think about tasks, which are easier to solve when using one type of representation rather than another.

For an experimental psychologist there is also one standard question to the experimenter: How do subjects understand the instructions? In psychology, we know of many situations in which the experimenter *wanted* subjects to understand the task in a certain way but the subjects understood it in another way. With Space Games, one has to clarify very carefully how subjects understand the task: 'Reproduce the same arrangement of objects *here*, which you have observed *there*.' The fact that certain cultures interpret such a task in their way (e.g., geocentric) does not exclude the possibility that they could also understand the other interpretation of that task (e.g., egocentric). Researchers might register the preferred mode of representation but should not conclude rashly this is the only mode.

Psychological research about the deep structure of perception shows the importance of criteria like intersubjectivity of space, age specificity, or gender specificity. With respect to Space Games it becomes evident that these instruments might, for example, help to elicit words used for certain directions or locations as well as actions related to spatial tasks. But Space Games cannot uncover space perception on the deep level. To do that, it is necessary to see how people appropriate their environment ('Aneignung'), how they turn space into place.

Chapter 15

Conducting Cognitive Tasks and Interpreting the Results: The Case of Spatial Inference Tasks

Thomas Widlok

Introduction

Cognitive anthropology, like cultural psychology, heavily relies on experiment-like tasks for understanding differences in the way in which members of different social groups conceptualize their environment. The tools of the trade are tasks which are directed at what anthropologists identify as 'common sense' – the concepts of just plain folks – but which put subjects under 'uncommon' circumstances, set-up by the researcher and isolated from the common social context. If 'common sense' is the outcome of a social process of concept generation (see Widlok 1998: 440), the interpretation of results gained through set tasks, which are uncommon for the social agents and which eclipse this social process of concept generation, is inherently problematic. The design of spatial inference tasks and the apparently mixed results gained with the help of Hai//om 'Bushmen' subjects in northern Namibia (see below) underlines this point. Despite being individually exposed to the uncommon tasks, the Hai//om orientation skills seem to rely heavily on common, everyday experiences that were socially shared. Inferential problem solving is facilitated by communication with fellow members of the group not because others may have superior knowledge but because intimate knowledge of the spatial behaviour and goals of others makes shared cognition particularly useful. This case of cognitive ethnography shows how to combine results gained through conducting cognitive tasks with the ethnography of local common sense. It suggests how to make use of ethnography when trying to interpret the results of tasks which were developed to distinguish different systems of spatial orientation.

Cognitive Tests

Although social scientists following Durkheim (1912) would by default be sceptical with regard to any 'explanation' of environmental skills in terms of 'instinct' (see Widlok 1997) and would want to account for cognition socially, that is supra-individually, they have had difficulties providing satisfactory comparative methods to demonstrate that their claims are true. In particular, the proposition that processes outside the individual mind are a critical part of cognition has met the opposition of neighbouring disciplines, most forcefully in experimental psychology (Cole 1985: 146). Those who regarded anthropological data as methodologically unsatisfactory felt pressed to collect 'anthropological data' meaning cross-cultural data, themselves (Eckensberger 1970: 34). As a consequence psychology and anthropology became divided by a common language (Jahoda 1993: 36), both notionally engaged in comparisons but practically working under two different paradigms. They drifted further apart when tests and experiments were introduced by psychology in an attempt to meet scientific standards. Cognitive anthropologists, partly out of conviction and partly out of a desire not to be ignored any longer, also started to subject their informants to experiment-like situations uncommon to them in order to elicit data that was considered acceptable by the other disciplines. Incorporating these hypothesis-testing devices seemed to be the prerequisite for challenging some of the views, established by neighbouring disciplines, about how human cognition works.

Cross-cultural comparisons of this kind were initially used to fulfil two functions, firstly to establish the relative identity of cognition within a group of subjects (however defined) and secondly to establish the relative difference of cognition between groups of subjects. 'Bushmen', too, had been subjected to spatial experiments. In parallel with the general context of an emerging experimental methodology, psychologists expressed their dissatisfaction with unsystematic anthropological statements – and other 'pre-psychological' reports – about the cognitive skills (or the lack of these skills) of the 'Bushmen' (see Reuning and Wortley 1973: 3). Porteus (1937), Winter (1976) and Reuning and Wortley (1973) designed and conducted a number of tests, several of which aimed to elucidate spatial skills such as recognition of tracks and estimation of distances. However, these *tests* differ critically from *tasks* because they measure good versus poor performance rather than different ways of solving problems. In sum, the overall performance rates were not strikingly different for the 'Bushmen' when compared to those of other subjects. Depending on the test, the performance was slightly better or slightly worse than that of other groups, a result that would not surprise ethnographers who have worked with southern African hunter-gatherers. The variation found was explained with reference to the degree to which the uncommon experiment-like situations resembled real-life problems encountered by the groups tested (Reuning and Wortley 1973:107). Environmental effects on performance were also targeted, with a rather narrow understanding of environment as being constituted by the natural environment. Here the most striking result was that the 'Bushmen', presented as living in an isolated, 'primitive' and homogeneous environment, were more heterogeneous in their responses than other African or European groups (ibid.: 108). In conclusion, the psychologists called upon the anthropologists to solve this puzzle and to find an explanation for the unexpected individual differences that were observed (ibid.: 96). Unfortunately, this was subsequently not taken up by the anthropologists who, instead, expressed their dissatisfaction with the psychological procedures in general for the very reasons that these procedures had been introduced by the psychologists who were dissatisfied with anthropological data: the fact that they are designed for comparability and not with reference to local perceptions, that they are formalized to increase reliability and are not culturally sensitive, and that they are amenable to statistical methods but difficult to link up with everyday life. Berland (1983) pointed out that while the tests may meet high standards of reliability and replicability and while the results are statistically rich, they are rather poor in explanatory power, difficult to grasp for the subjects and incapable of eliciting any cognitive processes that were not predicted beforehand.

As I will show in more detail below, anthropologists have recently preferred to design tasks that help to explore a cognitive field rather than to design tests that establish performance on a fixed scale (for example Boster 1985; Wassmann 1994). The spatial tasks which were developed in a collaborative effort by members of the Cognitive Anthropology Research Group in Nijmegen (to be described in detail below) try to avoid some of the methodological problems that had emerged in earlier work in cross-cultural psychology. The cognitive tasks of this series (see Baayen and Danziger 1994) are not testing performance but they allow at least two possible solutions to a spatial orientation problem which are equally correct, one being a relative response, the other being an absolute response in the experiment. However, these tasks maintain the experimental framework insofar as they try to keep a stimulus constant to see how the responses varied. For the comparisons envisaged, the stimulus and the experiment-like setting was the independent variable and the background of the people subjected to the test/task was the dependent variable.

Most of these experiments developed and carried out by anthropologists were primarily memory tasks in which a person is shown a spatial array, for instance a row of toy animals, asked to remember the way they are positioned, then turned around 180° and asked to rebuild the scene that he or she was shown before (see Widlok 1998). The stimulus is such that the objects are presented in a straight line which runs parallel to both the bodycentred axis of left and right (or in some variations the axis of front and back), relative to the body of the observer, and the geocentred axis of west and east (or north and south), 'absolute' with regard to the body of the observer. The consultants are brought into a situation where they have to chose between their body as a coordinate system (made up of the intersecting axes of left/right and front/back) or the wider world as a coordinate system (made up of the intersecting axes of west/east and north/south). Thus, the experimental design allows two equally valid but quite different responses to such a task. One is to focus on the spatial order of the animals relative to the observer, i.e. if the animals were positioned from my left to my right over there, they should be positioned from left to right over here (which is the *opposite* direction when viewed from above), when I have turned 180° (see figures and description by Senft in this volume). That would be coded as a relative response. Alternatively, one could focus on the spatial order of animals in some absolute frame, for instance with regard to the cardinal directions east/west/north/south. If the animals were positioned one after the other from west to east over there, they should also be positioned from west to east (which is the *same* direction when viewed from above) when I turn around over here. This strategy produces the opposite result of that in a relative framework and is coded as an absolute response in the experiment. This quasi-experiment,[1] which was carried out with speakers of very different languages and social backgrounds all over the world, shows that responses are patterned in the following way. Some communities seem to prefer the relative response, others the absolute response, and still others seem to be more interested in the fact that the animals are apparently following each other in a certain order, never mind which way they are heading (for a summary see Baayen and Danziger 1994; Pederson et al. 1998). The last-mentioned response is characteristic for the intrinsic frame of reference because it largely ignores the relative position of the observer as well as any larger absolute frame (Danziger 1996).

The results gained through this research refuted the received view in the field of spatial cognition, namely that a relative, body- or egocentred approach to space (in terms of left/right and front/back) was natural to humankind. There are cases, in which an alternative absolute or ground-centred system prevails, based, for instance, on cardinal directions (Levinson and Brown 1994; Wassmann 1994). While ostensibly relativizing Kant and Newton, these contributions called upon the cognitive scientists to pay more attention to cultural aspects, including language but also nonlinguistic aspects. They also called upon anthropologists to include carefully designed and selected uncommon experiment-like situations into their repertoire of ethnographic techniques. The quasi-experiments (described in some detail in Brown and Levinson 1993, and in Levinson 2003), turned out to be not only hypotheses-testing devices but also useful in detecting new problems. They were creating 'mandated', i.e. deliberately created, breakdowns of understanding as envisaged by Agar (1986: 25). They showed that the 'absolute' (allo-centric) system was a viable alternative to the 'relative' (egocentred) system, but also that there was another culturally realized possibility, that of an 'intrinsic' (orientation-free) system of conceptualizing spatial relations. At the same time it was suggested that repeated social experiences and forms of institutionalized social relations generate the conceptual frameworks elicited by the tasks (Danziger 1996: 79). However, how this integrated process worked could not be resolved on the basis of the experiments alone that had been conducted. In line with an overall modular view of cognition, the further experiments developed in this particular research programme were not designed to account for such a larger integration but were tailored to give answers to more specific aspects of the cognitive process, above all to separate linguistic from nonlinguistic aspects of cognition and to separate aspects to do with memory (the recall and reconstruction of earlier situations from a separate point in time) from those to do with inference (drawing conclusions from experiences joined into one situation). In other words, the quasi-experiments could show differences in orientation frames abstractly defined but could not elucidate the process whereby these frames are generated and maintained as working systems in practice.

Initially, the emphasis was put on the attempt to separate linguistic from nonlinguistic aspects of cognition. This raised a number of points with regard to the design of the quasi-experimental tasks – concerning the stimulus material, the linguistic input and memory effects, leading to other forms of research tools (see Widlok 1998 for details):

1. The more naturalistic the stimulus material was, the less likely was it to be confusing for the informants but the more likely was the introduction of uncontrolled variables by its specific form since details of form may influence the informants' decisions.
2. The more linguistic input went into the task the more likely were informants to understand the task but the less likely was their response independent of the surface structure of the language that is used.
3. The greater the time lapse between stimulus presentation and response elicitation, the less likely was an 'automatic' sensory response but the more likely was a response dependent on the way in which information is stored in memory rather than the way in which it is inferentially processed.

To some degree these trade-offs seem unavoidable, but when developing the task that I describe in the next section the attempt was made to preempt some of these criticisms by (1) keeping the stimulus simple and clear, by (2) minimizing the linguistic input and by (3) forcing inferential rather than memory skills.

In order to test whether the frames of reference described above were 'only' tools for memorizing spatial arrays or whether they were more generally employed, a number of inference tasks were developed. It was assumed that in orientation skills, wherever they are employed, processes of memorizing (and habitualizing) play a role but also, and maybe more fundamentally, processes of inferring spatial truths from the spatial information that is available in the situation. Pieces of spatial information are linked to one another and to the world through inferences. Euclidean geometry and other systems with strict rules of inference suggest that the process of everyday orientation (but also that of expert orientation) can be regarded as a series of logical inferences based on de-indexicalized spatial information. Indexical information itself can be of various formats. It may include cardinal items such as 'X is north of here' or relative items such as 'X is to my left'. In any case, this context dependent information needs to be de-indexicalized in order to make logical inferences (Kutschera and

Breitkopf 1979: 18; see also Küng 1993: 52). This applies as much to space as to any logical operation that requires a transformation of indicators (or tokens) such as 'here', 'there', 'I', 'you' into nontokens such as 'in the house', 'in the garden', 'Dr Jekyll', 'Mr Hyde'. This process seems central no matter what the sources of our information might be, whether it is a map, a compass reading, personal memory, or a route description. De-indexicalization is central no matter whether the information was coded in terms of an absolute or of a relative frame as identified above. The information may be recalled from memory, such as memorizing previous trips made, or it may be combined from simultaneously available sources, such as a map, a compass and the sight of a salient landmark. It seems to be widely assumed among cognitive scientists that even though the provision of information may be socially complex and culturally sensitive, the decisive process of de-indexicalizing may nevertheless be adequately presented and analyzed as an individual mental operation. As a corollary, understanding spatial inference is accessed through experiments in which the isolated individual is pushed to the limits in performing operations of inference. The 'transitivity task', which forms part of a battery of tasks developed by the Cognitive Anthropology Research Group and which I will discuss in some detail below, is an inference experiment of this kind. Individuals are asked to solve the task in isolation, but this does not necessarily warrant the assumption that the results gained are that of individual knowledge and cognition rather than of shared cognition. Only a strong cognitivist premise would need to attribute significantly divergent responses to a task across cultural environments to either differences in performance (see Porteus 1937) or as evidence for different mentalities (see Hallpike 1979).[2] Alternatively, and this is the approach taken here, disparate or mixed results may be regarded as a challenge to the underlying assumption that cognition is adequately grasped as a series of individual inference processes. What forms the 'testable' part of the cognitive process may be just that: a part of a larger cognitive process that can be targeted by tests and tasks but which may not constitute the decisive part of the process.

Spatial Inference Tasks

The research results that I have reported on so far are based either on experiment-like tests or at least on 'uncommon' tasks. 'Uncommon' tasks are those experimental tasks in which individuals are cut off from their everyday social life, the source of their common sense. Bringing together uncommon tasks with what may be called a 'cognitive ethnography of common sense' proves to be particularly important when spatial tasks produce 'mixed' results – not at all uncommon in non-Western settings. I suggest that making sense of mixed results in spatial cognition tasks forces us to shift our attention away from the isolated methodological question of the benefits and problems of applying formal tasks towards the underlying arguments about how the connection between cognition, culture and social life should be approached.

Many Hai//om people have outstanding skills of orienting themselves. As I have shown elsewhere (Widlok 1997) this skill emerges in the everyday practice of moving through the bush as well as in rather artificial pointing tasks set up by the anthropologist. Hai//om subjects were able to point correctly in the direction of places that were out of sight and at considerable distance. They surpassed European subjects by far. Given this expertise, we might expect that the Hai//om have brought to perfection and to particular sophistication one of the spatial cognition systems that have been identified by cognitive scientists. Two of these systems, or 'frames of reference', already mentioned lend themselves particularly to orientation, the third (the 'intrinsic' frame) being inherently orientation-free. The two orientation-bound systems have been variously labelled but are based on a distinction between egocentric (relative) and geocentric (absolute) orientation, i.e. relying either on the human body or on the landscape as frames.

Levinson, reviewing the extensive literature on human and animal orientation skills, points out that 'relative coordinate systems may favour "piloting", navigation by constant reference to familiar landmarks, while absolute coordinate systems favour true "dead-reckoning" types of navigation' (1996c: 23). Both systems can be distinguished by the 'merits' they have when solving real-life tasks. The absolute system, which has been likened to a compass, avoids specific viewpoints and therefore is logically superior when it comes to transitivity and fixing things spatially when the observer is mobile (e.g. turning or rotating). The relative system, by contrast, can rely on egocentric visual memory without the need to constantly maintain fixed bearings such as east/west and north/south. Some animals gain high proficiency in these systems, but, unlike humans, they do not learn to switch between different systems nor are they able to combine them (see Levinson 1996c: 23).

Figure 15.1 shows the setup of a task that was developed to find out which frame of reference is brought to bear in an inference problem. In this task the stimulus material consists of one large reference item (a yellow plastic block), which is shown first together with one other object (a blue cone) on table 1 and then together with another object (a red cylinder) on table 2. The two tables are placed at opposite ends of an axis on which investigator and consultant move, turning around between them. Both the red and the blue object are shown in relation to the yellow block, the reference item. Returning to the first table again, the reference item is not shown but the subjects have to infer the relation between the blue and the red item from what they have been shown before. Again, two solutions are envisaged, which are opposed to one another but which are both correct. One solution is to think of the relations of the blue and the red item to the yellow block in relative terms (left and right or front and back) and another is to think of them in absolute terms (e.g. in the east, further to the west or to the east).

For illustration purposes it is helpful to run through the task twice, once with a relative frame of reference in mind and once with an absolute frame of reference in mind (see Figure 15.1). In terms of relative relations the blue

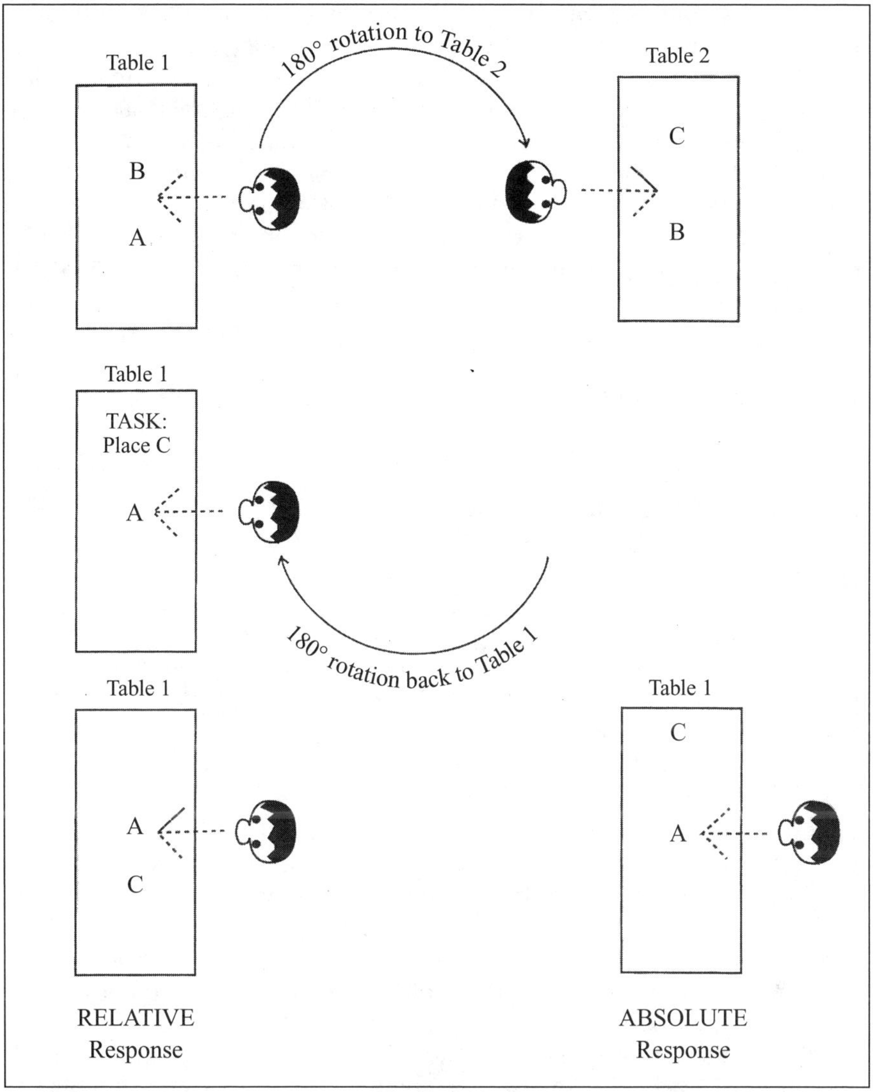

Figure 15.1: Setup of a task distinguishing strategies for solving spatial inference problems.

item A is seen as being left of B in the first step and the red item C is seen as being on the far left of block B in the second step. Without seeing B, when returning to the first table, one is inclined to say, C has to be to the left of A because it was further to the left of B than A was. In terms of absolute relations, by contrast, item A is seen to be north of B on table 1. Item C is seen to be to the far south of B on table 2. The inferred conclusion in this instance is: A has to be to the far north of C.

Since the task is not all that simple, two test runs were included. Firstly the subjects are familiarized with the idea that transitivity is involved in these tasks. This is done by using relations of containment between a trough, a fence and a toy pig and without rotation being involved. Secondly, several test runs (which are not counted in the results) are done to introduce the setup of the task and the objects involved. The main trials that follow are a randomized but systematic series of tasks that involve all possible constellations of left/right, front/back, east/west, north/south while moving between two tables positioned on one axis. Hai//om subjects solving the task in the first run (that is, tables positioned on the east/west axis) tended to use absolute coordinates rather than relative ones (see Table 15.1).

Table 15.1: Results of inference task, first run (tables on north/south axis).

Responses:	Relative	Untypable	Absolute
EW/LR axis (%):	8	17	75
Mean: 3.9 (high absolute response)			
NS/FB axis (%):	17	58	25
Mean: 2.6 (unexpectedly low absolute response)			

(N = 12, a larger sample of 18 subjects who, however, did not do both runs, shows the same tendency, even slightly more marked). The mean refers to a scale between 0.0 (extreme relative) and 5.0 (extreme absolute)

The results of the first run in this inference task confirms the Hai//om tendency to use absolute coordinates, which has been found in other nonlinguistic tasks as well as in language elicitation tasks (see Neumann and Widlok 1996; Widlok 1998). The descriptive statistics present the final results in two formats. Firstly, the percentage of all solutions chosen by all eighteen Hai//om subjects[3] according to their relative and absolute tendency (mixed responses are grouped together as 'untypable'). Secondly, a value for the tendency observed on a range between 0 (which denotes a 'pure' relative response) to 5 (which is an absolute response, in other words, at 5 out of 5 times the response was absolute). The value for the Hai//om sample is 3.9 on the across axis (that is, left to right, which in the first run is tantamount to north/south) and differs significantly from the clearly relative response of Dutch subjects (mean 0.2 on both axes). However, on the away axis (that is, front/back and in the first run east/west), the case is far less clear because a mean of 2.6 shows no clear tendency. In the Hai//om case the absolute tendency was much clearer on one axis than on the other and this difference of results is statistically significant, unlike any of the other of the tested relations (absolute tendency in comparison with sex, age, level of literacy). The question remained as to whether this was due to the unequal strength of axes in the absolute (cardinal) system or due to the unequal strength of axes in the relative (egocentred) system or indeed due to a combined effect in both these systems. Since the north/south (NS) and front/back (FB) axes fell together in this first run, a second run of 90° rotation was needed so that each axis was tested in relation to both of the corresponding axes, east/west (EW) and left/right (LR) in the other frame of reference (see Figure 15.2).

The first run of the spatial inference task, therefore, was not conclusive. However, since relative responses were always low (even when absolute responses were relatively low, they were still higher than the relative responses) a number of explanations can already be disregarded. The first run of the task shows that Hai//om speakers do not rely on an essentially

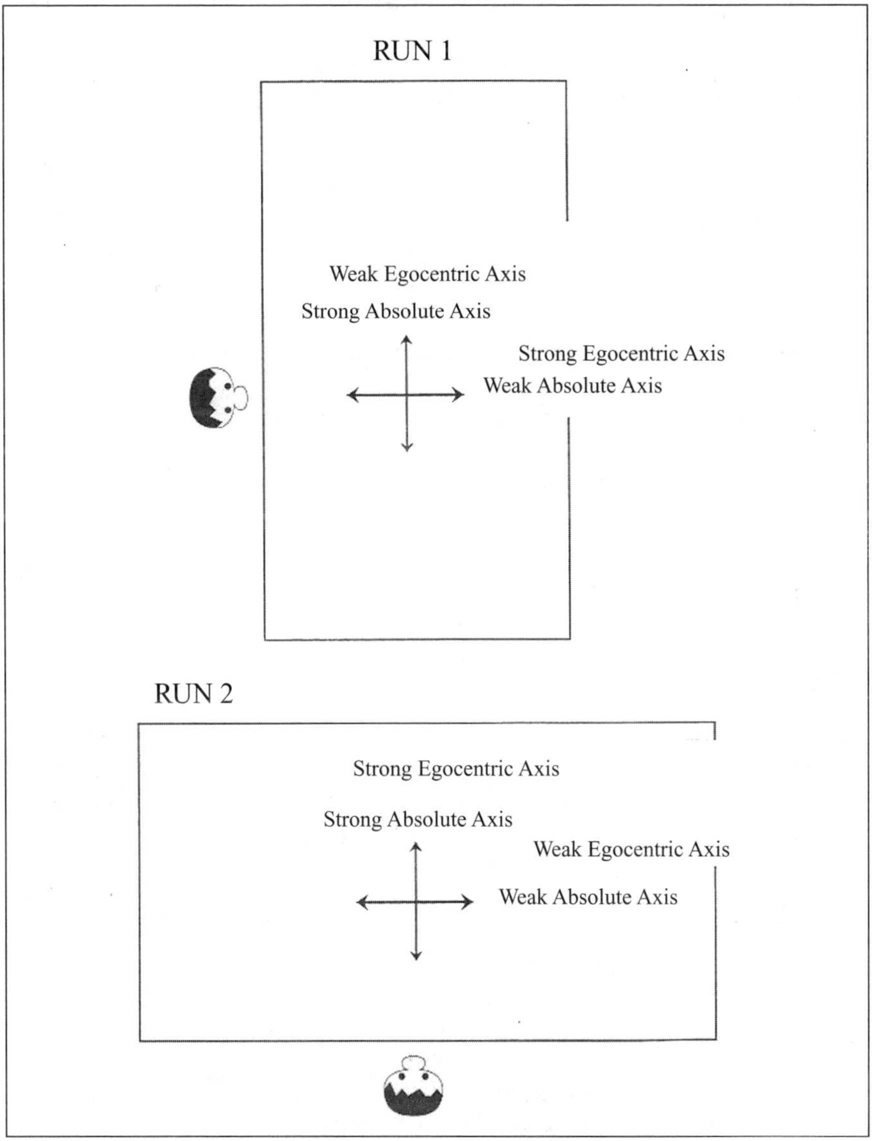

Figure 15.2: Spatial axes in the two runs of the transitive inference task.

relative/egocentric system. Since the overall tendency was always absolute, no matter which axes came together, it would not be valid to claim the existence of two frameworks which in some instances would collide. A number of logically possible hypotheses can be made to account for the mixed results gained in the first run of the task:

I The underlying absolute or ground-centred system is compromised by the fact that the north/south axis is weak compared to the east/west axis.

II The underlying absolute system is compromised by the fact that the front/back axis of the relative system is strong compared to the left/right axis.

III The underlying system is compromised by a combined effect of a weak north/south axis and a strong front/back axis.

The problem can be settled by statistical means when the same task is carried out with a shift of 90°, that is on the alternative axis, as Brown and Levinson (1993) have exemplified in a Mayan case study. A matrix can be constructed onto which the different hypotheses can then be mapped. The basic matrix (Table 15.2) summarizes the results of the first run, namely a high absolute response on one axis and a low absolute response, at least when considering the overall Hai//om tendency, on the other axis.

Table 15.2: Basic matrix for explaining results in transitive inference tasks.

Basic matrix	**Strong absolute axis (EW)**	**Weak absolute. axis (NS)**
Run 1 (EW = LR NS = FB)	High absolute response	Low absolute response
Run 2 (EW = FB NS = LR)	?	?

It is now possible to describe the expected results if one of the three hypotheses was to be verified with reference to this matrix and the two question marks in the left and the right quadrant.

Hypothesis I: It is Hai//om common sense in the field of spatial orientation to use an absolute frame of reference but on the weak (NS) axis of that system a relative framework may interfere. In this case the egocentred front/back responses in the first run would be an error in the sense of being wrong with regard to the dominant common sense as to what is a clever solution to the problem. On the left/right axis the 'foolish' or misguided responses may not have been as marked because they were matched by the strong east/west axis. After rotating 90° one would predict, following this hypothesis, that a similar rate of 'wrong' or foolish responses should occur on the north/south axis which is now paired with left/right. In the matrix we expect a high absolute response in the left quadrant and a low absolute response in the right quadrant.

Hypothesis II: It is Hai//om common sense in the field of spatial orientation to use an absolute frame of reference but on the strong (FB) axis of the relative framework may interfere with that tendency. In this case the effect of the first run should be reversed, i.e. the comparatively low absolute response should occur on the east/west axis which is now combined with the front/back axis, while the north/south axis (combined with the left/right axis) should reach a high mean for the absolute tendency. In the matrix we expect a low absolute response in the left and a high absolute response in the right quadrant.

Hypothesis III: It is Hai//om common sense in the field of spatial orientation to use an absolute frame of reference but this tendency can be compromised by a joint effect of the weak absolute (NS) axis and the strong (FB) axis. In this case the results of the second run should have high means for absolute responses in both cases since the weak absolute and the strong relative axis are no longer combined and therefore cannot produce the effect. In our matrix we expect high absolute responses in both quadrants.

There is a fourth logical possibility and that is a comparatively low absolute response on both axes in the second run (assuming still that the relative responses are not reaching a high level) and correspondingly a low absolute response in both quadrants of the matrix. This case would be matched by the hypothesis that only if the weak relative axis (left/right) is combined with the strong (east/west) axis, the Hai//om common sense of absolute spatial orientation would have an effect. However, as noted before, a comparatively low mean (i. e. around 2.5) on both axes, matching this possibility, would not positively indicate any tendency and is therefore indistinguishable from a result of a task that was not properly understood or simply badly administered by the researcher. Fortunately, the result of the second run did not produce low absolute responses in either case but quite to the contrary a high mean in both cases (see Table 15.3).

Table 15.3: Results of the inference task, second run (tables on east/west axis).

Responses:	**Relative**	**Untypable**	**Absolute**
EW/FB axis (%):	0	33	67
Mean: 4.1 (highest absolute response)			
NS/LR axis (%):	8	42	50
Mean: 3.8 (high absolute response)			

(N = 12, again a larger sample of 20 which includes subjects who did not do both runs, shows the same tendency.)

This means that hypothesis III – a joint compromising effect of a weak absolute and a strong relative axis on an overall absolute system – has been verified. Since the highest mean is reached when the east/west and the front/back axis come together there is a slight cumulative effect here, too.[4]

In the second run subjects gave few relative responses on either of the axes, that is to say the Hai//om consultants were more consistently absolute than in the first run. Learning effects may be safely disregarded because the two runs were carried out with almost half a year in between and because the two larger samples (including subjects who only did one run) show similar results. The statistical analysis settles the uncertainty that arose after the first run when I was not able to distinguish which aspects of the two frameworks had produced the mixed result. After the second run I was able to state that

- The two frames of orientation were not interchangeable alternatives.
- The weak north/south axis by itself was not a source for 'noncommonsensical' responses.
- The strong front/back axis by itself was not a source for 'noncommonsensical' responses.
- The joint effect of the weak absolute and the strong relative axis could produce apparently 'mixed' responses.

More generally, the experiment-like task and the statistical analysis allows a rather technical and formal description of the Hai//om spatial orientation system. The initial data now appear less messy than before but provide no insight beyond the assertion that a complex and apparently arbitrary system of spatial common sense is in place which has the following characteristics:

- The absolute frame at work is stronger than the relative frame. That is to say, when responding to the tasks, Hai//om informants tend to rely on an absolute frame of reference rather than a relative one.
- Within this absolute system, the east/west axis is stronger than the north/south axis, because it is used more often.
- Within the overall weak relative system the front/back axis is stronger than the left/right axis.
- The two frames of reference or systems interact in a specific way, namely the relatively strong front/back axis comes in when it coincides with the relatively weak north/south axis.

The statistical analysis within the experimental framework has made the mixed character of the Hai//om response more explicit but it still has not solved the underlying problem, namely to explain the mixed Hai//om system in the face of their outstanding skills of spatial orientation. Through the tasks we do not get any indication why the system developed that way and how such an apparently awkward combination could produce the very sophisticated orientation skills that have been documented for the group in question. More importantly, the explanatory power of the statistical analysis rests on the implicit assumption that the rules identified post facto are somehow followed by the individual as a guideline for the process of inference, either consciously or subconsciously. But to assume this is to commit a 'fallacy of the rule' (Bourdieu 1977). How can a common-sense structure be pervasive and convincing for those using it, if it relies on a rather complicated combination of underlying rules and systems? We have to face the possibility that the impression of a 'mixed' cognitive system used in common-sense orientation is in fact an analytical effect. What is worse, the detailed description given above suggests that the analytical effect is not one of bad experimental design but a consequence of the experimental design per se.

Comparative ethnography provides us with examples of the different ways in which people conceptualize their environment, the different systems of spatial terms they use and the different skills which they have developed according to the requirements of the environment they live in (navigation at sea, orientation in the bush, path-finding in urban settings etc.). What has been a matter of debate is whether, this diversity notwithstanding, there is a single underlying process of spatial cognition or whether differences in the cultural environment are effective enough to bring about a number of distinct cognitive systems (see Levinson 2003). The question is raised as a matter of cross-cultural comparison but it also emerges in case studies of apparently 'mixed' spatial systems like that of the Hai//om. In these cases we need to consider whether there is one cognitive process or several of them at work and how we can account for the divergence of processes.

It is part of the Western common sense to distinguish an absolute system from a relative system. In our everyday practice the two systems are rarely mixed. The relative frame is conventionally used for some activities and object relations (e.g. at the dinner table), the absolute frames for others (e.g. when navigating a ship). This is indeed a common-sense procedure in the strict anthropological sense because it convinces us that the system we commonly use consists of an exhaustive and well-structured set of descriptive categories to think and talk about spatial relations. But this impression we have may rely more on our way of segmenting the world (into areas that are preferably dealt with 'relatively' and others that are dealt with 'absolutely') than on any modular distinction in human cognition. To investigate this problem we need to look at the ways in which Hai//om common sense is constructed in the context of activities in space and how it becomes entrenched as a commonly accepted conceptual strategy – above all through the use of language and everyday routines.

Topographical Talk

Contrary to what one might expect, the Hai//om language neither possess a perfect geocentric system of cardinal terms nor a perfect egocentric system of spatial terms to guide speakers through space. Unlike systems used by some groups of Australian Aborigines, for instance, the Hai//om absolute reference system differs in many ways from a compass. Linguistically, the absolute frame of reference of the Hai//om is based on 'sunrise'/'sunset' for the east/west axis and on contingent landscape terms (such as !*Auib*, land of the river) for the north/south axis. For both axes alternative ways of expressing direction and location exist. The terms used are vague and variable; they do not always form a complete set (see Widlok 1998).

By the same token the Hai//om relative terms are not readily extended into other domains and remain undeveloped in comparison to strong relative systems like those employed in route descriptions (or in object manipulations) of speakers of English or other Indo-European languages. Hai//om seem not to routinely translate the information contained in mental maps into body-centred information in the form of indexical statements such as 'Turn left, here!'. Are these indications that Hai//om 'mix' and combine what appears to the cognitive scientist to be not only two linguistically but also logically (and presumably neurophysiologically) distinct systems? To begin answering this question we have to investigate the link between everyday spatial skills and cognitive systems or frames of reference in more detail.

Hai//om spatial activity combines piloting (wayfinding along known paths and landmarks) and dead-reckoning (navigating on a new route). Although most Hai//om today live in permanent settlements they are nevertheless a highly mobile group. Extended visiting, and moving between permanent settlements, or between permanent, semi-permanent and nonpermanent camps, is for the most part still done on foot and in small groups. The landscape is partly woodland, partly open bushland, so that it recommends itself to walking on established paths. The paths in the bush are in most cases not well-trodden and usually rather narrow. They only allow walking in line, which is in any case the preferred way of travelling. Often, though not always, the more senior people walk in front, especially if they know the route and if they know the clues they are looking for such as animal or human tracks. The people who follow may follow at some distance as they often gather fruits on the way or simply because their walking speed is different. If the person walking in front is out of sight, those behind also need to 'pilot' their way as they follow the fresh tracks of the person walking in front. Extending the experience of walking in line metaphorically into other fields is a well-established practice in Hai//om. People frequently refer to younger people as 'following' older ones and relative age is made explicit by presenting the younger person to 'walk in front' of the other person who 'is following behind'. Furthermore, on an even larger scale social and ethnic groups are considered to follow each other, with disadvantaged people like the 'Bushmen' being 'left behind'. The terms *aise* (in front) and *khause* (behind) themselves are clearly linked to the body parts *ais* (the face) and *khaub* (the behind). Hai//om do not extend the notions of left and right in a similar way, and not to the same extent that, for example, Indo-European languages do. Thus, with regard to relative terms it is only the front (*aise*) and back (*khause*) axis that features strongly in everyday practice and conversation with numerous extensions that help to familiarize children with the concepts and to entrench them across situations.

Apart from piloting their way on established paths between places, Hai//om frequently rely on their dead-reckoning skills, that is, finding a new way (e.g. a short cut) without retracing one's steps. They rely on good orientation skills because they frequently travel in the bush away from roads and paths searching for food and being driven by the distribution of resources in space – indexically, for instance, at the sight of the nearest fruit-bearing tree – rather than by a fixed route. Their exceptional skills at eventually returning to their point of departure take advantage of a set of landscape terms that contain a host of cross-domain information in combination with a good indication about the distance covered, i.e. the speed of travelling. Keeping a good record of both these sources of information in social interaction provides the necessary constraints required for successful orientation (Widlok 1997; see Hutchins 1995: 56–57). The landscape terms refer to the natural and social environment particularly as it changes from south to north following the climate and vegetation zones (far fewer changes occur from east to west), from arid, stony and bushy country dominated by the livestock enterprises of commercial farmers (and their Hai//om labourers) in the south to the sandy woodlands with limited agriculture and limited autonomy of communal farmers (and the Hai//om minority living among them) in the north. The ground, the vegetation and the people provide important spatial and directional information.

In these contexts Hai//om rely on the sun for the measurement of time. Sunset and sunrise provide the terms when referring to east (*sore≠âs*) and west (*sore≠oas*). The course of the sun not only indicates direction, it also indicates time and speed, important elements for orientation even when the direction is clear (due to a path). The north/south distinction, phrased in terms of adjoining landscapes, is also important for orientation purposes but it is not always nor necessarily implied in the east/west distinction as the two axes are somewhat independent. The east/west axis is used in some contexts in which north and south are irrelevant (when measuring time, when positioning huts and fireplaces). Correspondingly the north/south axis (when comparing landscapes, their people and resources) does not necessarily involve the east/west axis. Given the fact that several categories may be used to designate north and south (the nearest landscape or one that is further away, a geographically small or a bigger landscape, a place that stands for a landscape etc.) there are, at least in terms of conversation, several possible north/south axes that can be matched with the one cardinal east/west axis. Keeping these distinctions in mind, it is not surprising, therefore, that Hai//om conventional wisdom should see the combination of front/back with east/west as common sense and as 'natural', as Newton considered coordinating front/back with left/right, and east/west with north/south. By contrast, it is not commonsensical for Hai//om when orienting themselves in their environment to coordinate left/right with east/west or left/right with north/south. There is no field of activities and no domain of language use in which these combinations are established or entrenched.

There are, at least from a Western perspective, no conceptual gaps in the Hai//om way of spatial conceptualization. They can conceive and express both egocentric (relative) and geocentric (absolute) directions and relations, as can be shown systematically in language elicitation tasks (Widlok 1998). When some spatial relations seem to come more 'naturally' than others, this need not be attributed to 'weak' axes in the coordination systems which interact in certain ways with 'strong' axes. Rather, common sense may construe the coordinate systems differently from the way that the dominant European philosophical tradition does. Instead of superordinating the distinction between subject (relative/ego) and environment (absolute/geo), it conventionalizes an association between east/west and front/back, above all in the domain of walking, disregarding other possible associations.

A detailed analysis of the use of spatial language reported on elsewhere (Neumann and Widlok 1996) shows that two speakers of Hai//om are able to negotiate whether they use absolute or relative terms for spatial descriptions. Even though there is a general tendency to use absolute (or intrinsic) terms, the same situation may be interpreted either in terms of front/back/left/right or in terms of north/south/east/west. Once a course has been set up by one of the speakers, the interlocutors tend to stick to it. It is striking, however, that in these spatial descriptions, too, Hai//om speakers combine front/back with absolute terms (and east/west with relative terms) but that there is no combination of left/right with absolute terms (nor of north/south with relative terms).[5] This observation is in line with the results of the nonlinguistic task outlined above. Admittedly the evidence only covers a small section of Hai//om spatial discourse but I suggest that it indicates the more general way in which Hai//om entrench their commonsense combination of spatial axes in their communication.

To sum up the argument up to this point: The ethnography of communicated common sense not only matches the statistical results gained through 'uncommon' tasks, it also provides an explanation for these results and is therefore preferable. The responses to uncommon tasks no longer present themselves as arbitrary and as following rather complicated combinations of rules of preexisting systems (established ex post facto). Rather the results reflect the goal-oriented strategies adopted in common tasks, above all the common tasks of orienting oneself in space, or more precisely of piloting and dead-reckoning. Following this approach, we do not presuppose distinct (logically defined) and self-contained orientation systems that 'favour' either piloting or dead-reckoning. Taking cognitive ethnography as a starting point leads to an inversion of perspective. It is the social and natural environment in which piloting and dead-reckoning play an important role that helps to develop what we construe as distinct systems or as a combination of them. Clearly, all humans are potentially able to develop relative and absolute orientation strategies to some degree. But the embodied skill and disposition to use these systems are only acquired under specific conditions of development, which include common everyday tasks as much as the common sense conveyed to individuals by the social group. Thus, the initially puzzling Hai//om responses to uncommon tasks goes beyond a critique of the experimental methods. More fundamentally, it leads to a critique of cognitivist presuppositions and it suggests a more serious integration of environmental and developmental factors into cognitive sciences. In the final section of this paper I want to explore how such an integrated approach could help to solve the problem at hand.

Everyday Trust

In evolutionary and in autogenetic terms, Levinson (1996c: 22) has interpreted research with animals and with young children to indicate that an allocentric (i.e. any non-egocentric system) is more basic and only later overruled by egocentric systems. He attributes the primary allocentric system to the 'mammalian genes' and the secondary egocentric system to human culture. This begs the question as to how and why the egocentric system came into being. Despite the implicit but rigid nature (evolution) versus culture (history) division, Levinson hints at a way in which a gradual change might have taken place: the use of a language with absolute spatial coordinates, he argues, may have facilitated the emergence of an absolute system while the lack of such a language may have fostered the emergence of a relative system. He recognizes that it is 'hard to prove that language is the determinative factor' (1996c: 24). Indeed, it is much more convincing to regard language and other forms of social behaviour (including piloting and dead-reckoning) as activities that develop concurrently rather than sequentially and which prepare the ground, provide the context, in which the 'diverse capacities of human beings growing up in different surroundings' emerge (see Ingold 1994: 253). As such, the absolute and the relative system are both equally natural and cultural.

This is not a minor point because it undermines the established division into basic psychological processes and their cultural (or linguistic) expressions (see Mishra and Dasen in this volume). Instead of trying to separate out 'natural' (universalist) from 'cultural' (relativist) features this approach suggests that we look at spatial orientation skills as the outcome of complex developmental processes which may not be divided dualistically in this way. Human navigational skills, as has repeatedly been pointed out, are highly variable and humans are 'free to mix information from different systems' (Levinson 1996c: 22). This, however, is not an abstract freedom – nor is it only achieved by members of those language groups who have developed an egocentric system – but it is freedom from an innate cultural or natural programme which has been assumed to be the generative source of human behaviour. This freedom is limited by the conditions created over generations in which human cognitive and social activity takes place. It is not the cultural freedom from a natural endowment but rather it is cultural traditions and the nature of properties that make up 'a system' working together to constrain the direction of a developmental system.

In this developmental system the potential of individuals is unfolding as it is governed by an environment that favours certain directions of development. In the Hai//om case this environment includes above all a network of lively communicating and frequently travelling co-residents. The social network provides the individuals with an opportunity for keeping the orientation strategies going between individuals and across generations. It provides the supportive environment that is required to make use of the topographical gossip, the spatial information that is available in a distributed form among those who form part of the network. It is still possible to refer to the individual use of environmental stimulation and support as 'knowledge' but this is a far cry from an understanding of knowledge as a generative device that is implanted in the brain (or in the genes) and is only then filled with information gathered from the social environment, as in the dominant cognitivist view. Rather, it may only be called individual mastery or skill in the sense that it is a successful realization of the potential that is contained in these practices of making your way through the bush (and of talking about it). What is exhibited as individual responses to orientation tasks eclipses the supportive surrounding environment in which this skill has emerged. Given the fact that there is no prescriptive code that enforces the replication of a certain pattern, it is not surprising that 'mixed results' would occur.

There is one further implication in this which touches upon the very basic process of inference. When individuals can solve uncommon spatial tasks seemingly 'independently' and 'individually', they need not rely on an 'implanted' system, be it relative or absolute. Instead, the cognitive inferences which are worked out in the individual mind can be seen to rely heavily on the social and linguistic conditions that allow the individual mind to develop its skill in the first place. What appears (on the basis of artificial experiments and Western expectations) to be an individual process of de-indexicalization and inference may indeed be the outcome of complex and long-term social interactions. Some inferences required by an orientation task are not necessarily carried out in the individual mind at all. Maps and other orientation tools (including written or oral route descriptions) have to be regarded as complex sets of inferences that have been externalized from humans interacting with their environment and with each other – for the benefit of other individuals who can make use of them without reinventing or reexperiencing them. This aspect of 'shared cognition' has been illustrated with reference to complex spatial operations such as navigating a large modern ship safely into a harbour. As Hutchins (1995) has pointed out, to be able to carry out such an operation successfully relies on inference being summarized and fossilized in structures such as the positioning of co-workers on the bridge, the conventionalized procedures of reporting, and the design of trigonometrical tools. Neither the captain nor any of the individual navigators can achieve this cognitive feat single-handedly or single-mindedly. Some of the constitutive inferences in solving this spatial problem do not take place within individuals but between them, even between them and humans who are not present on the scene (e.g. those who designed the tools and the procedures for navigation and whom one may never encounter personally). But is this only true for complex cognition such as navigating a large ship? What about more restricted tasks that seem to require only an individual moving through his or her environment?

The process of map reading has been described above as one that involves de-indexicalizing, i.e. transforming context-dependent indicators such as 'here' and 'there' into a system that allows us to make inferences (see Gell 1995a). But navigating with a map differs considerably from knowing your way in familiar terrain. The stranger has to integrate his movements and perceptions 'from below' or 'vertically' with the map representation 'from above' their own body and mind (see Ingold 2000: 229). Those familiar with a terrain integrate 'laterally' as they connect places in 'a network of coming and going' (ibid.).[6] Knowledge can therefore be said to be regionally constructed, but it is also inherently socially shared because the 'network of coming and going' is made up of individual experiences with the movements of other people as well as of one's own movements.

Conclusion

In this paper I have shown how uncommon tasks may be used as a method for uncovering what might otherwise remain unnoticed. But making sense of the responses to uncommon tasks requires not only statistics, but above all the incorporation of 'habitual language use' (Pederson 1995) and of 'habitual cognition' (Danziger 1996: 79). The problem is not resolved by claiming that some questions require formal methods while others do not because differences of 'context' are always at the core of the problem posed to cognitive anthropology. Similarly, to single out context-bound cognition as 'everyday cognition' is evading the central question. Since *all* human cognition takes place in some socially defined context it is theoretically *and* empirically relevant that this context does not come into play as an add-on to internal cognition (as in the orthodox cognitivist view) but as the integral and basic ingredient as it emerges from cognitive ethnography. Tracking and orientation skills, I suggest, are best understood as a side-product of a more fundamental skill of making yourself at home in a region that outsiders have characterized as 'wild', 'remote' and 'hostile'. I have documented the cognitive dimension of the ways in which Hai//om 'Bushmen' in northern Namibia orient themselves in the region they inhabit. The key to their successful process of 'making themselves at home' is the sharing of knowledge across connected places in the region.

Hai//om, and others, who are moving through a familiar region do not need to put their trust into a map because they can trust in their engagement with others. Solving the problem of orienting oneself in the bush starts much earlier, strictly speaking, than the situation in which the problem of dead-reckoning (for instance) arises. Over a long period, in which the individual is engaged in multiple relations with fellow beings, spatial knowledge is generated. When being cut off from interactions in which journeys are discussed and reported on, any isolated individual would sooner or later lose the required orientation skills. But this knowledge is not only supra-individual because its source is social life and interaction. It is also the ease with which Hai//om and others solve orientation tasks which suggests that they do not need to run through a long linear processes of de-indexicalizing a huge quantity of spatial statements collected in a lifetime. It suggests that the stories (told and retold), the spatial language used in different domains and journeys made on the basis of earlier journeys, already contain summaries of spatial knowledge and of necessary inferences. In a system of spatial terms which is multidimensional (with reference to physical features, vegetation, resources, social relations), an indexical reference in one domain may have an implicit cross-link to some other domain and can be said to de-indexicalize itself thus saving the individual considerable effort. People have a variety of goals when they use spatial language, when they come and go. All these individual pursuits add to the growth of a common sense of movements and space. Individuals have access to a shared knowledge base but moreover they often also share goals. Goals can be routinely ascribed to well-known interlocutors on the basis of extensive interpersonal experience. In other words, indexical information linked to a certain person, his or her travels and sojourns, can without further processing be included.[7] In the limited universe of a face-to-face community an indexical statement uttered by a specific person may no longer need de-indexicalization. Indexicals may be amenable to inference when the 'hereabouts' and 'thereabouts' of known persons are well established not necessarily by individual mental effort in the mind but in entrenched public discourse. As individuals have a good background knowledge of living people and their position in a network of places and relations – as it is the case particularly in small-scale societies like that of most hunter-gatherers (Bird-David 1995) – individual de-indexicalizing may no longer be required. Two people who share a goal, for instance that of arriving somewhere, need to do less inferential work when they talk to each other and when they are in a relationship that is based on regular interactions. They index movements rather than places (see Ingold 2000: 226). 'There' is the same for two people who are basically on the same route, oriented similarly. The living people act as 'tools' in the cognitive process, because they enshrine spatial connections – just like tools (e.g. maps) include spatial cognition of other humans. Just as some material objects are more informative than others, namely when they have been worked on by fellow beings, some social partners are more informative than others. The more informative ones are not those who *know more* than others but those who *are better known*. Their utterances and activities are therefore easier to incorporate, they take over cognitive functions for the individuals who know them well. It is therefore more appropriate to speak of mutual engagement than simply of interaction. Social engagement is valued for its own sake and spatial orientation skills can be seen as a protracted spin-off of this mutual engagement. People do not engage so that they can orient themselves spatially but these skills grow as mobile people engage with one another in and with a region diversified into different landscapes. There is no special 'certified' knowledge or special trust required since everyday mutual commitment in the sense of presupposed familiarity and trust forms a sufficient shared base.

Social partners with whom everyday tasks are shared are not just sources of information. As complex personalities, the indexical information they provide also includes goal orientations which are partly shared between speaker and listener. Topographical gossip and other forms of interaction create links which may not be intended or controlled by individual speakers. These links of pieces of information do not need to be followed through by individual listeners in order to be useful to them. In order to know where places are, individuals do not need to do all the travelling themselves. Nor do they need to do all the de-indexicalizing and inferencing of spatial statements themselves. For wayfinding they do not need the omnipresence conventionally ascribed to the 'bird's eye view' but a 'sufficiently extended set of paths' is in fact sufficient for their practical orientation purposes (Ingold 2000: 227). Although they may be able to display spatial skills in uncommon tasks put to them in isolation, their capacity to orient themselves is developed only in the social context in which they have been brought up and continue to live. Furthermore, the specific characteristic strategies which can be elicited through uncommon tasks (like the one described above) can be traced back to characteristics of the social and natural environment which allowed the individuals to develop their skills and at the same time committed them to a common-sense way of solving these tasks.

Where does this leave the issue of methods in psychology and anthropology with which this contribution (and this volume as a whole) began? It is true that the established disciplinary boundaries still highlight quantitative versus qualitative methods or experimental rigorousness versus sensitivity towards cultural contexts (see Mishra and Dasen in this volume). With marriages between 'cultural psychologists' and 'cognitive anthropologists' these boundaries become fuzzier and the issue itself may soon seem somewhat anachronistic. However, a new dividing line emerges and here nothing less is at stake than our understanding of what cognition is, namely either a basically universal natural process with cultural overtones or a developmental process that involves individual brains and bodies embedded in relations with the wider natural and social environment that resists being split into nature and culture (see Ingold

2000: 391). This new split may not follow the established disciplinary boundaries and it is not easily bridged by complementing one's canon of methods, for instance by adopting cognitive tasks. It is a fundamental theoretical divide informing the ways in which results are being interpreted and the direction in which new methods are developed for the future.

Acknowledgements

The results presented here were gained through my involvement with the Max Planck Institute for Psycholinguistics in Nijmegen. I am grateful to Stephen Levinson and the members of the Cognitive Anthropology Research Group for their support. My sincere thanks also go to the ≠Akhoe Hai//om of Mangetti-West who participated in doing these tasks, in many cases not because they gave any importance to them but simply as an act of friendship towards me. For more information on the ≠Akhoe Hai//om, who are for short called Hai//om in this contribution, and for the use of click symbols in transcribing their language see Widlok (1999).

Notes

1. Strictly speaking, it is more appropriate to talk of quasi-experiments since they are set-up far less rigidly than experiments in the natural sciences usually are.
2. This is in parallel to Hutchins's observation that 'if one believes that technology is the consequence of cognitive capabilities, and if one further believes that the only place to look for the sources of cognitive capabilities is inside individual minds, then observed differences in level of technology … will inevitably be seen as evidence of advanced and primitive minds' (Hutchins 1995: 355).
3. Due to the small number of potential subjects there, no 'random' selection was possible. However, the sample included men and women, younger and older subjects (all adult, none literate).
4. A very recent restudy (in 2005) with the same community involved a number of tasks carried out by Daniel Haun, a psychologist of the Max Planck Institute for Psycholinguistics in Nijmegen, which follows the basic design of rotation memory tasks outlined above (see Levinson 2003: 132). Haun's data supports the results presented here,
5. In a comparative paper (Widlok 1998) I have shown that the Bantu-speaking Himba of Namibia are a contrastive case in which common sense combines the left/right axis with the east/west axis, both linguistically as well as in terms of social behaviour towards the ancestors in a Himba homestead (see Crandall 1996).
6. Ingold maintains that ultimately even situations like that of map reading are examples of lateral integration which becomes transformed in the representations of science as a 'vertical' interplay between universal theories and particular instantations (2000: 229).
7. Such inferences may be called abductions (see Gell 1998).

Notes on the Contributors

Bettina Beer is a Professor at the Institute of Anthropology at Heidelberg University. She wrote her 'Habilitation' on *Körperkonzepte, interethnische Beziehungen und Rassismustheorien* (Berlin, 2002). Her most recent publications are *Methoden und Techniken der Feldforschung* (ed.) (Berlin, 2006), *Ethnologie. Eine Einführung*, edited with Hans Fischer (Berlin, 2003). Her fieldwork areas are in the Philippines (Visaya region) and in Papua New Guinea.

Martin Brandt is a Senior Lecturer of Cognitive and Biological Psychology at the Department of Psychology, University of Mannheim, Germany. He wrote his doctoral thesis on 'Modellbasierte Analyse globaler Gedächtnismodelle' (Lengerich, 2001). His most recent publications are 'Gedächtniskonzeptionen und Wissensrepräsentationen' (in *Allgemeine Psychologie* (ed.), Heidelberg, 2002) and 'The Principle of Multiple Memory Systems' (in *Principles of Learning and Memory* (ed.), Cambridge MA, 2003), both with Axel Buchner. His research interests include human memory and mathematical models of cognitive processes.

Pierre R. Dasen is Professor of Anthropology of Education at the Faculty of Psychology and Education of the University of Geneva. He studied developmental psychology in Geneva, was an assistant to J. Piaget, and received a Ph.D. from the Australian National University. He studied the cognitive development of Aboriginal children in Australia, Inuit in Canada, Baoulé in Côte d'Ivoire, and Kikuyu in Kenya; he has also contributed to research in cognitive anthropology among the Yupno of Papua New Guinea, and in Bali. His research topics have included visual perception, the development of sensori-motor intelligence, the causes and effects of malnutrition, the development of concrete operations as a function of eco-cultural variables and daily activities, definitions of intelligence, number systems, and spatial orientation. His interests are in everyday cognition, informal education and parental ethnotheories. He is currently completing a study of spatial cognitive development in Bali, India and Nepal. P. Dasen is the co-author or co-editor of several volumes and textbooks on cross-cultural psychology and intercultural education. His most recent publications are *Cross-cultural Psychology. Research and Applications* (second revised edition, with co-authors J.W. Berry, Y.H. Poortinga, and M.H. Segall, Cambridge, 2002) and *Pédagogies et pédagogues du Sud* (edited with A. Akkari, Paris, 2004).

Edgar Erdfelder is a Professor of Experimental and Differential Psychology at the University of Mannheim, Germany. He studied psychology, German literature, linguistics, and history at the Georg August University in

Göttingen, Germany. Edgar Erdfelder received a diploma in Psychology in Göttingen in 1980 and earned his doctorate in Behavioural Science at the University of Trier, Germany, in 1986. His most recent publications include *Allgemeine Psychologie und deduktivistische Methodologie* (Göttingen, 2004) with Joachim Funke, 'Power Analysis for Categorical Methods' (in *Encyclopedia of Statistics in Behavioral Science*, London 2005) with Franz Faul and Axel Buchner, 'Word Frequency of Irrelevant Speech Distractors Affects Serial Recall' (in *Memory & Cognition*, 2005) with Axel Buchner, and 'Experimental Methods of Psychological Assessment' (in *Handbook of Multimethod Measurement in Psychology*, Washington, D.C., 2006), together with Jochen Musch. From 2000 to 2002, Edgar Erdfelder was Consulting Editor of the *Journal of Experimental Psychology: Learning, Memory, and Cognition*. For several years, he has served on the editorial boards of the journals *Experimental Psychology*, *Methodology*, *Methods of Psychological Research – online*, and *Psychology Science*. His research interests include stochastic modelling and measurement of cognitive processes, human memory and judgement, and cognitive aging.

Joachim Funke is a Professor of Theoretical, Experimental and Cognitive Psychology and teaches at the Institute of Psychology, University of Heidelberg. He wrote his 'Habilitation' on Knowledge Representation (Berlin, 1992). His most recent publications are *Problemlösendes Denken* (Stuttgart, 2003), the encyclopedic edition *Denken und Problemlösen* (Göttingen, 2006), the reader *Allgemeine Psychologie und deduktivistische Methodologie* edited with Edgar Erdfelder (Göttingen, 2004) and *Handbuch der Allgemeinen Psychologie – Kognition* edited with Peter Frensch (Göttingen, 2006). His research interests are human thinking, decision making, and problem solving.

Werner Hennings is a Professor at the Faculty of Sociology, University of Bielefeld, teaching Social Geography. His research topics are social and spatial aspects of development processes, sustainability and 'narrative spaces' (space as a text). His fieldwork areas are in the South Pacific, predominantly in Samoa with six field trips between 1982 and 2006. Recent publications are 'Welche Entwicklung für die Verlierer der Globalisierung? Ein Plädoyer für die Subsistenzperspektive' (in *Lokal verankert – weltweit vernetzt. 52. Deutscher Geographentag, Tagungsbericht und Abhandlungen*, Stuttgart, 2000), 'Nachhaltige Stadtentwicklung – nur eine Vision oder messbar' (in *Stadt und Region. Dynamik von Lebenswelten. 53. Deutscher Geographentag, Tagungsbericht und Abhandlungen*, Leipzig 2002), 'Wallensteins Welt. Die räumlichen Erzählungen des kaiserlichen Generalissimus, Herzog von Friedland' (in *Die Kunst sich einzumischen. Festschrift für Tilman Rhode-Jüchtern*, Berlin, 2005).

Verena Keck is Lecturer and Research Fellow at the Institute of Anthropology, University of Heidelberg. She has carried out field research among the Yupno people in Papua New Guinea, in Bali, Indonesia and in Guam, Micronesia. Her research interests include medical anthropology, identity and aging as well as contemporary Pacific cultures topics. Currently she is working on her 'Habilitation' about a neurodegenerative disease in Guam. Her writings include the edited *Common Worlds and Single Lives: Constituting Knowledge in Pacific Societies* (London, 1998), *Historical Atlas of Ethnic and Linguistic Groups in Papua New Guinea*, Volume 1 (Basel, 1994), and *Social Discord and Bodily Disorders: Healing among the Yupno of Papua New Guinea* (Durham, NC, 2005) as well as articles in international journals.

Susanne Kuehling has received her Ph.D. from the Australian National University and was Assistant Professor at Heidelberg University (Institute of Anthropology). She currently holds a research grant from the Deutsche Forschungsgemeinschaft and serves as a board member at the European Society for Oceanists. Her main research areas are Melanesia (Dobu Island, Papua New Guinea) and Micronesia (Central Caroline Islands). She is currently working on a book-length manuscript on the cultural meanings of space. Her book, *Dobu: Ethics of Exchange on a Massim Island, Papua New Guinea* has recently been published (Honolulu, 2005). Future projects include a multisited study on the kula exchange (with Prof. S. Montague).

Stephen C. Leavitt is Associate Professor of Anthropology at Union College (Schenectady, NY). He was trained in psychological anthropology at the University of California, San Diego and has done field research in Papua New Guinea and more recently in Fiji. He has written on Bumbita Arapesh (PNG) adolescence, sexuality, mourning, and the narrative construction of religious experience. His most recent articles are '"We Are Not Straight": Bumbita Arapesh Strategies for Self-Reflection in the Face of Images of Western Superiority' in *The Making of Global and Local Modernities in Melanesia* (2005) and 'From "Cult" to Religious Conviction: The Case for Making Cargo Personal' in *Cargo, Cult, and Culture Critique* (Honolulu, 2004).

Peter Meusburger is Professor at the Department of Geography, University of Heidelberg. His chair for economic and social geography concentrates on geography of knowledge and education, as well as on the geography of labour markets. His research interests include the topics knowledge and space, spatial disparities of educational achievement, educational attainment of ethnic minorities, spatial distribution of jobs for highly skilled decision makers, and the transformation process in Hungary. His most recent publications are 'The Spatial Concentration of Knowledge. Some Theoretical Considerations' (in *Erdkunde*, 2000), 'The Role of Knowledge in the Socio-Economic Transformation of Hungary in the 1990's' (in his *Transformations in Hungary. Essays in Economy and Society*, 2001), 'Wissen als Erklärungsvariable in den Mensch-Umwelt-Beziehungen' (in his *Humanökologie. Ansätze zur Überwindung der Natur-Kultur-Dichotomie*, 2003) and 'Culture, pouvoir et scolarité dans les États-nations multi-ethniques. Contextes culturels et géographie du savoir' (in *Géographie et Cultures*, 2003).

Ramesh C. Mishra is a Professor in the Department of Psychology, Banaras Hindu University, in Varanasi, India. He received his Ph.D. from the

University of Allahabad under the supervision of Professor D. Sinha and has spent two years of postdoctoral studies at Queen's University, Kingston, Canada with Professor J.W. Berry. He is the author or co-author of more than seventy five papers and chapters, and of several books, on various aspects of developmental psychology, cognition and social psychology in India. His most recent publications are *Psychology in Human and Social Development: Lessons from Diverse Cultures* (edited with J.W. Berry and R.C. Tripathi, New Delhi, 2002) and forthcoming *Cultural and Cognitive Dimensions of Tribal Education* (New Delhi).

Katja Neves-Graça is an Assistant Professor at the Department of Sociology and Anthropology, Concordia University (Montreal, Canada). There she also currently holds a Junior Professor-Researcher grant from the Fonds Québecois de Recherche sur la Société et la Culture. Her main research interests pertain to human-environmental relations, ecological learning and aesthetics, and the sociology of techno-science. Recent publications include (2004) 'Revisiting the Tragedy of the Commons: Whale Watching in the Azores and Its Ecological Dilemmas' (*Human Organisation*, 63 (3): 289–300) and (2005) 'Chasing Whales with Bateson and Daniel' (*Ecological Humanities Corner of the Australian Humanities Review Journal*, 35).

Ton Otto is Professor of Anthropology and Ethnography at the University of Aarhus, Denmark. He has conducted long-term fieldwork in Papua New Guinea (particularly in Manus Province). He is interested in issues of social and cultural change, which include religious conversion, cargo cults, political leadership, warfare, tradition, identity and the use of resources. He also engages in methodological and epistemological questions as well as in material and visual culture. His most recent publications include 'Work, Wealth and Knowledge: Enigmas of Cargoist Identifications' in Holger Jebens (ed.) *Cargo, Cult and Culture Critique* (Honolulu, 2004) and two co-edited volumes: *Tradition and Agency: Tracing Cultural Continuity and Invention* (with Poul Pedersen, Aarhus, 2005) and *Warfare and Society – Archaeological and Social Anthropological Perspectives* (with Henrik Thrane and Helle Vandkilde, Aarhus, 2006).

Margaret C. Rodman is a Professor at York University, at the Department of Anthropology. A fascination with the social and cultural construction of 'place' links Professor Rodman's research in the South Pacific to her fieldwork on Canadian housing. She has a long-standing research interest in the Pacific islands. The topics she has studied in Vanuatu (formerly the New Hebrides) include customary land tenure, colonial history, and development issues. Her publications include books on Vanuatu land tenure, *Masters of Tradition* (1987), and fisheries, *Deep Water* (1989). She co-authored *New Neighbours* on Toronto housing cooperatives (with M. Cooper, 1992). More recently she is the author of *Houses Far From Home* (2001), co-editor with Jan Rensel of *Home in the Islands* (1997), and co-editor of the collective book, *House-Girls Remember: Domestic Workers in Vanuatu* (University of Hawai'i Press 2007).

Gunter Senft is Senior Research fellow at the Max Planck Institute for Psycholinguistics in Nijmegen and extraordinary professor of general linguistics at the University of Cologne. He has been studying the language and the culture of the Trobriand Islanders since 1982. His main research interests include Austronesian and Papuan languages, anthropological linguistics, and the interface between language, culture and cognition. For further information, including a list of his publications, consult the following website: http://www.mpi.nl/Members/GunterSenft. His email address is: gunter.senft@mpi.nl.

Katharina Stockhaus is a Ph.D. Candidate at the Institute of Anthropology, University of Heidelberg, where she studies the life and work of indigenous Pacific authors and has taught courses on Maori language, the Pacific diaspora in New Zealand and Pacific literature. Before, she studied Languages and Cultures of Austronesia at the University of Hamburg and wrote her master's thesis about 'Maori Mythology in Contemporary New Zealand Literature', which developed during several stays at the University of Auckland (New Zealand).

Prof. **Andrew Strathern** and Dr. **Pamela J. Stewart** are a husband-and-wife research team, situated in the Anthropology Department, University of Pittsburgh, USA, and they are, respectively, Visiting Professor and Visiting Research Fellow, Department of Anthropology, University of Durham, England. They have published widely in books and articles on their research in the Pacific, Europe and Asia on a range of topics, such as. violence, gender, expressive genres, kinship, medical anthropology, cultural history, and contemporary change. Their most recent co-authored books are *Witchcraft, Sorcery, Rumors, and Gossip* (Cambridge, 2004) and *Empowering the Past, Confronting the Future: The Duna People of Papua New Guinea* (London, 2004). Their most recent co-edited books include *Expressive Genres and Historical Change: Indonesia, Papua New Guinea, and Taiwan* (Aldershot, 2005); and *Terror and Violence: The Imagination and The Unimaginable* (London, 2006).

Gisela Trommsdorff is Professor and holds the Chair of Developmental and Cross-Cultural Psychology at the University of Konstanz. Her research topics are cross-cultural studies on socio-emotional development, control and value orientation, intergenerational relations. Her most recent publications are *Enzyklopädie der kulturvergleichenden Psychologie*, volumes I and II (edited with H.J.Kornadt, Göttingen, 2005), *Positionen der Kulturanthropologie* (edited with A. Assmann and U. Gaier, Frankfurt, 2004), and *Challenges of the 21st century in Japan and Germany* (edited. with U. Teichler, Lengerich, 2002).

Jürg Wassmann is Professor of Anthropology and Head of the Institute of Anthropology, University of Heidelberg. His field area is Papua New Guinea, where he has carried out fieldwork among the Iatmul and the Yupno, and Bali, Indonesia. His main research interests are culture and cognition, memory, knowledge systems and concepts of personhood. He is author of *The Song to the Flying Fox* (Boroko, 1991), *Das Ideal des leicht gebeugten*

Menschen (Berlin, 1993), *Historical Atlas of Ethnic and Linguistic Groups in Papua New Guinea*, Volume 3 (Basel, 1994), has edited *Pacific Answers to Western Hegemony: Cultural Practices of Identity Construction* (London, 1998), co-edited *Alltagswissen. Les savoirs quotidiens. Everyday Cognition.* (Fribourg, 1993), and has published numerous articles in international journals. In preparation is the (co-edited) *Over the Fence. Anthropology – An Undisciplined Discipline?*

Thomas Widlok is a Senior Researcher at the Max Planck Institute for Psycholinguistics in Nijmegen and Privatdozent at the University of Cologne. He has carried out anthropological research on the relation between language and thought, on property and inequality, personhood and corporate organization and on local knowledge. He has published books and articles based on his field research in Africa and in Australia as well as comparative work. Forthcoming publications include 'Where Settlements and the Landscape Merge: Towards an Integrated Approach to the Spatial Dimension of Social Relations' (in *The Anthropology of Landscape*, edited by Michael Bollig and Nina Grundkowski) and 'Local Experts – Expert Locals. A Comparative Perspective on Australia and Namibia' (in *Culture and the Changing Environment*, edited by M. Casimir, Oxford).

References

Abu-Lughod, L. 1991. 'Writing against Culture', in *Recapturing Anthropology: Working in the Present*, ed. R.G. Fox. Santa Fe, NM: School of American Research Press, pp. 137–62.

_____. 1993. *Writing Women's Worlds: Bedouin Stories*. Berkeley: University of California Press.

Abu-Lughod, L. and C. Lutz 1990. 'Introduction: Emotion, Discourse, and the Politics of Everyday Life', in *Language and the Politics of Emotion*, ed. C. Lutz and L. Abu-Lughod. Cambridge: Cambridge University Press, pp. 1–23.

Agar, M. 1980. 'Stories, Background Knowledge and Themes: Problems in the Analysis of Life History Narrative', *American Ethnologist* 7(2): 223–39.

_____. 1986. *Speaking of Ethnography*, Vol. 2, Qualitative Research Methods Series. Beverley Hills: Sage.

Agnew, J. 1999. 'Mapping Political Power beyond State Boundaries: Territory, Identity, and Movement in World Politics', *Millenium* 28: 499–521.

Agrawal, A. 2005. *Envionmentality: Technologies of Government and the Making of Subjects*. Durham, London: Duke University Press.

Almagor, U. 1978. *Pastoral Partners. Affinity and Bond Partnership Among the Dassanetch of South-West Ethiopia*. Manchester: Manchester University Press.

_____. 1987. 'The Cycle and Stagnation of Smells: Pastoralists–Fishermen Relationships in an East African Society', RES: *Anthropology and Aesthetics* 14: 106–21.

_____. 1990. 'Odors and Private Language: Observations on the Phenomenology of Scent', *Human Studies* 13: 253–74.

Ames, S.L. and A.W. Stacy. 1998. 'Implicit Cognition in the Prediction of Substance Use among Drug Offenders', *Psychology of Addictive Behaviors* 12: 272–81.

Anderson, J.R. 1985. *Cognitive Psychology and Its Implications*, 2nd edn. New York: Freeman.

_____. 1995. *Learning and Memory. An Integrated Approach*. New York, Chichester, Bribane: John Wiley & Sons.

Appadurai, A. 1988. 'Putting Hierarchy in Its Place', *Cultural Anthropology* 3: 36–49.

_____. 1996. *Modernity at Large: Cultural Dimensions of Globalization*. Minneapolis: University of Minnesota Press.

Apter, A. 1999. 'The Subvention of Tradition: A Genealogy of the Nigerian Durbar', in *State/Culture: State-Formation after the Cultural Turn*, ed. G. Steinmetz. Ithaca and London: Cornell University Press.

Asad, T. 1991. 'From the History of Colonial Anthropology to the Anthropology of Western Hegemony', in *Colonial Situations: Essays on the Contextualization of Ethnographic Knowledge*, ed. G. Stocking. Madison: University of Wisconsin Press, pp. 314–24.

Assmann, A. 2002. 'Vier Formen des Gedächtnisses [Four types of memory]', *Erwägen, Wissen, Ethik* 13: 183–95.

Atkinson, R.C. and R.M. Shiffrin 1968. 'Human Memory. A Proposed System and Its Control Processes', in *The Psychology of Learning and Motivation*, Vol. 2, ed. K.W. Spence and J.T. Spence. London: Academic Press, pp. 89–195.

Augé, M. 1994. *Orte und Nicht-Orte*. Frankfurt am Main: S. Fischer Verlag.

———. 1995. Non-Places. Introduction to an Anthropology of Supermodernity. London: Pluto.

Axel, B. 2002. *From the Margins: Historical Anthropology and Its Futures*. Durham: Duke University Press.

Ayabe-Kanamura, S., S. Saito, H. Distel, M. Martínez-Gómez and R. Hudson. 1998. 'Differences and Similarities in the Perception of Everyday Odors: A Japanese–German Cross-Cultural Study', in *Olfaction and Taste XII*, ed. C. Murphy, pp. 694–700.

Azuma, H. 1984. 'Secondary Control as a Heterogeneous Category', *American Psychologist* 39(9): 970–71.

———. 1986. 'Why Study Child Development in Japan?', in *Child Development and Education in Japan*, ed. H.W. Stevenson, H. Azuma and K. Hakuta. New York: Freeman, pp. 3–12.

Baayen, H. and E. Danziger 1994. 1993 *Annual Report*. Max Planck Institute for Psycholinguistics.

Babadzan, A. 1988. '*Kastom* and Nation Building in the South Pacific', in *Ethnicities and Nations: Processes of Interethnic Relations in Latin America, Southeast Asia, and the Pacific*, ed. R. Guidieri, F. Pellizzi and S.J. Tambiah. Houston: The Rothko Chapel & University of Texas Press, pp. 199–228.

———. (ed.). 1999. 'Les politiques de la tradition: Identités culturelles et identités nationales dans le Pacifique. Numéro spécial', *Journal de la Société des Océanistes* 109(2).

———. 2000. 'Anthropology, Nationalism and "the Invention of Tradition"', *Anthropological Forum* 10: 131–55.

Baddeley, A.D. 1997. *Human Memory. Theory and Practice*, revised edition. Hove, UK: Psychology Press.

Bailly, A. 1998. 'Marginality and Centrality', *Semiotica* 122: 291–95.

Bakhtin, M.M. 1981. 'Discourse in the Novel', in *The Dialogic Imagination: Four Essays by M.M. Bakhtin*, ed. Michael Holquist, transl. C. Emerson and M Holquist. Austin: University of Texas Press, pp. 259–422.

Baltes, P.B. 1997. 'On the Incomplete Architecture of Human Ontogeny: Selection, Optimization, and Compensation as Foundation of

Developmental Theory', *American Psychologist* 52(4): 366–80.

Bamford, S. 1998. 'Humanized Landscapes, Embodied Worlds. Land and the Construction of Intergenerational Continuity among the Kamea of Papua New Guinea', *Social Analysis* 42(3): 28–54.

Bandura, A. 2001. 'Social Cognitive Theory: A Genetic Perspective', *Annual Review of Psychology* 52(1): 1–26.

Bandura, A. and E.A. Locke 2003. 'Negative Self-Efficacy and Goal Effects Revisited', *Journal of Applied Psychology* 88(1): 87–99.

Bargatzky, T. and R. Kuschel 1994. *The Invention of Nature*. Frankfurt am Main: Peter Lang Publishing.

Barlow, L. 1995. 'Achieving Womanhood and the Achievements of Women in Murik Society', in *Gender Rituals. Female Initiation in Melanesia*, ed. N. Lutkehaus and P. Roscoe. New York: Routledge, pp. 85–112.

Barry III, H. 1980. 'Description and Uses of the Human Relations Area Files', in *Handbook of Cross-Cultural Psycholog*. Vol. 2., *Methodology*, ed. H. Triandis and J.W. Berry. Boston: Allyn & Bacon, pp. 445–78.

Barth, F. 1975. *Ritual and Knowledge among the Baktaman of New Guinea*. New Haven: Yale University Press.

_____. 1987. *Cosmologies in the Making: A Generative Approach to Cultural Variation in Inner New Guinea*. Cambridge: Cambridge University Press.

_____. 1993. *Balinese Worlds*. Chicago: University of Chicago Press.

_____. 2002. 'An Anthropology of Knowledge', *Current Anthropology* 43: 1–18.

Bartlett, F.C. 1932. *Remembering: A Study in Experimental and Social Psychology*. Cambridge: Cambridge University Press.

Batchelder, W.H. and D.M. Riefer 1986. 'The Statistical Analysis of a Model for Storage and Retrieval Processes in Human Memory', *British Journal of Mathematical and Statistical Psychology* 39: 129–49.

_____. 1990. 'Multinomial Processing Models of Source Monitoring', *Psychological Review* 97: 548–64.

Bateson, C. 2005 [originally printed in 1972, 1991]. *Our Own Metaphor: A Personal Account of a Conference on the Effects of Conscious Purpose on Human Adaptation*. Cresskill, NJ: Hampton Press Inc.

Bateson, G. 1958. *Naven: A Survey of the Problems Suggested by a Composite Picture of the Culture of a New Guinea Tribe Drawn from Three Points of View*. Stanford: Stanford University Press.

_____. 1972. *Steps to an Ecology of Mind*. New York: Ballantine Books.

_____. 1978. 'The Pattern Which Connects', *The CoEvolution Quarterly* Summer: 5–15.

_____. 1991. *Sacred Unity: Further Steps to an Ecology of Mind*, ed. R. Donaldson. New York: Harper Collins.

Bateson, G. and M. Bateson 1987. *Angels Fear: Towards an Epistemology of the Sacred*. Toronto, New York, London, Sydney, Auckland: Bantam Books.

Baudrillard, J. 1995. 'Die Stadt und der Hass. Über die "kritische Masse" und ihre Gewalt', *Frankfurter Rundschau* (September 30, 1995: Dokumentationsseite).

Bauman, Z. 1992. *Intimation of Postmodernity*. London: Routledge.

Bayen, U.J., K. Murnane, K. and E. Erdfelder 1996. 'Source Discrimination, Item Detection, and Multinomial Models of Source Monitoring', *Journal of Experimental Psychology: Learning, Memory, and Cognition* 22: 197–215.

Bayen, U.J., G.V. Nakamura, S. Depuis, and C.L. Yang 2000. 'The Use of Schematic Knowledge about Sources in Source Monitoring', *Memory & Cognition* 28: 480–500.

Bechtel, W. and A. Abrahamsen 2001. *Connectionism and the Mind. Parallel Processing, Dynamics, and Evolution in Networks*. Oxford: Blackwell Publishers.

Beek, W.E.A. van. 1992. 'The Dirty Smith: Smell as a Social Frontier among the Kapsiki/Higi of North Cameroon and North-Eastern Nigeria', *Africa* 62: 38–58.

Beer, B. 2000a. 'Im Reich der Sinne – Plädoyer für eine Ethnologie der kulturellen Wahrnehmungsweisen', *Ethnoscripts. Analysen und Informationen aus dem Institut für Ethnologie der Universität Hamburg* 2: 6–15.

———. 2000b. 'Geruch und Differenz – Körpergeruch als Kennzeichen konstruierter "rassischer" Grenzen', *Paideuma* 46: 207–30.

———. 2001. 'Geschmack', in *NetzkunstWörterBuch*, ed. A. Eske and K. Alsleben, www.hfbk.uni-hamburg.de/netz/kunst.html, Hamburg: Edition Kuecocokue, pp. 185–89.

———. 2002. *Körperkonzepte, interethnische Beziehungen und Rassismustheorien. Eine kulturvergleichende Untersuchung*. (Habilitationsschrift). Berlin: Dietrich Reimer Verlag.

Behar, R. 1990. 'Rage and Redemption: Reading the Life Story of a Mexican Marketing Woman', *Feminist Studies* 16(2): 223–58.

———. (ed.) 1996. *Women Writing Culture*. Berkeley: University of California Press.

Bell, R.Q. 1979. 'Parent, Child, and Reciprocal Influences', *American Psychologist* 34(10): 821–26.

Bender, B. 1993. 'Stonehenge – Contested Landscapes (Medieval to Present-Day)', in *Landscape: Politics and Perspectives*, ed. B. Bender. Oxford: Berg, pp. 245–79.

———. 1995. 'Introduction: Landscape – Meaning and Action', in *Landscape: Politics and Perspectives*, ed. B. Bender. Oxford: Berg, pp. 1–17.

———. 1996. 'Landscape', in *Encyclopaedia of Social and Cultural Anthropology*, ed. Alan Barnard and Jonathan Spencer. London and New York: Routledge, pp. 323–24.

———. 1999. *Stonehenge: Making Space*. Oxford & New York: Berg.

———. (ed.) 1995. *Landscape: Politics and Perspectives*. Oxford: Berg Publishers.

Bender, B. and M. Winer (eds) 2001. *Contested Landscapes: Movement, Exile and Place*. Oxford: Berg.

Bennardo, G. (ed.) 2002. *Representing Space in Oceania. Culture in Language and Mind*. Canberra: Pacific Linguistics.

Berger, P.L., and T. Luckmann. 1967. *The Social Construction of Reality: A Treatise in the Sociology of Knowledge*. Harmondsworth: Penguin Books.

Berkes, F. 1999. *Sacred Ecology: Traditional Ecological Knowledge and Resource Management*. Philadelphia: Taylor and Francis.

Berkes, F. and C. Folke 2000. *Linking Social and Ecological Systems: Management and Social Mechanisms for Building Resilience*. Cambridge: Cambridge University Press.

Berland, J. 1983. 'Dress Rehearsals for Psychological Performance', in *Human Assessment and Cultural Factors*, ed. S. Irvine and J. Berry. New York: Plenum, pp. 139–54.

Berleant, A. 1997. *Living in the Landscape. Toward an Aesthetics of Environment*. Lawrence, Kansas: University Press of Kansas.

Berlin, B. and P. Kay 1969. *Basic Color Terms: Their Universality and Evolution*. Berkeley: University of California Press.

Bernard, H. 2000. *Social Research Methods: Qualitative and Quantitative Approaches*. Thousand Oaks: Sage.

Berreman, G. 1981. *The Politics of Truth: Essays in Critical Anthropology*. Madras: South Asian.

Berry, J.W. 1976. *Human Ecology and Cognitive Style: Comparative Studies in Cultural and Psychological Adaptation*. New York: Sage/Halsted.

_____. 1980. 'Ecological Analyses for Cross-Cultural Psychology', in *Studies in Cross-Cultural Psychology*, Vol. 2, ed. N. Warren. New York: Academic Press, pp. 157–89.

_____. 1984. 'Toward a Universal Psychology of Cognitive Competence', *International Journal of Psychology* 19: 335–61.

_____. 1993. 'The Way Ahead: From Indigenous Psychologies to a Universal Psychology', in *Indigenous Psychologies*, ed. U. Kim and J.W. Berry. Newbury Park, CA: Sage, pp. 277–80.

Berry, J. W. and P.R. Dasen 1974. *Culture and Cognition: Readings in Cross-Cultural Psychology*. London: Methuen.

Berry, J.W., Y.H. Poortinga, M.H. Segall and P.R. Dasen 2002. *Cross-Cultural Psychology. Research and Applications*, second, revised edition. Cambridge: Cambridge University Press.

Bhabha, Homi K. 1994. *The Location of Culture*. New York: Routledge.

Bicker, A., P. Sillitoe and J. Pottier (eds) 2004. *Development and Local Knowledge. New Approaches to Issues in Natural Resources Management, Conservation and Agriculture*. London and New York: Routledge.

Biersack, A. 1999. 'Introduction: From the "New Ecology" to the New Ecologies', *American Anthropologist* 101(1): 5–18.

_____. (ed.) 1991. *Clio in Oceania: Toward a Historical Anthropology*. Washington: Smithsonian Institute Press.

Bird-David, N. 1995. 'Hunter-Gatherers' Kinship Organization: Implicit Roles and Rules', in *Social Intelligence and Interaction*, ed. E. Goody, Cambridge: Cambridge University Press, pp. 68–84.

_____. 2004. 'No Past, No Present: A Critical-Nayaka Perspective on Cultural Remembering', *American Ethnologist* 31(3): 406–21.

Bloch, M. 1991. 'Language, Anthropology, and Cognitive Science', *Man* 26: 183–98.

_____. 1998. *How We Think They Think*. Oxford: Westview.

Blommaert , J. and C. Bulcaen 2000. 'Critical Discourse Analysis', *Annual Review of Anthropology* 29: 447–66.

Bohnemeyer, J. and M. Caelen 2001. 'The ECOM Clips. A Stimulus for the Linguistic Coding of Event Complexity', in *Manual for the Field Season 2001. Language and Cognition Group*, ed. S.C. Levinson and N. Enfield. Nijmegen: Max Planck Institute for Psycholinguistics, pp. 168–75.

Bolton, L. 1999. 'Women, Place and Practice in Vanuatu: A View from Ambae', *Oceania* 70(1): 43–55.

Bonnemaison, J. 1986. *L'Arbre et la Pirogue*. Paris: Edition de L'Orston.

Borchgrevink, A. 1999. *Clean and Green: Morality and Knowledge in a Philippine Farming Community*. (Dissertations & Theses No. 6/99). Oslo: Centre for Development and the Environment (SUM).

———. 2002. 'Clean and Green: Indigenous Knowledge and Cultural Models in a Philippine Community', *Ethnos* 67: 223–44.

Boroditsky, L. and M. Ramscar 2002. 'The Roles of Body and Mind in Abstract Thought', *Psychological Science* 13: 185–89.

Borofsky, R. 1994. 'On the Knowledge and Knowing of Cultural Activities', in *Assessing Cultural Anthropology*, ed. R. Borofsky. New York: McGraw-Hill, pp. 331–47.

———. 2000. *Rememberance of Pacific Pasts: An Invitation to Remake History*. Honolulu: University of Hawai'i Press.

Boster, J. 1985. 'Requiem for the Omniscient Informant: There's Life in the Old Girl Yet', in *Directions in Cognitive Anthropology*, ed. J. Dougherty. Urbana: University of Illinois Press, pp. 177–197.

Bourdieu, P. 1977. *Outline of a Theory of Practice*, trans. R. Nice. Cambridge: Cambridge University Press.

———. 1990. *The Logic of Practice*. Stanford, CA: Stanford University Press.

———. 1991. *Language and Symbolic Power*. Cambridge: Polity Press.

Bowlby, J. 1969. *Attachment and Loss*: Vol. 1. *Attachment*. New York: Basic Books.

Boyer, P. 1990. *Tradition as Truth and Communication*. Cambridge: Cambridge University Press.

Bratman, M.E. 1999. *Faces of Intention*. Cambridge: Cambridge University Press.

Bravo, M.T. 1996. *The Accuracy of Ethnoscience: A Study of Inuit Cartography and Cross-cultural Commensurability*. Manchester: University of Manchester.

Bredenkamp, J. and E. Erdfelder 1996. 'Methoden der Gedächtnispsychologie' [Methods of the Psychology of Memory], in *Gedächtnis. Enzyklopädie der Psychologie* (Themenbereich C, Serie II, Band 4), ed. D. Albert and K.H. Stapf. Göttingen: Hogrefe, pp. 1–94.

Briggs, C.L. 1996a. *Disorderly Discourse: Narrative, Conflict and Inequality*. New York: Oxford University Press.

———. 1996b. 'The Politics of Discursive Authority in Research on the "Invention of Tradition"', *Cultural Anthropology* 11: 435–69.

———. 2005. 'Communicability, Racial Discourse, and Disease', *Annual Review of Anthropology* 34: 269–91.

Brislin, R.W., W. Lonner and R. Thorndike 1973. *Cross-Cultural Research Methods*. New York: John Wiley.

Bronfenbrenner, U. 1979. *The Ecology of Human Development: Experiments by Nature and Design*. Cambridge, MA: Harvard University Press.

Bronfenbrenner, U. and S.J. Ceci 1994. 'Nature-Nurture Reconceptualized in Developmental Perspective: A Bioecological Model', *Psychological Review* 101(4): 568–86.

Brosius, P. 1999. 'Analyses and Interventions: Anthropological Engagements with Environmentalism', *Current Anthropology* 40(3): 277–309.

Brown, M. 1998. 'Can Culture Be Copyrighted?', *Current Anthropology* 39: 193–222.

Brown, P. and S.C. Levinson 1993. 'Linguistic and Nonlinguistic Coding of Spatial Arrays: Explorations in Mayan Cognition', CARG *Working Paper No*. 24. Nijmegen: Mimeo.

Brown, P. and A. Ploeg (eds). 1997. 'Change and Conflict in Papua New Guinea Land and Resource Rights', Special Issue, *Anthropological Forum* 7(4): 507–690.

Brown, R.H. 1993. 'Modern Science: Institutionalization of Knowledge and Rationalization of Power', *The Sociological Quarterly* 34: 153–68.

Bruner, E. (ed.) 1986. *The Anthropology of Experience*. Urbana: University of Illinois Press.

Bruner, J. 1994. 'The "Remembered" Self', in *The Remembering Self: Construction and Accuracy in the Self-Narrative*, ed. U. Neisser and R. Fivush. Cambridge: Cambridge University Press, pp. 41–54.

_____. 1996. *The Culture of Education*. Cambridge, MA: Harvard University Press.

Brunton, R. 1996. 'The Hindmarsh Island Bridge and the Credibility of Australian Anthropology', *Anthropology Today* 12(4): 2–7.

Bubandt, N. 1998. 'The Odour of Things: Smell and the Cultural Elaboration of Disgust in Eastern Indonesia', *Ethnos* 63: 48–80.

_____. 2005. 'On the Genealogy of Sasi: Transformations of an Imagined Tradition in Eastern Indonesia', in *Tradition and Agency: Tracing Cultural Continuity and Invention*, ed. T. Otto and P. Pedersen. Aarhus: Aarhus University Press, pp. 193–232.

Buchner, A. and M. Brandt 2003. 'The Principle of Multiple Memory Systems', in *Principles of Learning and Memory*, ed. G. Lueer, R.H. Kluwe and F. Roesler. Cambridge, MA: Birkhaeuser, pp. 93–111.

Buchner, A. and W. Wippich 2000. 'On the Reliability of Implicit and Explicit Memory Measures', *Cognitive Psychology* 40: 227–59.

Buchner, A., E. Erdfelder and B. Vaterrodt Plünnecke 1995. 'Toward Unbiased Measurement of Conscious and Unconscious Memory Processes within the Process Dissociation Framework', *Journal of Experimental Psychology: General* 124: 137–60.

Bühler, K. 1934 (1978). *Sprachtheorie. Die Darstellungsfunktion der Sprache*. Jena: Gustav Fischer.

Butalia, U. 1998. *The Other Side of Silence: Voices from the Partition of India*. Delhi: Penguin.

Camilleri, C. and G. Vinsonneau 1996. *Psychologie et Culture: Concepts et Méthodes*. Paris: Armand Colin/Masson.

Campbell, D.T. and D.W. Fiske 1959. 'Convergent and Discriminant Validation by the Multitrait-Multimethod Matrix', *Psychological Bulletin* 56: 81–105.

Campbell, J.R., K.A. Heller and A.X. Feng 2004. 'Restructuring Attribution Research with Cross-Cultural Data', in *Cross-Cultural Research: Basic Issues, Dilemmas, and Strategies*, ed. J.R. Campbell, K. Tirri, P. Ruohotie and H. Walberg. Hämeenlinna: Research Centre for Vocational Education, University of Tampere, pp. 61–79.

Carrier, J. (ed.). 2004. *Confronting Environments: Local Understanding in a Globalized World*. Walnut Creek, Toronto, Oxford: AltaMira Press.

Carrier, J.G. and A.H. Carrier 1989. *Wage, Trade and Exchange in Melanesia: A Manus Society in the Modern State*. Berkeley: University of California Press.

Carry, L. 1999. 'Unexpected Stories: Life History and the Limits of Representation', *Qualitative Inquiry* 5(3): 411–27.

Casey, E.S. 1997. *The Fate of Place: a Philosophical Study*. Berkeley: University of California Press.

Chase, S.E. 1995. 'Taking Narratives Seriously: Consequences for Method and Theory in Interview Studies', in *Interpreting Experience: The Narrative Study of Lives*, ed. R. Josselson and A. Leiblich. Thousand Oaks, CA: Sage, pp. 1–26.

Chevallier, D. 2005. 'The Cow, the Cheese and the Anthropologist: The Revival of Traditional Practices in Rural France', in *Tradition and Agency: Tracing Cultural Continuity and Invention*, ed. T. Otto and P. Pedersen. Aarhus: Aarhus University Press, pp. 119–35.

Chiu, C.Y., M.W. Morris, Y.Y. Hong and T. Menon 2000. 'Motivated Cultural Cognition: The Impact of Implicit Cultural Theories on Dispositional Attribution Varies as a Function of Need for Closure', *Journal of Personality and Social Psychology* 78(2): 247–59.

Chodorow, N.J. 1999. *The Power of Feelings*. New Haven: Yale University Press.

Choi, I. and Y. Choi 2002. 'Culture and Self-Concept Flexibility', *Personality & Social Psychology Bulletin* 28(11): 1508–17.

Chomsky, N. 1965. *Aspects of the Theory of Syntax*. Cambridge, MA: The Massachusetts Institute of Technology.

Chu, S. and J.J. Downes 2000. 'Odour-evoked Autobiographical Memories: Psychological Investigations of Proustian Phenomena', *Chemical Senses* 25: 11–116.

Church, A.T. 2000. 'Culture and Personality: Toward an Integrated Cultural Trait Psychology', *Journal of Personality* 68(4): 651–703.

Clark, A. 1993. *Sensory Qualities*. Oxford: Clarendon Press.

Clark, H.H. and D. Wilkes-Gibbs 1986. 'Referring as a Collaborative Process', *Cognition* 11: 1– 39.

Classen, C. 1990. 'Sweet Colors, Fragrant Songs: Sensory Models of the Andes and the Amazon', *American Ethnologist* 17: 722–35.

———. 1991. 'The Sensory Orders of "Wild Children"', in *The Varieties of Sensory Experience*, ed. D. Howes. Toronto: University of Toronto Press, pp. 47–60.

———. 1993. *Worlds of Sense. Exploring the Senses in History and across Cultures*. London, New York: Routledge.

———. 1998. 'Introduction', in his *The Color of Angels: Cosmology, Gender and the Aesthetic Imagination*. London, New York: Routledge, pp. 1–9. <http://alcor.concordia.ca/~senses/Classen.htm> [1.6.2000].

Classen, C., D. Howes and A. Synnott 1994. *Aroma. The Cultural History of Smell*. London, New York: Routledge.

Clifford, J. 1982. *Person and Myth: Maurice Leenhardt in the Melanesian World*. Berkeley: University of California Press.

———. 1988. *The Predicament of Culture: Twentieth Century Ethnography, Literature, and Art*. Cambridge, MA: Harvard University Press.

———. 2001. 'Indigenous Articulations', *The Contemporary Pacific* 13: 468–90.

Clifford, J. and G. Marcus (eds). 1986. *Writing Culture: The Poetics and Politics of Ethnography*. Berkeley: University of California Press.

Codrington, R.H. 1972. *Melanesians: Studies in Their Anthropology and Folk-Lore*. New York: Dover Publications.

Cohen, A. (ed.). 2000. *Signifying Identities: Anthropological Perspectives on Boundaries and Contested Values*. London and New York: Routledge.

Cohen, E. 1988. 'The Broken Cycle: Smell in a Bangkok Soi (Lane)', *Ethnos* 53: 37–49.

Cohen, R. 1997. *Global Diasporas. An Introduction*. London: University College London Press.

Cole, M. 1985. 'The Zone of Proximal Development: Where Culture and Cognition Create Each Other', in *Culture, Communication and Cognition. Vygotskian Perspectives*, ed. J. Wertsch. Cambridge: Cambridge University Press, pp. 187–214.

Cole, M. and S. Scribner 1974. *Culture and Thought: A Psychological Introduction*. New York: John Wiley.

———. 1975. 'Theorizing about Socialization of Cognition', *Ethos* 3: 241–68.

Cole, M., J. Gay, J.A. Glick and D.W. Sharp 1971. *The Cultural Context of Learning and Thinking: An Exploration in Experimental Anthropology*. New York: Basic Books.

Colombo Dougoud, R. 2005. '"Ol I kam long hul bilong Wotñana" (They come from the hole of Wotñana): How a Papua New Guinean Artefact Became Traditional', in *Tradition and Agency: Tracing Cultural Continuity and Invention*, ed. T. Otto and P. Pedersen. Aarhus: Aarhus University Press, pp. 235–66.

Comaroff, J. and J. Comaroff 1992. *Ethnography and the Historical Imagination*. Westview: Boulder.

Connerton, P. 1989. *How Societies Remember*. Cambridge: Cambridge University Press.

Côté, J.E. 1994. *Adolescent Storm and Stress. An Evaluation of the Mead/Freeman Controversy*. Hillside, NJ: Lawrence Erlbaum.

Crandall, D. 1996. 'Female over Male or Left over Right: Solving a Classificatory Puzzle among the Ovahimba', *Africa* 66: 327–48.

Crapanzano, V. 1984. 'Life Histories: A Review Essay', *American Anthropologist*, 86: 953–60.

Crumley, C. 2001. *New Directions in Anthropology and Environment: Intersections*. Walnut Creek: AltaMira Press.

Csordas, T. (ed.). 1994. *Embodiment and Experience: The Existential Ground of Culture and Self*. Cambridge: Cambridge University Press.

D'Andrade, R. 1992. 'Schemas and Motivation', in *Human Motives and Cultural Models*, ed. R. D'Andrade and C. Strauss. Cambridge: Cambridge University Press, pp. 23–44.

_____. 1995. *The Development of Cognitive Anthropology*. Cambridge: Cambridge University Press.

_____. 2005. 'Some Methods for Studying Cultural Cognitive Structures', in *Finding Culture in Talk: A Collection of Methods*, ed. N. Quinn. New York: Palgrave Macmillan, pp. 83–104.

D'Andrade, R. and C. Strauss (eds). 1992. *Human Motives and Cultural Models*. Cambridge: Cambridge University Press.

Dalgleish, T. and M.J. Power (eds). 1999. *Handbook of Cognition and Emotion*. Chichester: Wiley Verlag.

Danziger, E. 1996. 'Parts and Their Counterparts: Spatial and Social Relationships in Mopan Maya', *Journal of the Royal Anthropological Institute* 2: 67–82.

_____. (ed.). 1993. *Cognition and Space Kit of the Cognitive Anthropology Research Group*. Nijmegen: MPI for Psycholinguistics.

Dasen, P.R. 1975. 'Concrete Operational Development in Three Cultures', *Journal of Cross-Cultural Psychology* 6: 156–72.

Dasen, P.R. and G. Jahoda 1986. 'Preface. Cross-Cultural Human Development', Special Issue of *International Journal of Behavioural Development* 9(4): 413–16.

de León, L. 1991. 'Space Games in Tzotzil: Creating a Context for Spatial Reference'. CARG *Working Paper* No. 4. Nijmegen: Mimeo.

De Wet, C. (ed.). 2005. *Development-Induced Displacement. Problems, Policies and People*. Oxford: Berghahn.

Deci, E.L. and R.M. Ryan 1995. 'Human Autonomy: The Basis for True Self-Esteem', in *Efficacy, Agency, and Self Esteem. Plenum Series in Social/clinical Psychology*, ed. M.H. Kernis. New York: Plenum Press, pp. 31–49.

Delamont, S. 2003. *Feminist Sociology*. London: Sage.

Denzin, N. 1989. *Interpretive Biography*. Beverly Hills: Sage University Paper Series on Qualitative Research Methods.

Denzin, N. and Y. Lincoln (eds). 1998. *Collecting and Interpreting Qualitative Materials*. Thousand Oaks, London, New Delhi: Sage.

DeWalt, K. and B. DeWalt 2002. *Participant Observation*. Walnut Creek: AltaMira Press.

Dirks, N. 1992. *Colonialism and Culture*. Ann Arbor: Chicago University Press.

Dubois, D. and C. Rouby 2002. 'Names and Categories for Odors: The Veridical Label', in *Olfaction, Taste, and Cognition*, ed. C. Rouby, B. Schaal, D. Dubois et al. Cambridge: Cambridge University Press, pp. 47–66.

Duelke, B. 2005. '"No matter that modern world" – An Aboriginal Approach to a Polychronic Past: A Case Study from Northern Australia', in *Tradition and Agency: Tracing Cultural Continuity and Invention*, ed. T. Otto and P. Pedersen. Aarhus: Aarhus University Press, pp. 267–90.

Durkheim, E. 1912. *Les Formes Élémentaires de la Vie Religieuse*. Paris: Alcan.

Durkheim, E. and M. Mauss 1963. *Primitive Classification*. Chicago: University of Chicago Press. (1903. *De Quelques Formes Primitives de Classification*.)

Dweck, C.S. 1999. *Self-theories: Their Role in Motivation, Personality, and Development*. Philadelphia: Psychology Press.

Ebbinghaus, H. 1885. *Über das Gedächtnis: Untersuchungen zur experimentellen Psychologie*. Leipzig: Dunker und Humbold. Translated by H.A. Ruger and C.E. Bussenius 1913. *Memory. A Contribution to Experimental Psychology*. New York: Teachers College, Columbia University.

Eckensberger, L. 1970. *Methodenprobleme der kulturvergleichenden Psychologie*. Saarbrücken: Sozialwissenschaftlicher Studienkreis für Internationale Probleme.

Eliade, M. 1969. *Images and Symbols*. New York, Princeton: Princeton University Press. (reprint 1991).

Engen, T. 1982. *The Perception of Odors*. New York, London, Paris: Academic Press.

_____. 1987. 'Remembering Odors and their Names', *American Scientist* 75: 497–503.

_____. 1991. *Odor Sensation and Memory*. New York, Westport, Connecticut, London: Praeger.

Erdfelder, E. and J. Bredenkamp 1998. 'Recognition of Script-Typical versus Script-Atypical Information: Effects of Cognitive Elaboration', *Memory & Cognition* 26: 922–38.

Erdfelder, E. and A. Buchner 2003. 'Prozessdissoziationsprozedur: Quo vadis?' [Process dissociation procedure: Quo vadis?], *Zeitschrift für Psychologie* 211: 17–25.

Errington, F. and D. Gewertz. 1996. 'The Individuation of Tradition in a Papua New Guinean Modernity', *American Anthropologist* 98: 114–26.

Errington, S. 1989. 'Fragile Traditions and Contested Meanings', *Public Culture* 1(2): 49–59.

Essau, C.A. and G. Trommsdorff 1996. 'Coping with University-Related Problems: A Cross-Cultural Comparison', *Journal of Cross-Cultural Psychology* 27(3): 315–28.

Eves, R. 1998. *The Magical Body. Power and Meaning in a Melanesian Society*. Singapore: Harwood Academic Publishers.

Ewing, K.P. 1990. 'The Illusion of Wholeness: Culture, Self, and the Experience of Inconsistency', *Ethos* 18: 251–78.

———. 1997. *Arguing Sainthood: Psychoanalysis, Modernity, and Islam*, Durham, NC: Duke University Press.

Eysenck, M. and M. Keane 1997. *Cognitive Psychology*. Hove: Psychology Press.

Fabian, J. 1983. *Time and the Other*. New York: Columbia University Press.

———. 1990. *History from Below*. Amsterdam: Benjamins.

Fairclought, N. and R. Wodak 1995. 'Critical Discourse Analysis', in *Discourse as Social Interaction*, ed. N. Fairclought and R. Wodak. London: Sage, pp. 258–84.

Fajans, J. 1985. 'The Person in Social Context. The Social Character of Baining "Psychology"', in *Person, Self and Experience. Exploring Pacific Ethnopsychologies*, ed. G. White and J. Kirkpatrick. Berkeley, Los Angeles, London: University of California Press, pp. 367–400.

Feld, S. 1990 (1982). *Sound and Sentiment: Birds, Weeping, Poetics and Song in Kaluli Expression*. Philadelphia: University of Pennsylvania Press.

———. 1991. 'Sound as a Symbolic System: The Kaluli Drum', in *The Varieties of Sensory Experience*, ed. D. Howes. Toronto: University of Toronto Press, pp. 79–99.

———. 1996. 'Waterfalls of Song: An Acoustemology of Place Resounding in Bosavi, Papua New Guinea', in *Senses of Place*, ed. S. Feld and K. Basso. Santa Fe, NM: School of American Research Press, pp. 91–136.

Feld, S. and K. Basso (eds). 1996. *Senses of Place*. Santa Fe, NM: School of American Research Press.

Feldman, S.S. and T. Quatman 1988. 'Factors Influencing Age Expectations for Adolescent Autonomy: A Study of Early Adolescents and Parents', *Journal of Early Adolescence* 8(4): 325–43.

Feldman, S.S. and D.A. Rosenthal 1990. 'The Acculturation of Autonomy Expectations in Chinese High Schoolers Residing in Two Western Nations', *International Journal of Psychology* 25(3): 259–81.

Feldman, St. P. 1997. 'The Revolt against Cultural Authority: Power/Knowledge as an Assumption in Organization Theory', *Human Relations* 50: 937–55.

Finger, S. 1994. *Origins of Neuroscience. A History of Explorations into Brain Function*. New York: Oxford University Press.

Finger, T. E., W.L. Silver and D. Restrepo (eds). 2000. *The Neurobiology of Taste and Smell*, 2nd edition. New York, Chichester, Weinheim: Wiley.

Finnegan, R. 1992. *Oral Traditions and the Verbal Arts: A Guide to Research Practices*. London and New York: Routledge.

Fisher, J. 1997. 'Relational Sense: Towards a Haptic Aesthetics', *Parachute* 87: 4–11, <http://alcor.concordia.ca/~senses/Fisher.htm> [1.6.2000].

Flinn, J. L. Marshall and J. Armstrong 1998. Fieldwork and Families: Constructing New Models for Ethnographic Research. Honolulu: University of Hawai'i Press.

Flyvbjerg, B. 1998. *Rationality and Power. Democracy in Practice*. Chicago, London: University of Chicago Press.

Fodor, J. 2001. *The Mind Doesn't Work That Way*. Cambridge, MA: MIT-Press.

Fonagy, P., G. Gergely, E.L. Jurist and M. Target 2002. *Affect Regulation, Mentalization, and the Development of the Self*. New York: Other Press.

Foote, K., A. Tóth and A. Árvay 2000. 'Hungary after 1989: Inscribing a New Past on Place', *The Geographical Review* 90: 301–34.

Foster, R.J. 1991. 'Making National Cultures in the Global Ecumene', *Annual Review of Anthropology* 20: 235–60.

_____. 1995. *Social Reproduction and History in Melanesia: Mortuary Ritual, Gift Exchange, and Custom in the Tanga Islands*. Cambridge: Cambridge University Press.

_____. (ed.). 1995b. *Nation Making: Emergent Identities in Postcolonial Melanesia*. Ann Arbor: University of Michigan Press.

Foucault, M. 1972. *The Archaeology of Knowledge*, New York: Pantheon.

_____. 1978. *The History of Sexuality*, Vol. I: *An Introduction*. New York: Pantheon.

_____. 1980. *Power/Knowledge. Selected Interviews and other Writings 1972–1977*, ed. C. Gordon. Brighton: Harvester.

_____. 1990. 'Andere Räume', *Zeitmitschrift* 1: 5–15.

_____. 1994. *The Order of Things : An Archaeology of Human Sciences*. New York: Vintage (reissue edition).

Fox, J. (ed.). 1993. *Inside the Austronesian Houses: Perspectives on Domestic Designs for Living*. Canberra: Australian National University Press.

_____. (ed.). 1997. *The Poetic Power of Place: Comparative Perspectives on Austronesian Ideas of Locality*. Canberra: Research School of Pacific and Asian Studies, Australian National University.

Fox, J., P. Bellwood and D. Tyron (eds). 1995. *The Austronesians: Historical and Comparative Perspectives*. Canberra: The Australian National University Press.

Franck, G. 1998. *Ökonomie der Aufmerksamkeit. Ein Entwurf*. München, Wien: Hanser.

Freeman, D. 1983. *Margaret Mead and Samoa. The Making and Unmaking of an Anthropological Myth*. Cambridge, MA: Harvard University Press.

Freksa, C. 2003. 'Understanding Spatial Cognition through Synthesis and Analysis'. Paper presented at the Symposium *Space Games: New Research Approaches to Spatial Orientation, Spatial Perception, and Spatial Cognition*, 30.-31.3.2003, Internationales Wissenschaftsforum Heidelberg, Germany.

Frensch, P.A. and J. Funke 1995. *Complex Problem Solving. The European Perspective*. Hillsdale: Lawrence Erlbaum Associates.

Friedman, J. 1993. 'Will the Real Hawaiian Please Stand: Anthropologists and Natives in the Global Struggle for Identity', *Bijdragen Tot de Taal, Land- en Volkenkunde* 149: 737–67.

_____. 1994. *Cultural Identity and Global Processes*. London: Sage.

_____. 1998. 'Knowing Oceania or Oceanian Knowing. Identifying Actors and Acting Identities in Turbulent Times', in *Pacific Answers to Western Hegemony*, ed. J. Wassmann. Oxford: Berg, pp. 36–66.

Frykman, J., and O. Löfgren. 1996. 'The Study of Swedish Customs and Habits', in *Force of Habit: Exploring Everyday Culture*, ed. J. Frykman and O. Löfgren. Lund: Lund University Press, pp. 5–19.

Fung, H.H., R.P. Abeles and L.L. Carstensen 1999. 'Psychological Control in Later Life: Implications for Life-Span Development', in *Action and Development: Origins and Functions of Intentional Self Development*, ed. M. Lerner and J. Brandtstädter. Hillsdale: Sage Publications, pp. 345–72.

Geertz, C. 1960. *The Religion of Java*. Chicago: University of Chicago Press.

———. 1973. *The Interpretation of Cultures*. London: Basic Books.

———. 1975. 'On the Nature of Anthropological Understanding', *American Scientist* 63(1): 47–53.

———. 1990. '"Popular Art" and the Javanese Tradition', *Indonesia* 50: 77–94.

Gell, A. 1975. *Metamorphosis of the Cassowaries. Umeda Society, Language and Ritual*. London, New Jersey: Athlone Press.

———. 1977. 'Magic, Perfume, Dream …', in *Symbols and Sentiments*, ed. I.M. Lewis. London: Academic Press, pp. 25–38.

———. 1995a. 'How to Read a Map', *Man* 20: 271–86.

———. 1995b. 'The Language of the Forest: Landscape and Phonological Iconism in Umeda', in *The Anthropology of Landscape. Perspectives on Place and Space*, ed. E. Hirsch and M. O'Hanlon. Oxford: Clarendon Press, pp. 232–54.

———. 1998. *Art and Agency. An Anthropological Theory*. Oxford: Clarendon Press.

Gelman, R. and T.K.-F. Au (eds). 1996. *Perceptual and Cognitive Development*. San Diego, New York, Boston: Academic Press.

Gergen, K.J. 1990. 'Social Understanding and the Inscription of Self', in *Cultural Psychology: Essays on Comparative Human Development*, ed. J.W. Stigler, R.A. Shweder and G. Herdt. Cambridge: Cambridge University Press, pp. 569–606.

Geurts, K.L. 2002. *Culture and the Senses. Bodily Ways of Knowing in an African Community*. Berkeley, Los Angeles, London: University of California Press.

Gewertz, D. and F. Errington 1991. *Twisted Histories, Altered Contexts*. Cambridge: Cambridge University Press.

Giddens, A. 1984. *The Constitution of Society. Outline of the Theory of Structuration*. Cambridge: Polity Press.

Gilbert, D.T. and P.S. Malone 1995. 'The Correspondence Bias', *Psychological Bulletin* 117(1): 21–38.

Gilgun, J. 2004. 'Fictionalizing Life Stories: Yukee the Wine Thief', *Qualitative Inquiry* 10(5): 691–705.

Gladwin, T. 1971. *East Is a Big Bird*. Cambridge, MA: Harvard University Press.

Godden, D. and A.D. Baddeley 1975. 'Context-Dependent Memories in Two Natural Environments: On Land and under Water', *British Journal of Psychology* 66: 325–31.

Goldstein, E.B. 1999. *Sensation and Perception*. Albany, Belmont: Pacific Grove.

———. 2002. *Sensation and Perception*, sixth edn. Belmont, CA: Wadsworth.

Gorgendiere, L. de la (ed.). 1996. *Ethnicity in Africa: Roots, Meanings and Implications*. Edinburgh: Centre of African Studies.

Gottmann, J. 1980. 'Confronting Centre and Periphery', in *Centre and Periphery. Spatial Variation in Politics*, ed. J. Gottmann. Beverly Hills, London: Sage, pp. 11–25.

Gould, S.J. 1999. 'A Critique of Heckhausen and Schulz's (1995) Life-Span Theory of Control From a Cross-Cultural Perspective', *Psychological Review* 106(3): 597–604.

Gow, P. 1995. 'Land, People and Paper in Western Amazonia', in *The Anthropology of Landscape*, ed. E. Hirsch and M. O'Hanlon. Oxford: Clarendon Press, pp. 43–62.

Graumann, C.-F. 2002. 'The Phenomenological Approach to People-Environment Studies', in *Handbook of Environmental Psychology*, ed. R.B. Bechtel and A. Churman. New York: Wiley, pp. 95–113.

Gray, J.N. 2000. *At Home in the Hills: Senses of Place in the Scottish Borders*. New York: Berghahn Books.

Greenfield, P.M. 1997. 'Culture as Process: Empirical Methods for Cultural Psychology', in *Handbook of Cross-Cultural Psychology*, 2nd edition. Volume 1: *Theory and Method*, ed. J.W. Berry, Y.H. Poortinga and J. Pandey. Boston: Allyn & Bacon, pp. 301–46.

Gregg, G.S. 1991. *Self-Representation: Life Narrative Studies in Identity and Ideology*. New York: Greenwood Press.

Gruber, H. 2001. 'Acquisition of Expertise', in *International Encyclopedia of the Social & Behavioral Sciences* 8, ed. N.J. Smelser and P.B. Baltes. Amsterdam: Elsevier, pp. 5145–50.

Grusec, J.E. and J.J. Goodnow 1994. 'Impact of Parental Discipline Methods on the Child's Internalization of Values: A Reconceptualization of Current Points of View', *Developmental Psychology* 30(1): 4–19.

Guha, R. 1989. 'Dominance without Hegemony and Its Historiography', in *Subaltern Studies 6: Writings on South Asian History and Society*, ed. R. Guha. Dehli: Oxford University Press, pp. 210–309.

_____. (ed.). 1982. *Subaltern Studies: I. Writings on South Asian History and Society*. Delhi: Oxford University Press.

Guisinger, S. and S.J. Blatt 1994. 'Individuality and Relatedness: Evolution of a Fundamental Dialectic', *American Psychologist* 49(2): 104–11.

Gulliver, P.H and Silverman, M. 1995. *Merchants and Shopkeepers: A Historical Anthropology of an Irish Market Town, 1200–1991*, Toronto: University of Toronto Press.

Gumperz, J. and S. Levinson (eds). 1996. *Rethinking Linguistic Relativity*. Cambridge: Cambridge University Press.

Gunderson, L. and C. Holling (eds). 2002. *Panarchy: Understanding Transformations in Human and Natural Systems*. Washington: Island Press.

Gupta, A. and J. Ferguson 1992. 'Beyond "Culture": Space, Identity, and the Politics of Identity', *Cultural Anthropology* 7: 6–23.

Guyer, J. 1996. 'Traditions of Invention in Equatorial Africa', *African Studies Review* 39(4): 1–28.

Hagen, J. 1999. 'Reckoning Kinship in Maneo (Seram, Indonesia)', *American Ethnologist*, 26(1): 173–95.

Hägerstrand, T. 1970. 'What About People in Regional Science?', *Papers of the Regional Science Association* XXIV: 7–21.

———. 1975. 'Space, Time and Human Conditions', in *Dynamic Allocation of Urban Space*, ed. A. Karlquist, L. Lunquist and F. Snickar. Farnborough: Saxon House, pp. 3–14.

———. 1982. 'Diorama, Path and Project', *Tijdschrift voor Economische en Sociale Geographie* (73): 323–56.

Hallpike, C.R. 1979. *The Foundations of Primitive Thought*. Oxford: Clarendon Press.

Hamelin, C., and E. Wittersheim (eds). 2002. *La Tradition et L'état: Églises, Pouvoirs et Politiques Culturelles dans le Pacifique*. Paris: L'Harmattan.

Handler, R., and J. Linnekin. 1984. 'Tradition, Genuine or Spurious', *Journal of American Folklore* 97: 273–90.

Hannerz, U. 1996. *Transnational Connections. Culture, People, Places*. London: Routledge.

Hanson, A. 1989. 'The Making of the Maori: Culture Invention and Its Logic', *American Anthropologist* 91: 890–902.

———. 1991. 'Reply to Langdon, Levine, and Linnekin', *American Anthropologist* 93: 449–50.

Hardwig, J. 1991. 'The Role of Trust in Knowledge', *The Journal of Philosophy* 88: 693–708.

Harneit-Sievers, A. (ed.). 2002. *A Place in the World: New Local Historiographies from Africa and South-Asia*. Leiden: Brill.

Harper, R.G. 1985. 'Power, Dominance, and Nonverbal Behavior: An Overview', in *Power, Dominance, and Nonverbal Behavior*, ed. S.L. Ellyson and J.F Dovidio. New York: Springer Verlag, pp. 29–48.

Harries-Jones, P. 1995. *A Recursive Vision: Ecological Understanding and Gregory Bateson*. Toronto: University of Toronto.

———. 2005. 'Gregory Bateson's "Uncovery" of Ecological Aesthetics', Paper Presented on 28 August 2005 at the Bateson Symposium: 'Bateson and the Epistemology of the Sacred – The Science-Religion Pattern'.

Harris, G. 1989. 'Concepts of Individual, Self, and Person in Description and Analysis', *American Anthropologist* 91(3): 599–612.

Harrison, S. 2000. 'From Prestige Goods to Legacies: Property and the Objectification of Culture in Melanesia', *Comparative Studies in Society and History* 42: 662–79.

———. 2001. 'Smoke Rising from the Villages of the Dead', *The Journal of the Royal Anthropological Institute* 7: 257–74.

———. 2006. *Fracturing Resemblances. Identity and Mimetic Conflict in Melanesia and the West*. Oxford: Berghahn.

Hastrup, K. 1992. *Other Histories*. London: Routledge.

Harvey, D. 1989. *The Conditions of Postmodernity*. Oxford: Blackwell.

Hatt, H. 1990. 'Physiologie des Riechens und Schmeckens', in *Vom Reiz der Sinne*, ed. A. Maelicke. Weinheim, New York, Basel, Cambridge: VCH, pp. 93–127.

Hau'ofa, E. 1994 (1993). 'Our Sea of Islands', *The Contemporary Pacific* 6(1): 148–61.

Hays, W.L. 1972. *Statistics for the Social Sciences* (2nd edition). London: Holt, Rinehart and Winston.

Heider, F. 1958. *The Psychology of Interpersonal Relations*. Oxford: John Wiley.

Heine, S.J., D.R. Lehman, H.R. Markus and S. Kitayama 1999. 'Is There a Universal Need for Positive Self-Regard?', *Psychological Review* 106(4): 766–94.

Hermans, H. 2002. 'The Dialogical Self. Toward a Theory of Personal and Cultural Positioning', Culture and Psychology 7(3): 243–281.

Herz, R.S. 2002. 'Olfaction and Affective Cognition', in *Olfaction, Taste, and Cognition*, ed. C. Rouby, B. Schaal, D. Dubois et al. Cambridge: Cambridge University Press, pp. 160–77.

Hilgard, E.R. 1952. 'Methods and Procedures in the Study of Learning', in *Handbook of Experimental Psychology*, ed. S.S. Stevens. New York: Wiley, pp. 517–67.

Hill, D. (ed.). 1993. *Space Stimuli Kit 1 of the Cognitive Anthropology Research Group*. Nijmegen: MPI for Psycholinguistics.

Hirsch, E. 1995. 'Landscape: Between Place and Space', in *The Anthropology of Landscape*, ed. E. Hirsch and M. O'Hanlon. Oxford: Clarendon Press, pp. 1–30.

Hirsch, E. and M. O'Hanlon (eds.). 1995. *The Anthropology of Landscape: Perspectives on Place and Space*. Oxford: Oxford University Press.

Hirsch, E. and M. Strathern (eds). 2005. *Transactions and Creations. Property Debates and the Stimulus of Melanesia*. Oxford: Berghahn.

Hobsbawm, E. 1983. 'Introduction: Inventing Traditions', in *The Invention of Tradition*, ed. E. Hobsbawm and T. Ranger. Cambridge: Cambridge University Press, pp. 1–14.

Hobsbawm, E. and T. Ranger (eds). 1983. *The Invention of Tradition*. Cambridge: Cambridge University Press.

Hodder, I. 1998. 'The Interpretation of Documents and Material Culture', in *Collecting and Interpreting Qualitative Materials*, ed. N.Y. Hodder. Thousand Oaks, London, New Delhi: Sage, pp. 110–29.

Hofstede, G.H. 2001. *Culture's Consequences: Comparing Values, Behaviors, Institutions and Organizations Across Nations*, 2nd edition. Thousand Oaks: Sage Publications.

Højbjerg, C.K. 2005. 'Tradition Invented and Inherited in West Africa: A Study of Loma Cultural Identity', in *Tradition and Agency: Tracing Cultural Continuity and Invention*, ed. T. Otto and P. Pedersen. Aarhus: Aarhus University Press, pp. 85–117.

Hollan, D. 1994. 'Suffering and the Work of Culture: A Case of Magical Poisoning in Toraja', *American Ethnologist* 21: 74–87.

Holy, L. and M. Stuchlik. 1983. *Actions, Norms and Representations: Foundations of Anthropological Enquiry*. Cambridge: Cambridge University Press.

Holzkamp, K. 1972. 'Zum Problem der Relevanz psychologischer Forschung für die Praxis', in his *Kritische Psychologie. Vorbereitende Arbeiten*. Frankfurt a.M.: Fischer, pp. 9–34.

Hoskins, J. 1987. 'The Headhunter as Hero: Local Traditions and Their Reinterpretation in National History', *American Ethnologist*, 14(4): 605–22.

Howes, D. 1987. 'Olfaction and Transition: An Essay on the Ritual Uses of Smell', *The Canadian Review of Sociology and Anthropology* 24: 398–416.

_____. 1988. 'On the Odour of the Soul: Spatial Representation and Olfactory Classification in Eastern Indonesia and Western Melanesia', *Bijdragen Tot de Taal-, Land-en Volkenkunde* 144: 84–113.

_____. 1992. 'Neuguinea. Eine olfaktorische Ethnographie', *dragoco report* 2: 71–81.

_____. 2000. *Body Signs: Body Decoration and Sensory Symbolism in South America*. <http://alcor.concordia.ca/~senses/Howes.htm> [accessed 1.6.2000].

_____. 2003. *Sensual Relations. Engaging the Senses in Culture and Social Theory*. Ann Arbor: University of Michigan Press.

_____. (ed.). 1991. *The Varieties of Sensory Experience: A Sourcebook in the Anthropology of the Senses*. Toronto: University of Toronto Press.

Hsu, F.L.K. 1985. 'The Self in Cross-Cultural Perspective, in *Culture and Self: Asian and Western Perspectives*, ed. A. Marsella, G. DeVos and F.L. Hsu. London: Tavistock, pp. 24–55.

Huizinga, J. 1956. *Homo Ludens. Vom Ursprung der Kultur im Spiel*. Hamburg: Rowohlt.

Hutchins, E. 1995. *Cognition in the Wild*. Cambridge, MA: MIT Press.

Hviding, E. 1996. *Guardians of Marovo Lagoon: Practice, Place, and Politics in Maritime Melanesia*. Honolulu: University of Hawai'i Press.

Ibarra H. and S.B. Andrews 1993. 'Power, Social Influence, and Sense Making: Effects of Network Centrality and Proximity on Employee Perceptions', *Administrative Science Quarterly* 38: 277–303.

Inglehart, R. 1997. *Modernization and Postmodernization: Cultural, Economic, and Political Change in 43 Societies*. Princeton: Princeton University Press.

Ingold, T. 1994. '"People Like Us": The Concept of the Anatomically Modern Human', in *Ape, Man, Apeman: Changing Views since 1600, Evaluative Proceedings of the Pithecanthropus Centennial Congress*, ed. R. Corbey and B. Theunissen. Leiden: Department of Prehistory, pp. 241–62.

_____. 2000. *Perception of the Environment: Essays in Livelihood, Dwelling and Skill*. New York and London: Routledge.

Issanchou, S, D. Valentin, C. Sulmont,. et al. 2002. 'Testing Odor Memory: Incidental versus Intentional Learning, Implicit versus Explicit Memory', in *Olfaction, Taste, and Cognition*, ed. C. Rouby, B. Schaal, D. Dubois et al. Cambridge: Cambridge University Press, pp. 211–29.

Jaarsma, S. (ed.). 2002. *Handle with Care. Ownership and Control of Ethnographic Materials*. Pittsburgh: University of Pittsburgh Press.

Jacoby, L.L. 1991. 'A Process Dissociation Framework: Separating Automatic from Intentional Uses of Memory', *Journal of Memory and Language* 30: 513–41.

———. 1998. 'Invariance in Automatic Influences of Memory: Toward a User's Guide for the Process-Dissociation Procedure', *Journal of Experimental Psychology: Learning, Memory, and Cognition* 24: 3–26.

Jahoda, G. 1982. *Psychology and Anthropology: A Psychological Perspective*. London: Academic Press.

———. 1993. 'Themes in the History of Culture and Cognition', in *Alltagswissen. Les Savoirs Quotidiens. Everyday Cognition*. ed. J. Wassmann and P. Dasen. Freiburg: Universitätsverlag, pp. 19–41.

Jahoda, G. and B. Krewer 1997. 'History of Cross-Cultural and Cultural Psychology', in *Handbook of Cross-Cultural Psychology*, 2nd edition, *Volume 1: Theory and Method*, ed. J.W. Berry, Y. H. Poortinga and J. Pandey. Boston: Allyn & Bacon, pp. 1–42.

Jay, M. 1993. *Downcast Eyes. The Denigration of Vision in Twentieth-Century French Thought*. Berkeley, Los Angeles, London: University of California Press.

Johnson, M. 1987. *The Body in the Mind: The Bodily Basis of Meaning, Imagination, and Reason*. Chicago, London: University of Chicago Press.

Johnson, M.K. and L. Hasher 1987. 'Human Learning and Memory', *Annual Review of Psychology* 38: 631–68.

Jolly, M. 1992. 'Specters of Inauthenticity', *The Contemporary Pacific* 4: 49–72.

Jolly, M. and N. Thomas (eds). 1992. 'The Politics of Tradition in the Pacific. Special Issue', *Oceania* 62(4): 241–354.

Jones, E.E. and V.A. Harris 1967. 'The Attribution of Attitudes', *Journal of Experimental Social Psychology* 3(1): 1–24.

Jones, K. 2001. 'Trust: Philosophical Aspects', in *International Encyclopedia of the Social & Behavioral Sciences*, Vol. 23, ed. N.J. Smelser and P.B. Baltes. Amsterdam: Elsevier, pp. 15917–22.

Jorgensen, D. 1998. *Participant Observation: A Methodology for Human Studies*. Oakland: Sage.

Kagan, J. and R.E. Klein 1973. 'Cross-Cultural Perspectives on Early Development', *American Psychologist* (November): 947–61.

Kagitcibasi, C. 1996. *Family and Human Development Across Cultures: A View From the Other Side*. Mahwah: Erlbaum.

Kahn, M. 1990. 'Stone-Faced Ancestors: The Spatial Anchoring of Myth in Wamira, Papua New Guinea', *Ethnology* 29(1): 51–66.

———. 1996. 'Your Place and Mine: Sharing Emotional Landscapes in Wamira, Papua New Guinea', in *Senses of Place*, ed. S. Feld and K.H. Basso, Santa Fe: School of American Research Press, pp. 167–96.

———. 2004. 'Placing Tahitian Identities. Rooted in Land and Enmeshed in Representations', in *Globalization and Culture Change in the Pacific Islands*, ed. V.S. Lockwood. Upper Saddle River, NJ: Pearson Prentice Hall, pp. 285–306.

Kant, I. (1768) 1991. 'Von dem ersten Grunde des Unterschiedes der Gegenden im Raume', in *The Philosophy of Right and Left*, ed. J. van Cleve and R. Frederck. Dordrecht: Kluwer, pp. 27–33.

Kardiner, A. and R, Linton 1945. *The Individual and His Society*. New York: Columbia University Press.

Karkelar, M. 1991. *Voices from Within: Early Personal Narratives of Bengali Women*. Delhi: Oxford University Press.

Karstedt, L. 2003. 'Ethnologie und Schematheorie. Das Konzept der "kognitiven Schemata" als methodische Grundlage für eine ethnologische Untersuchung der Sinne', *Zeitschrift für Ethnologie* 128: 293–310.

Keck, V. 2003. 'Interdisziplinäre Projekte und Teamarbeit', in *Methoden und Techniken der Feldforschung*, ed. B. Beer. Berlin: Reimer, pp. 203–23.

_____. 2005. *Social Discord and Bodily Disorders. Healing among the Yupno of Papua New Guinea*. Durham: Carolina Academic Press.

_____. (ed.). 1998. *Common Worlds and Single Lives. Constituting Knowledge in Pacific Societies*. Oxford, New York: Berg.

Keeler, W. 1987. *Javanese Shadow Plays, Javanese Selves*. Princeton: Princeton University Press.

Keesing, R.M. 1982. 'Kastom in Melanesia: An Overview', *Mankind* 13: 297–301.

_____. 1989. 'Creating the Past: Custom and Identity in the Contemporary Pacific', *The Contemporary Pacific* 1: 19–42.

_____. 1991. 'Reply to Trask', *The Contemporary Pacific* 3: 168–71.

_____. 1992. *Custom and Confrontation: The Kwaio Struggle for Cultural Autonomy*. Chicago: University of Chicago Press.

Keesing, R.M. and R. Tonkinson (eds). 1982. 'Reinventing Traditional Culture: The Politics of Kastom in Island Melanesia. Special Issue', *Mankind* 13(4): 297–399.

Keifenheim, B. 2000. *Wege der Sinne. Wahrnehmung und Kunst bei den Kashinawa-Indianern Amazoniens*. Frankfurt am Main: Campus Verlag.

Kemp, P. 1976. *Oxford Companion to Ships and the Sea*. New York: Oxford.

King, R. and J. Connell 1999. *Small Worlds, Global Lives. Islands and Migration*. London and New York: Pinter.

Kirk, J. and M.L. Miller 1986. *Reliability and Validity in Qualitative Research*, Vol. 1. Beverly Hills, CA: Sage.

Kirkpatrick, J. and G.M. White 1985. 'Exploring Ethnopsychologies', in *Person, Self, and Experience. Exploring Pacific Ethnopsychologies*, ed. G. White and J. Kirkpatrick. Berkeley, Los Angeles and London: University of California Press, pp. 3–32.

Klein, W. 1982. 'Local Deixis in Route Description', in *Speech, Place and Action*, ed. R.J. Jarvella and W. Klein. Chinchester: John Wiley, pp. 161–75.

_____. 1991. 'Raumausdrücke', *Linguistische Berichte* 132: 77–114.

Kluwe, R.H., G. Lüer and F. Rösler (eds). 2003. *Principles of Learning and Memory*. Basel, Boston, Berlin: Birkhäuser Verlag.

Knox, P. and S. Marston 1998. *Places and Regions in Global Context – Human Geography*. Saddle River, NJ: Prentice Hall, Inc.

_____. 2001. *Humangeographie*. Heidelberg/Bern: Spektrum Akademischer Verlag.

Koentjaraningrat, R.M. 1985. *Javanese Culture*. Singapore: Oxford University Press.

Kojima, H. 1984. 'A Significant Stride Toward the Comparative Study of Control', *American Psychologist* 39(9): 972–73.

Kolig, E. 2005. 'The Politics of Indigenous – or Ingenious – Tradition: Some Thoughts on the Australian and New Zealand Situation', in *Tradition and Agency: Tracing Cultural Continuity and Invention*, ed. T. Otto and P. Pedersen. Aarhus: Aarhus University Press, 293–326.

Konrád G. and I. Szelényi 1978. *Die Intelligenz auf dem Weg zur Klassenmacht*. Frankfurt a.M.: Suhrkamp.

Kornadt, H.-J. 2007. 'Motivation [Motivation]', in *Enzyklopädie der Psychologie: Themenbereich C Theorie und Forschung, Serie VII Kulturvergleichende Psychologie. Band 2: Kulturelle Determinanten des Erlebens und Verhaltens* [Encyclopaedia of Psychology: Subject Area C. Theory and Research: Serial VII. Cross-Cultural Psychology: Vol. 2. Cultural Determinants of Experience and Behavior], ed. G. Trommsdorff and H.-J. Kornadt. Göttingen: Hogrefe.

Kornadt, H.-J., L.H. Eckensberger and W.B. Emminghaus 1980. 'Cross-Cultural Research on Motivation and Its Contribution to a General Theory of Motivation', in *Handbook of Cross-Cultural Psychology*, Vol. 3: *Basic Processes*, ed. H.C. Triandis and W.J. Lonner. Boston: Allyn & Bacon, pp. 223–321.

Köster, E.P. 2002. 'The Specific Characteristics of the Sense of Smell', in *Olfaction, Taste, and Cognition*, ed. C. Rouby, B. Schaal, D. Dubois et al. Cambridge: Cambridge University Press, pp. 27–43.

Kottak, C. 1999. 'The New Ecological Anthropology', *American Anthropologist* 101(1): 23–35.

Krogstad, A. 1989. 'The Treasure-House of Smell: From an "Unsensing" to "Sensual" Anthropology', *Folk* 31: 87–103.

Küchler, S. 1993. 'Landscape as Memory: The Mapping of Process and Its Representation in a Melanesian Society', in *Landscape: Politics and Perspectives*, ed. Barbara Bender. Oxford: Berg, pp. 85–106.

Kuczynski, L. 2003. 'Beyond Bidirectionality: Bilateral Conceptual Frameworks for Understanding Dynamics in Parent-Child Relations', in *Handbook of Dynamics in Parent-Child Relations*, ed. L. Kuczynski. Thousand Oaks: Sage Publications, pp. 1–24.

Kuehling, S. 2001. 'Two Places in One'. Paper delivered at the ARIC conference, Geneva, September 2001.

———. 2003. 'Kula: Myth and Magic in the Trobriand Islands', *The Contemporary Pacific* 15(2): 512–16.

———. 2005. *Dobu. Ethics of Exchange on a Massim Island, Papua New Guinea*. Honolulu: University of Hawai'i Press.

Kuipers, J.C. 1991. 'Matters of Taste in Weyewa', in *The Varieties of Sensory Experience*, ed. D. Howes. Toronto: University of Toronto Press, pp. 111–27.

Küng, G. 1993. 'Zum weiteren Umfeld der Kognitiven Wissenschaft: Bemerkungen eines Philosophen', in *Alltagswissen. Les Savoirs Quotidiens. Everyday Cognition*, ed. J. Wassmann and P. Dasen. Freiburg: Universitätsverlag, pp. 43–61.

Kutschera, F. v. and Breitkopf, A. 1979. *Einführung in die moderne Logik*. Freiburg: Alber.

Lakoff, G. 1987. *Women, Fire, and Dangerous Things. What Categories Reveal about the Mind*. Chicago, London: University of Chicago Press.

Lakoff, G. and M. Johnson 1980. *Metaphors We Live By*. Chicago: University of Chicago Press.

Lam, A.G. and N.W.S Zane. 2004. 'Ethnic Differences in Coping with Interpersonal Stressors: A Test of Self-Construals as Cultural Mediators', *Journal of Cross-Cultural Psychology* 35(4): 446–59.

Larsson, M. 2002. 'Odor Memory: A Memory Systems Approach', in *Olfaction, Taste, and Cognition*, ed. C. Rouby, B. Schaal, D. Dubois et al. Cambridge: Cambridge University Press, pp. 231–45.

Lave, J. 1988. *Cognition in Practice. Mind, Mathematics, and Culture in Everyday Life*. Cambridge: Cambridge University Press.

Law, J. 2004. *After Method: Mess in Social Science Research*. London, New York: Routledge.

Lawrence, D.L. and S.M. Low 1990. 'The Built Environment and Spatial Form', *Annual Review of Anthropology* 19: 453–505.

Lawson, S. 1997. 'The Tyranny of Tradition: Critical Reflections on Nationalist Narratives in the South Pacific', in *Narratives of Nation in the South Pacific*, ed. T. Otto and N. Thomas. Amsterdam: Harwood Academic Publishers, pp. 15–31.

Leach, E. 1967. 'An Anthropologist's Reflections on a Social Survey', in *Anthropologists in the Field*, ed. P.C.W. Gutkind and D.G. Jongman. New York: Humanities Press.

Leach, J. 2003a. *Creative Land. Place and Procreation on the Rai Coast of Papua New Guinea*. Oxford: Berghahn.

———. 2003b. 'Owning Creativity. Cultural Property and the Efficacy of Custom on the Rai Coast of Papua New Guinea', *Journal of Material Culture* 8(2): 123–43.

Leavitt, S.C. 1995. 'Suppressed Meanings in Narratives about Suffering: A Case from Papua New Guinea', *Anthropology and Humanism* 20(2): 1–20.

———. 2005. '"We are not straight": Bumbita Arapesh Strategies for Self-Reflection in the Face of Images of Western Superiority', in *The Making of Global and Local Modernities in Melanesia: Humiliation, Transformation and the Nature of Cultural Change*. ed. J. Robbins and H. Wardlow. Burlington VT: Ashgate, pp. 73–84.

Leavitt, S.C. and G. Herdt (eds). 1998. *Adolescence in Pacific Societies*. Pittsburgh: University of Pittsburgh Press.

Lebra, T.S. 1994. 'Mother and Child in Japanese Socialization: A Japan-U.S. Comparison', in *Cross-Cultural Roots of Minority Child Development*, ed. P.M. Greenfield and R.R. Cocking. Hillsdale: Erlbaum, pp. 259–74.

Leenhardt, M. (1947) 1984. *Do Kamo. Die Person und der Mythos in der melanesischen Welt*. Frankfurt a.M., Berlin, Wien: Ullstein. [trans. B.M. Gulati 1979 *Do Kamo: Person and Myth in the Melanesian World*. Chicago: University of Chicago.]

Lefcourt, H.M. 1973. 'The Function of the Illusions of Control and Freedom', *American Psychologist* 28(5): 417–25.

Lepowsky, M. 1994. *Fruit of the Motherland. Gender in an Egalitarian Society*. New York: Columbia University Press.

Lerner, R.M. 1982. 'Children and Adolescents as Producers of Their Own Development', *Developmental Review* 2(4): 342–70.

Levelt, P. 1989. *Speaking. From Intention to Articulation*. Cambridge, MA: MIT Press.

Levinson, D. and M.J. Malone 1981. *Toward Explaining Human Culture: A Critical Review of the Findings of Worldwide Cross-Cultural Research*. New Haven, CT: HRAF Press.

Levinson, S.C. 1992. 'Primer for the Field Investigation of Spatial Description and Conception', *Pragmatics* 2: 5–47.

______. 1996a. 'Language and Space', *Annual Review of Anthropology* 25: 353–82.

______. 1996b. 'Frames of Reference and Molyneaux's Question: Cross-Linguistic Evidence', in *Language and Space*, ed. P. Bloom, M. A. Peterson, L. Nadel and M. Garrett. Cambridge, MA: MIT Press, pp. 109–69.

______. 1996c. *The Role of Language in Everyday Human Navigation*. Max Planck Institute for Psycholinguistics.

______. 2003. *Space in Language and Cognition. Explorations in Cognitive Diversity*. Cambride: Cambridge University Press.

Levinson, S.C. and P. Brown 1994. 'Immanuel Kant among the Tenejapans. Anthropology as Empirical Philosophy', *Ethos* 22: 3–41.

Lewis, D. 1976. 'Observations on Route Finding and Spatial Orientation among the Aboriginal Peoples of the Western Desert Region of Central Australia', *Oceania* 46(4): 249–82.

Lillard, A. 1998. 'Ethnopsychologies: Cultural Variations in Theories of Mind', *Psychological Bulletin* 123(1): 3–32.

Linde, C. 1993. *Life Stories: The Creation of Coherence*. New York: Oxford University Press.

Lindenbaum, S. and M. Lock (eds). 1993. *Knowledge, Power, and Practice: The Anthropology of Medicine and Everyday Life*. Berkeley: University of California Press.

Linnekin, J. 1991. 'Cultural Invention and the Dilemma of Authenticity', *American Anthropologist* 93: 446–49.

______. 1992. 'On the Theory and Politics of Cultural Construction in the Pacific', *Oceania* 62: 249–63.

Little, P. 1999. 'Environments and Environmentalism in Anthropological Research: Facing a New Millennium', *Annual Review of Anthropology* 28, 253–84.

Livingstone, D. 2002. 'Knowledge, Space and the Geographies of Science', *Hettner Lectures* 5, Heidelberg: Department of Geography, University of Heidelberg, pp. 7–40.

Lockwood, V.S. (ed.) 2004. *Globalization and Culture Change in the Pacific Islands*. Upper Saddle River: Pearson Prentice Hall.

Lonner, W.J. and J.W. Berry (eds). 1986. *Field Methods in Cross-Cultural Research*. Boston: Allyn and Bacon.

Löw, M. 2001. *Raumsoziologie*. Frankfurt am Main: Suhrkamp Verlag.

Lowe, A. and P. Willis 1980. *Culture, Media, Language*. London: Hutchinson.

Luhmann, N. 1997. *Die Gesellschaft der Gesellschaft*. Frankfurt am Main: Suhrkamp Verlag.

Lutz, C. 1988. *Unnatural Emotions: Everyday Sentiments on a Micronesian Atoll and Their Challenge to Western Theory*. Chicago: University of Chicago Press.

MacDonald, S., U. Kimberly and H. Hayne 2000. 'Cross-Cultural and Gender Differences in Childhood Amnesia', *Memory* 8: 365–76.

Macintyre, M. 1993. 'Fictive Kinship or Mistaken Identity? Fieldwork on Tubetube Island, Papua New Guinea', in *Gendered Fields. Women, Men and Ethnography*, ed. D. Bell, P. Caplan and W.J. Karim. London: Routledge, pp. 44–62.

Macmillan, N.A. and C.D. Creelman 1991. *Detection Theory: A User's Guide*. Cambridge: Cambridge University Press.

Maehr, M.L. and J.G. Nicholls 1980. 'Culture and Achievement Motivation: A Second Look', in *Studies in Cross-Cultural Psychology*, Vol. 2, ed. C. Warren. New York: Academic Press, pp. 221–67.

Mageo, J.M. 1998. *Theorizing Self in Samoa: Emotions, Genders, and Sexualities*. Ann Arbor: University of Michigan.

———. (ed.). 2002. *Power and the Self*. Cambridge: Cambridge University Press.

———. 2003. *Dreaming and the Self: New Perspectives on Subjectivity, Identity and Emotion*. Albany: SUNY.

Magnis-Suseno, F. v. 1989. *Neue Schwingen für Garuda. Indonesien zwischen Tradition und Moderne*. München: Kindt Verlag

Magnusson, D. 1990. 'Personality Research: Challenges for the Future', *European Journal of Personality* 4(1): 1–17.

Malkki, L. 1995. 'Refugees and Exile: From "Refugee Studies" to the National Order of Things', *Annual Review of Anthropology* 24: 495–523.

Manning, P. and B. Cullum-Sawn 1998. 'Narrative, Content, and Semiotic Analysis', in *Collecting and Interpreting Qualitative Materials*, ed. N. Y. Hodder. Thousand Oaks, London, New Delhi: Sage, pp. 246–73.

Marcus, G.E. 2002. 'Beyond Malinowski and after Writing Culture: on the Future of Cultural Anthropology and the Predicament of Ethnography', TAJA 13(2): 191–99.

Marcus, G.E. and M.J. Fischer 1986. *Anthropology as Cultural Critique: An Experimental Moment in the Human Sciences*. Chicago: Chicago University Press.

Markham, A. 2005. '"Go Ugly Early": Fragmented Narrative and Bricolage as Interpretive Method', *Qualitative Inquiry* 11(6): 813–39.

Markus, H.R. and S. Kitayama 1991. 'Culture and the Self: Implications for Cognition, Emotion, and Motivation', *Psychological Review* 98(2): 224–53.

———. 2002. 'Models of Agency: Sociocultural Diversity in the Construction of Action', in *The 49th Annual Nebraska Symposium on Motivation: Cross-*

Cultural Differences in Perspectives on Self, ed. V. Murphy-Berman and J. Berman Lincoln. Lincoln: University of Nebraska Press, pp. 1–57.

Marshall, M. 2004. *Namoluk Beyond the Reef. The Transformation of a Micronesian Community*. Boulder: Westview Press.

Massey, D. 1999. 'Philosophy and Politics of Spatiality: Some Considerations', *Hettner Lectures* 2, Heidelberg: Department of Geography, University of Heidelberg, pp. 27–42.

Mathur, S. 2000. 'History and Anthropology in South Asia: Rethinking the Archive', *Annual Review of Anthropology* 29: 89–106.

Matsumoto, D. 1999. 'Culture and Self: An Empirical Assessment of Markus and Kitayama's Theory of Independent and Interdependent Self Construal', *Asian Journal of Social Psychology* 2(3): 289–310.

May, T. 2001. *Social Research: Issues, Methods and Processes*. Buckingham, Philadelphia: Open University Press.

Mazur, J.E. 2002. *Learning and Behavior*, 5th edition. Upper Saddle River, NJ: Prentice Hall.

McClintock, Ch.G. 1988. 'Evolution, Systems of Interdependence, and Social Values', *Behavioral Science* 33: 59–76.

McCloskey, M. 1991. 'Networks and Theories: the Place of Connectionism in Cognitive Science', *Psychological Science* 2: 387–95.

McCrae, R.R. and P.T.J. Costa 1996. 'Toward a New Generation of Personality Theories: Theoretical Contexts for the Five-Factor Model', in *Five-Factor Model of Personality: Theoretical Perspectives*, ed. J.S. Wiggins. New York: Guilford Press, pp. 51–87.

McCulloch, G. 2004. *Documentary Research: In Education, History and the Social Sciences*. London, New York: Routledge Falmer.

McKean Skaff, M. and P. Gardiner 2003. 'Cultural Variations in Meaning of Control', in *Personal Control in Social and Life Course Contexts*, ed. S.H. Zarit, L.I. Pearlin and K.W. Schaie. New York: Springer Publishing Co., pp. 83–105.

McLuhan, M. (1964) 1994. *Understanding Media. The Extensions of Man*. London: Routledge.

Mead, M. 1928. *Coming of Age in Samoa: A Psychological Study of Primitive Youth for Western Civilization*. New York: Morrow Quill Paperbacks.

Meijl, T. van 2000. 'The Politics of Ethnography in New Zealand', in *Ethnographic Artifacts: Challenges to a Reflexive Anthropology*, ed. S.R. Jaarsma and M.A. Rohatynskyj. Honolulu: University of Hawai'i Press, pp. 86–103.

Meltzoff, A.N. and R. Brooks 2001. '"Like Me" as a Building Block for Understanding Other Minds: Bodily Acts, Attention, and Intention', in *Intentions and Intentionality: Foundations of Social Cognition*, ed. B.F. Malle, L.J. Moses and D.A. Baldwin. Cambridge MA: MIT Press, pp. 171–91.

Merlan, F. 2001. 'An Assessment of Von Doussa on Anthropology', 11 pages. Retrieved 7 August 2002, from the World Wide Web: http://aas.asn.au/hindmarsh.htm.

Meusburger, P. 1998. *Bildungsgeographie. Wissen und Ausbildung in der räumlichen Dimension*. Heidelberg: Spektrum Akademischer Verlag.

———. 2000. 'The Spatial Concentration of Knowledge. Some Theoretical Consideration', *Erdkunde* 54(4): 352–64.

———. 2005. 'Sachwissen und symbolisches Wissen als Machtinstrument und Konfliktfeld. Zur Bedeutung von Worten, Bildern und Orten bei der Manipulation des Wissens', *Geographische Zeitschrift* 93: 148–164.

———. 2006. 'Wissen und Raum – ein subtiles Beziehungsgeflecht', in *Bildung und Wissensgesellschaft*, ed. K. Kempter and P. Meusburger. Heidelberger Jahrbücher 49, Berlin, Heidelberg: Springer-Verlag, pp. 269–308.

Miles, M. and A.M. Huberman, 1994. *Qualitative Data Analysis: An Expanded Sourcebook*. Thousand Oaks, CA: Sage.

Miller, G. and P. Johnson-Laird 1976. *Language Perception*. Cambridge: Cambridge University Press.

Miller, J.G. 1984. 'Culture and the Development of Everyday Social Explanation', *Journal of Personality and Social Psychology* 46(2): 961–78.

———. 2003. 'Culture and Agency: Implications for Psychological Theories of Motivation and Social Development', in *Cross-Cultural Difference in Perspectives on the Self*, ed. V. Murphy-Berman and J. Berman. Lincoln: University of Nebraska Press, pp. 59–99.

Milner, B. 1965. 'Memory Disturbances after Bilateral Hippocampal Lesions', in *Cognitive Processes and the Brain*, ed. P. Milner and S. Glickman. Princeton, NJ: D. Van Nostrand Co. Inc., pp. 104–5.

Milton, K. 2002. *Loving Nature: Towards an Ecology of Emotion*. London, New York: Routledge.

Minsky, M. 1975. 'A Framework for Representing Knowledge', in *The Psychology of Computer Vision*, ed. P. Winston. New York: McGraw Hill, pp. 211–77.

Mintzberg, H. 1979. *The Structuring of Organisations. A Synthesis of the Research*. Englewood Cliffs N. J.: Prentice Hall.

Mirowsky, J. and C.E. Ross 1984. 'Mexican Culture and Its Emotional Contradictions', *Journal of Health & Social Behavior* 25(1): 2–13.

Mischel, W. 2004. 'Toward an Integrative Science of the Person', *Annual Review of Psychology* 55: 1–22.

Mishra, R.C., P.R. Dasen and S. Niraula 2003. 'Ecology, Language, and Performance on Spatial Cognitive Tasks', *International Journal of Psychology* 38: 366–83.

Mishra, R.C., D. Sinha and J.W. Berry 1996. *Ecology, Acculturation and Psychological Adaptation: A Study of Adivasis in Bihar*. New Delhi: Sage.

Mishra, R.C. and P.R. Dasen 2005. 'Spatial Language and Cognitive Development in India: An Urban/Rural Comparison', in *Culture and Human Development: The Importance of Cross-Cultural Research to the Social Sciences*, ed. W. Friedlmeier, P. Chakkarath and B. Schwarz. Hove, UK: Psychology Press, pp. 153–79.

Modell, J. (ed.) 2002. *Constructing Moral Communities. Pacific Islander Strategies for Settling in New Places*. Special Issue of *Pacific Studies* 25 (1/2).

Montello, D. R. 2001. 'Spatial Cognition', in *International Encyclopedia of the Social & Behavioral Sciences*, ed. N. J. Smelser and P. B. Baltes. Amsterdam: Elsevier, pp. 14771–75.

Moore, S.F. 1987. 'Explaining the Present. Theoretical Dilemmas in Processual Ethnography', *American Ethnologist* 14(4): 727–36.

Moral, B. and F. Escobar 2004. 'Time Geography Paradigm to Faciliate Visualisation of Gender Mobility Differences in Chuuk, Micronesia: a Research Proposal', in *Micronesia. Visiones desde Europa*, ed. B. Moral. Madrid: Ediciones Gondo, pp. 141–165.

Morling, B. and S.T. Fiske 1999. 'Defining and Measuring Harmony Control', *Journal of Research in Personality* 33(4): 379–414.

Morphy, H. (1993) 1995. 'Colonialism, History and the Construction of Place. The Politics of Landscape in Northern Australia', in *Landscape: Politics and Perspective*, ed. B. Bender. Oxford: Berg, pp. 205–44.

Morris, B. 1994. *Anthropology of the Self: The Individual in Critical Perspective*. Boulder, CO: Pluto Press.

Morris, M.W. and K. Peng. 1994. 'Culture and Cause: American and Chinese Attributions for Social and Physical Events', *Journal of Personality and Social Psychology Bulletin* 67(6): 949–71.

Morris, M.W., T. Menon and D. Ames 2001. 'Culturally Conferred Conceptions of Agency: A Key to Social Perception of Persons, Groups, and Other Actors', *Personality and Social Psychology Review* 5(2): 169–82.

Morton Lee, H. 2003. *Tongans Overseas. Between two Shores*. Honolulu: University of Hawai'i Press.

Mulder, N. 1992. 'Contemporary Cultural Dynamics in Yogyakarta, Central Java', *Orientierungen. Zeitschrift zur Kultur Asiens*, Sonderheft: 53–63.

Munroe, R.L. and R.H Munroe. 1997. 'A Comparative Anthropological Perspective', in *Handbook of Cross-Cultural Psychology*, 2nd edition, Volume 1: *Theory and Method*, ed. J.W. Berry, Y.H. Poortinga and J. Pandey. Boston: Allyn & Bacon, pp. 171–214.

Myers, C.S. 1899. 'Observations on Hearing, Taste, Smell, Reaction, Time, etc.', *Journal of the Anthropological Institute* 29: 222.

_____. 1901a. 'Olfactory Acuity and Discrimination of Odour-Strengths', in *Reports of the Cambridge Anthropological Expedition*. Volume II: *Physiology and Psychology*, ed. A.C. Haddon. New York, London: Cambridge University Press, pp. 169–85.

_____. 1901b. 'Taste. Taste-Names in Murray Island', in *Reports of the Cambridge Anthropological Expedition*. Volume II: *Physiology and Psychology*, ed. A.C. Haddon. New York, London: Cambridge University Press, pp. 186–88.

Myers, F.R. 1986. *Pintupi Country, Pintupi Self: Sentiment, Place, and Politics among Western Desert Aborigines*. Washington D.C.: Smithsonian Institution Press.

Naroll, R., G.L. Michik and F. Naroll 1980. *Holocultural Research Methods*. Boston: Allyn & Bacon.

Neumann, S. and T. Widlok 1996. 'Rethinking Some Universals of Spatial Language Using Controlled Comparison', in *The Construal of Space in Language and Thought*, ed. R. Dirven and M. Pütz. Berlin: de Gruyter, pp. 345–69.

Neves-Graça, K. 2002. *'A Whale of a Thing': Transformations from Whale Hunting to Whale Watching in Lajes do Pico, Azores*. Doctoral Dissertation, York University, Toronto. Dissertation Abstracts.

_____. 2004. 'Revisiting the Tragedy of the Commons: Ecological Dilemmas of Whale Watching in the Azores', *Human Organization* 63(3): 289–300.

_____. 2005. *Chasing Whales with Bateson and Daniel*. Australian Humanities Review: Ecological Humanities Corner, 35.

_____. 2006. 'Politics of Environmentalism and Ecological Knowledge at the Intersection of Local and Global Processes', *Journal of Ecological Anthropology*, 10 (Spring).

Nihill, M. 2000. 'Gift Exchange as Sensory Experience Among the Anganen', *Oceania* 71: 110–28.

Nisbett, R.E., K. Peng, I. Choi and A. Norenzayan. 2001. 'Culture and Systems of Thought: Holistic versus Analytic Cognition', *Psychological Review* 108(2): 291–310.

Norenzayan, A., I. Choi, and R.E. Nisbett. 2002. 'Cultural Similarities and Differences in Social Inference: Evidence from Behavioral Predictions and Lay Theories of Behavior', *Personality and Social Psychology Bulletin* 28(1): 109–20.

Nuckolls, C. 1995. 'The Misplaced Legacy of Gregory Bateson: Toward a Cultural Dialectic of Knowledge and Desire', *Cultural Anthropology* 10(3): 367–94.

Nüse, R. 1996. 'Über die Realitätsadäquatheit der phänomenalen Welt oder: Things look as they do because they are what they are', *Gestalttheorie* 18: 52–67 and 86–114.

Obeyesekere, G. 1985. 'Depression, Buddhism and the Work of Culture in Sri Lanka', in *Culture and Depression*, ed. A. Kleinman and B. Good. Berkeley: University of California Press, pp. 134–52.

Ochs, E. and L. Capps. 1996. 'Narrating the Self', *Annual Reviews of Anthropology* 25: 19–43.

Olson, D. R. 1970. 'Language and Thought: Aspects of a Cognitive Theory of Semantics', *Psychological Review*, 77: 257–73.

Olwig, K.F. and K. Hastrup (eds). 1997. *Siting Culture. The Shifting Anthropological Object*. London: Routledge.

Oosterhout, D. 2001. 'The Scent of Sweat: Notions of Witchcraft and Morality in Inanwatan', in *Humors and Substances*, ed. A. Strathern und P. Stewart. Westport CT, London: Bergin & Garvey, pp. 23–50.

Otto, T. 1991. *The Politics of Tradition in Baluan: Social Change and the Construction of the Past in a Manus Society*. Nijmegen: Centre for Pacific Studies.

_____. 1992a. 'The Paliau Movement in Manus and the Objectification of Tradition', *History and Anthropology* 5: 427–54.

———. 1992b. 'The Ways of *Kastam*: Tradition as Category and Practice in a Manus Village', *Oceania* 62.

———. 2002. 'Chefs, big men et bureaucrates: Weber et les politiques de la tradition à Baluan (Papouasie-Nouvelle-Guinée)', in *La Tradition et l'État: Églises, Pouvoirs et Politiques Culturelles dans le Pacifique*, ed. C. Hamelin and E. Wittersheim. Paris: L'Harmattan, pp. 103–29.

Otto, T., and A. Borsboom (eds). 1997. *Cultural Dynamics of Religious Change in Oceania*. Leiden: KITLV Press.

Otto, T., and N. Thomas (eds). 1997. *Narratives of Nation in the South Pacific*. Amsterdam: Harwood Academic Publishers.

Otto, T., and P. Pedersen (eds). 2000. 'Anthropology and the Revival of Tradition: Between Cultural Continuity and Invention, Special Issue', *FOLK: Journal of the Danish Ethnographic Society* 42.

———. 2005. *Tradition and Agency: Tracing Cultural Continuity and Invention*. Aarhus: Aarhus University Press.

Otto, T., and R.J. Verloop. 1996. 'The Asaro Mudmen: Local Property, Public Culture?', *The Contemporary Pacific* 8: 349–86.

Oyserman, D., H.M. Coon, and M. Kemmelmeier 2002. 'Rethinking Individualism and Collectivism: Evaluation of Theoretical Assumptions and Meta-Analyses', *Psychological Bulletin*. 128(1): 3–72.

Pacific Islands Monthly, 'The Stephens Family of Urelapa', June 1948: 39.

Pandya, V. 1990. 'Movement and Space: Andamanese Cartography', *American Ethnologist* 17: 775–97.

Paul, R.A. 1990. 'What Does Anybody Want? Desire, Purpose, and the Acting Subject in the Study of Culture', *Cultural Anthropology* 5: 431–51.

Peacock, J. and D.Holland. 1993. 'The Narrated Self: Life Stories in Process', *Ethnos* 21(4): 367–83.

Pederson, E. 1993. 'Geographic and Manipulable Space in Two Tamil Linguistic Systems', in *Spatial Information Theory. A Theoretical Basis for GIS*, eds. A.U. Frank and I. Campari. Berlin: Springer, pp. 294–311.

———. 1995. 'Language as Context, Language as Means: Spatial Cognition and Habitual Language Use', *Cognitive Linguistics* 6(1): 33–62.

Pederson, E. and B. Schmitt. 1993. 'Eric's Maze Task', in *Cognition and Space – Kit of the Cognitive Anthropology Research Group*, ed. E. Danziger. Nijmegen: Mimeo, pp. 71–76.

Pederson, E., E. Danziger, D. Wilkins, S.C. Levinson, S. Kita, and G. Senft. 1998. 'Semantic Typology and Spatial Conceptualisation', *Language* 74: 557–89.

Pelly, David F. 1991. 'How Inuit Find their Way in the Trackless Arctic', *Canadian Geographic* Aug./Sept.: 58–64.

Petit, P. 2005. '"Be proud to be Bwile: It is your tribe": Ethnicity, Political Jubilees and Traditions of Origins among the Bwile of Zambia', in *Tradition and Agency: Tracing Cultural Continuity and Invention*, ed. T. Otto and P. Pedersen. Aarhus: Aarhus University Press, pp. 51–82.

Piaget, J. and B. Inhelder. 1967. *The Child's Conception of Space*. New York: Norton (1947).

Pinker, S. 1994. *The Language Instinct*. New York: Harper Collins.

Poewe, K. 1996. 'Writing Culture and Writing Fieldwork: The Proliferation of Experimental and Experiential Ethnographies', *Ethnos* 61: 177–206.

Polt, R. 1999. *Heidegger: An Introduction*. Ithaca: Cornell University Press.

Pomponio, A. 1992. *Seagulls Don't Fly into the Bush. Cultural Identity and Development in Melanesia*. Belmont: Wadsworth.

Poortinga, Y.H. 1983. 'Psychometric Approaches to Intergroup Comparison: The Problem of Equivalence', in *Human Assessment and Cultural Factors*, ed. S.H. Irvine and J.W. Berry. New York: Plenum Press, pp. 237–57.

_____. 1997. 'Towards Convergence?', in *Handbook of Cross-Cultural Psychology*, 2nd edition, Volume 1: *Theory and Method*, ed. J.W. Berry, Y.H. Poortinga and J. Pandey. Boston: Allyn & Bacon, pp. 347–87.

Porteus, S.D. 1937. *Primitive Intelligence and Environment*. New York: Macmillan.

Pourtois, J.-P. and H. Desmet. 1997. *Epistémologie et Instrumentation en Sciences Humaines*, 2nd edition. Liège, Bruxelles: Pierre Mardaga.

Prasetyo, E.D. 1998. *The White Banyan*. Yogyakarta: Self-published.

Prior, L. 2004. 'Following in Foucault's Footsteps: Text and Context in Qualitative Research', in *Approaches to Qualitative Research: a Reader on Theory and Practice*, ed. S. Hesse-Biber and P. Leavy. New York, Oxford: Oxford University Press, pp. 317–33.

Puff, C.R. (ed.). 1982. *Handbook of Research Methods in Human Memory and Cognition*. New York: Academic Press.

Quinn, N. 1992. 'The Motivational Force of Self-Understanding', in *Human Motives and Cultural Models*, ed. R.D. D'Andrade and C. Strauss. Cambridge: Cambridge University Press, pp. 90–126.

Ranger, T. 1983. 'The Invention of Tradition in Colonial Africa', in *The Invention of Tradition*, ed. E. Hobsbawm and T. Ranger. Cambridge: Cambridge University Press, pp. 211–62.

_____. 1993. 'The Invention of Tradition Revisited: The Case of Colonial Africa', in *Legitimacy and the State in Twentieth-Century Africa*, ed. T. Ranger and O. Vaughan. London: Macmillan Press, pp. 62–111.

_____. 1999. 'Concluding Comments', in *Ethnicity and Nationalism in Africa: Constructivist Reflections and Contemporary Politics*, ed. P. Yeros. London: Macmillan, pp. 133–44.

_____. 2005. 'Postscript', in *Tradition and Agency: Tracing Cultural Continuity and Invention*, ed. T.Otto and P. Pedersen. Aarhus: Aarhus University Press, pp. 327–38.

Rappaport, R. 1979. *Ecology, Meaning and Religion*. Richmond, CA: North Atlantic Books.

_____. 1999. *Ritual and Religion in the Making of Humanity*. Cambridge: Cambridge University Press.

Rasmussen, S. 1999. 'Making Better "Scents" in Anthropology: Aroma in Tuareg Sociocultural Systems and the Shaping of Ethnography', *Anthropological Quarterly* 72: 55–73.

Rawson, N.E. 2000. 'Human Olfaction', in *The Neurobiology of Taste and Smell*, ed. T. E. Finger, W.L. Silver and D. Restrepo. New York, Chichester, Weinheim: Wiley-Liss, pp. 257–84.

Reber, A.S. 1967. 'Implicit Learning of Artificial Grammars', *Journal of Verbal Learning & Verbal Behavior* 6: 855–63.

_____. 1989. 'Implicit Learning and Tacit Knowledge', *Journal of Experimental Psychology: General* 118: 219–35.

_____. 1992. 'The Cognitive Unconscious: An Evolutionary Perspective', *Consciousness and Cognition* 1: 93–133.

Reichel-Dolmatoff, G. 1971. *The Sexual and Religious Symbolism of Tukano Indians*. London: University of Chicago Press.

_____. 1978. *Hallucinatory Imagery of the Tukano Indians*. Los Angeles: UCLA Latin America Center Publ.

. 1985a. *Basketry as Metaphor: Arts and Crafts of the Desana Indians of the Northwest Amazon*. Los Angeles: Occasional Papers of the Museum of Cultural History.

_____. 1985b. 'Tapir Avoidance in the Colombian Northwest Amazon', in *Animal Myths and Metaphors in South America*, ed. G. Urton. Salt Lake City: University of Utah Press, pp. 107–43.

_____. 1989. *Desana Texts and Contexts*. (*Acta Ethnologica et Linguistica*, Nr. 62). Wien: Föhrenau.

Reimann, R. 1998. *Der Schamane sieht eine Hexe – der Ethnologe sieht nichts. Menschliche Informationsverarbeitung und ethnologische Forschung*. Frankfurt: Campus.

Rensel, J. and M. Rodman. (eds). 1997. *Home in the Islands. Housing and Social Change in the Pacific*. Honolulu: University of Hawai'i Press.

Reuning, H. and W. Wortley. 1973. *Psychological Studies of the Bushmen. Psychologia Africana Monograph Supplement* 7. Johannesburg: National Institute for Personnel Research.

Reyna, S.P. 2002. *Connections. Brain, Mind, and Culture in a Social Anthropology*. London, New York: Routledge.

Richardson, J.T.E. and G.M. Zucco. 1989. 'Cognition and Olfaction: A Review', *Psychological Bulletin* 105: 352–60.

Richardson-Klavehn, A. and R.A. Bjork. 1988. 'Measures of Memory', *Annual Review of Psychology* 39: 475–543.

Riebe, I. 1987. 'Kalam Witchcraft: A Historical Perspective', in *Sorcerer and Witch in Melanesia*, ed. Michele Stephen. New Brunswick, NJ: Rutgers University Press, pp. 211–45.

Riefer, D.M. and W.H. Batchelder. 1995. 'A Multinomial Modeling Analysis of the Recognition-Failure Paradigm', *Memory & Cognition* 23: 611–30.

Rivers, W.H.R. 1914. *Kinship and Social Organization*. London: Constable.

Roberson, D., I. Davies, and J. Davidoff. 2000. 'Color Categories Are Not Universal: Replications and New Evidence from a Stone-Age Culture', *Journal of Experimental Psychology: General* 129: 369–98.

Rodman, M.C. 1992.'Empowering Place: Multilocality and Multivocality', *American Anthropologist* 94: 640–56.

———. 2001 *Houses Far From Home: British Colonial Space in the New Hebrides*. Honolulu: University of Hawai'i Press

———. 2003. 'The Heart in the Archives: Colonial Contestation of Desire and Fear in the New Hebrides, 1933', *The Journal of Pacific History* 38(3): 291–312.

———. 2004. 'Traveling Stories, Colonial Intimacies, and Women's Histories in Vanuatu', *The Contemporary Pacific* 16(2): 233–257.

Roediger, H.L. 1990. 'Implicit Memory: Retention without Remembering', *American Psychologist* 45: 1043–56.

Roepstorff, A. and A. Simoniukstyte. 2005. 'Cherishing the Nation's Time and Space: Lithuanian Identity and the Maintenance of Tradition', in *Tradition and Agency: Tracing Cultural Continuity and Invention*, ed. T. Otto and P. Pedersen. Aarhus: Aarhus University Press, pp. 157–92.

Rosaldo, M. Z. 1980. *Knowledge and Passion: Ilongot Notions of Self and Social Life*. Cambridge: Cambridge University Press.

———. 1984. 'Toward an Anthropology of Self and Feeling', in *Culture Theory: Essays on Mind, Self, and Emotions*, ed. R.A. Schweder and R.A. Levine. Cambridge: Cambridge University Press, pp. 137–57.

Rosaldo, R. 1980. *Ilongot Headhunting 1883–1974: A Study in Society and History*. Stanford: Stanford University Press.

———. 1989. 'Grief and a Headhunter's Rage', in *Culture and Truth: The Remaking of Social Analysis*. Boston: Beacon, pp. 1–21.

Rosch Heider, E.H. 1972. 'Universals in Color Naming and Memory', *Journal of Experimental Psychology* 93: 10–20.

Rosch, E.H. 1973. 'Natural Categories', *Cognitive Psychology* 4: 328–50.

Rosenwald, G.C. and R.L. Ochberg (eds). 1992. *Storied Lives: The Cultural Politics of Self-Understanding*. New Haven: Yale University Press.

Ross, L. 1977. 'The Intuitive Psychologist and His Shortcomings', in *Advances in Experimental Social Psychology*, vol. 10, ed. L. Berkowitz. New York: Academic Press, pp. 173–220.

Rothbaum, F. and G. Trommsdorff. 2007. 'Do Roots and Wings Complement or Oppose One Another? The Socialization of Relatedness and Autonomy in Cultural Context', in *The Handbook of Socialization*, ed. J.E. Grusec and P. Hastings. New York: Guilford Press, pp. 461–489.

Rothbaum, F., J.R. Weisz and S.S. Snyder. 1982. 'Changing the World and Changing the Self: A Two-Process Model of Perceived Control', *Journal of Personality and Social Psychology* 42(1): 5–37.

Rothbaum, F., J.R. Weisz, M. Pott, K. Miyake, and G. Morelli. 2000. 'Attachment and Culture: Security in the United States and Japan', *American Psychologist* 55, (10): 1093–1104.

Rotter, J.B. 1966. 'Generalized Expectancies for Internal Versus External Control of Reinforcement', *Psychological Monographs* 80(1): 1–28.

Rouby, C., B. Schaal, D. Dubois, et al. (eds). 2002. *Olfaction, Taste, and Cognition*. Cambridge: Cambridge University Press.

Rumelhart, D.E. and J.L. McClelland (eds) 1986. *Parallel Distributed Processing. Explorations in the Microstructure of Cognition*. Cambridge, MA: MIT Press.

Rumsey, A. and J.F. Weiner. (eds). 2001. *Emplaced Myth: Space, Narrative, and Knowledge in Aboriginal Australia and Papua New Guinea.* Honolulu: University of Hawai'i Press.

Rutherford, D. 1996. 'Of Birds and Gifts: Reviving Tradition on an Indonesian Frontier', *Cultural Anthropology* 11: 577–616.

Rutter, M. and J. Silberg. 2002. 'Gene-Environment Interplay in Relation to Emotional and Behavioral Disturbance', *Annual Review of Psychology* 53(1): 463–90.

Ryan, R.M. and E.L. Deci. 2000. 'Self-Determination Theory and the Facilitation of Intrinsic Motivation, Social Development, and Well-Being', *American Psychologist* 55(1): 68–78.

Ryle, G. 1949. *The Concept of Mind.* London: Hutchinson & Company.

Saegert, J., E. Hayaman, and H. Ahmar. H. 1975. 'Memory for Language of Input in Polyglots', *Journal of Experimental Psychology: Human Learning and Memory* 5: 607–13.

Sahlins, M. 1985. *Islands of History.* Chicago: University of Chicago Press.

_____. 1999. 'Two or Three Things That I Know About Culture', *Journal of the Royal Anthropological Institute* (N.S.) 5: 399–421.

Said, E. 1979. *Orientalism.* New York: Vintage.

Sampson, E.E. 1988. 'The Debate on Individualism: Indigenous Psychologies of the Individual and Their Role in Personal and Societal Functioning', *American Psychologist* 43(1): 15–22.

Saussure, F. de 1916. *Cours de Linguistique Générale.* Paris: Payot.

Scarduelli, P. 2005. 'Padanian Identity: Ethnogenesis as a Political Strategy', in *Tradition and Agency: Tracing Cultural Continuity and Invention*, ed. T. Otto and P. Pedersen. Aarhus: Aarhus University Press, pp. 137–55.

Schaal, B., R. Soussignan, and L. Marlier. 2002. 'Olfactory Cognition at the Start of Life: The Perinatal Shaping of Selective Odor Responsiveness', in *Olfaction, Taste, and Cognition*, ed. C. Rouby, B. Schaal, D. Dubois et al. Cambridge: Cambridge University Press, pp. 421–40.

Schacter, D.L. 1987. 'Implicit Memory: History and Current Status', *Journal of Experimental Psychology: Learning, Memory, and Cognition* 13: 501–18.

_____. 1992. 'Understanding Implicit Memory. A Cognitive Neuroscience Approach', *American Psychologist* 47: 559–69.

_____. 1999. 'The Seven Sins of Memory: Insights from Psychology and Cognitive Neuroscience', *American Psychologist* 54: 182–203.

_____. 2001. *The Seven Sins of Memory. How the Mind Forgets and Remembers.* New York: Houghton Mifflin.

Schank, R. and R. Abelson, 1977. *Scripts, Plans, Goals and Understanding: An Inquiry into Human Knowledge Structures.* Hillsdale, N.J.: Lawrence Erlbaum Associates.

Schensul, S., J. Schensul and M. LeCompte. 1999. *Essential Ethnographic Methods: Observations, Interviews, and Questionnaires* (Ethnographer's Toolkit, Vol. 2). Walnut Creek: AltaMira Press.

Schlehe, J. 1998. *Die Meeresgöttin des Südens, Ratu Kidul. Geisterpolitik im javanischen Alltag.* Berlin: Dietrich Reimer Verlag.

Schleidt, M. 1987. 'Was riecht man rund um die Welt? Humanethologische Aspekte des Geruchs', *dragoco report* 5: 135–49.

Schleidt, M., P. Neumann, and H. Morishita 1988. 'Pleasure and Disgust: Memories and Associations of Pleasant and Unpleasant Odours in Germany and Japan', *Chemical Senses* 13: 279–93.

Schrauf, R.W. 2000. 'Bilingual Autobiographical Memory: Experimental Studies and Clinical Cases', *Culture and Psychology* 6: 387–417.

Schrauf, R.W. and D.C. Rubin 1998. 'Bilingual Autobiographical Memory in Older Adult Immigrants: A Test of Cognitive Explanations of the Reminiscence Bump and the Linguistic Encoding of Memories, *Journal of Memory and Language* 39: 437–57.

———. 2000. 'Internal Languages of Retrieval: The Bilingual Encoding of Memories for the Personal Past', *Memory and Cognition* 28: 616–23.

Schulz, R. and J. Heckhausen. 1999. 'Aging, Culture and Control: Setting a New Research Agenda', *Journal of Gerontology: Psychological Sciences* 54B(3): 139–45.

Schwarz, B., E. Schäfermeier, and G. Trommsdorff 2005. 'The Relationships Between Value Orientation, Child-Rearing Goals, and Child-Rearing Behavior: A Comparison of Korean and German Mothers', in *Culture and Human Development: The Importance of Cross-Cultural Research to the Social Sciences*, ed. W. Friedlmeier, P. Chakkarath and B. Schwarz. Hove: Psychology Press, pp. 203–30.

Schwartz, S.H. 1994. 'Beyond Individualism/Collectivism: New Cultural Dimensions of Values', in *Individualism and Collectivism: Theory, Methods, and Applications*, ed. U. Kim, H.C. Triandis, C. Kagitcibasi, S.-C. Choi and G. Yoon. Thousand Oaks: Sage Publications, pp. 85–119.

Schwarzkopf, L. and H.-U. Schminke. 2000. 'Die Vulkanwelt Indonesiens', *Geographische Rundschau*, 4: 4–12.

Seeger, A. 1981. *Nature and Society in Central Brazil: The Suya Indians of Mato Grosso*. Cambridge, MA: Harvard University Press.

———. 1988. 'Anthropology and Odor: From Manhattan to Mato Grosso', *Perfumer & Flavorist* 13: 41–48.

Segall, M.H. 1989. 'Le Système HRAF (Human Relations Area Files) au Service de la Psychologie Interculturelle', in *La Recherche Interculturelle*, Vol. 1, ed. J. Retschitzki, M. Bossel-Lagos and P.R. Dasen. Paris: L'Harmattan, pp. 271–79.

Segall, M.H., P.R. Dasen, J.W. Berry,and Y.H. Poortinga 1999. *Human Behavior in Global Perspective: An Introduction to Cross-Cultural Psychology*, revised 2nd edition. Boston: Allyn & Bacon.

Sekuler, R. and R. Blake. 1990. *Perception*. New York, St. Louis, San Francisco: McGraw-Hill.

Senft, G. 1994a. 'Ein Vorschlag, wie man standardisiert Daten zum Thema "Sprache, Kognition und Konzepte des Raumes" in verschiedenen Kulturen erheben kann', *Linguistische Berichte* 154: 413–29.

———. 1994b. 'Spatial Reference in Kilivila: The Tinkertoy Matching Games – A Case Study', *Language and Linguistics in Melanesia* 25: 55–93.

_____. 1995. 'Elicitation', in *Handbook of Pragmatics, Manual*, ed. J. Blommaert, J-O. Östman and J. Verschueren. Amsterdam: John Benjamins, pp. 577–81.

_____. 1997. 'Introduction', in *Referring to Space – Studies in Austronesian and Papuan Languages*, ed. G. Senft. Oxford: Clarendon Press, pp. 1–38.

_____. 1999. 'ENTER and EXIT in Kilivila', *Studies in Language* 23:1–23.

_____. 2000. 'COME and GO in Kilivila', in SICOL. *Proceedings of the Second International Conference on Oceanic Linguistics: Vol. 2, Historical and Descriptive Studies*, eds. B. Palmer and P. Geraghty. Canberra: Pacific Linguistics, pp. 105–36.

_____. 2001. 'Frames of Spatial Reference in Kilivila', *Studies in Language* 25: 521–55.

_____. 2003. 'Ethnographic Methods', in *Handbook for Psycholinguistics*, ed. G. Rickheit, T. Hermann and W. Deutsch. Berlin: Walter de Gruyter, pp. 106–14.

Serpell, R. 1979. 'How Specific Are Perceptual Skills', *British Journal of Psychology*, 70: 365–80.

Sethi, A., W. Mischel, J.L. Aber, Y. Shoda and M.L. Rodriguez 2000. 'The Role of Strategic Attention Deployment in Development of Self-Regulation: Predicting Preschoolers' Delay of Gratification from Mother-Toddler Interactions', *Developmental Psychology* 36(6): 776–7.

Shepard, R. N. and J. Metzler, 1971. 'Mental Rotation of Three-Dimensional Objects', *Science*, 171: 701–03.

Shils, E. 1971. 'Tradition', *Comparative Studies in Society and History* 13: 122–59.

Shore, B. 1991. 'Twice-Born, Once Conceived: Meaning Construction and Cultural Cognition', *American Anthropologist* 93: 9–27.

_____. 1996. *Culture in Mind: Cognition, Culture, and the Problem of Meaning*. New York: Oxford University Press.

Shotter, J. 1975. *Image of Man in Psychological Research*. London: Methuen.

Shweder, R.A. 1982. 'Beyond Self-Constructed Knowledge: The Study of Culture and Morality', *Merrill-Palmer Quarterly* 28(1): 41–69.

_____. 1991. *Thinking Through Cultures: Expeditions in Cultural Psychology*. Cambridge, M.A.: Harvard University Press.

Shweder, R.A. and E.J. Bourne 1984. 'Does the Concept of the Person Vary Cross-Culturally?', in *Culture Theory*, ed. R.A. Shweder and R.A. LeVine. Cambridge: Cambridge University Press, pp. 158–99.

Sillitoe, P. 1983. *Roots of the Earth: the Cultivation and Classifications of Crops in the Papua New Guinea Highlands*, Manchester: Manchester University Press.

_____. 1996. *A Place Against Time: Land and Environment in the Papua New Guinea Highlands*. Amsterdam: Harwood.

_____. 2003. *Managing Animals in New Guinea: Preying the Game in the Highlands*. London and New York: Routledge.

Sillitoe, P., A. Bicker and J. Pottier (eds). 2002. *Participating in Development: Approaches to Indigenous Knowledge*. London and New York: Routledge.

Silverman, M. 2001. *An Irish Working Class: Explorations in Political Economy and Hegemony*, 1800–1950. Toronto: Toronto University Press.

Simpson, C. 1962. *Plumes and Arrows*. Sydney: Angus and Robertson.

Sissons, J. 1997. 'Nation or Desti-Nation? Cook Islands Nationalism since 1965', in *Narratives of Nation in the South Pacific*. eds. T. Otto and N. Thomas. Amsterdam: Harwood Academic Publishers, pp. 163–88.

Skinner, E.A. 1996. 'A Guide to Constructs of Control', *Journal of Personality & Social Psychology* 71(3): 549–70.

Small, C. 1997. *Voyages. From Tongan Villages to American Suburbs*. Ithaca: Cornell University Press.

Smith, P.B., F. Trompenaars and S. Dugan 1995. 'The Rotter Locus of Control Scale in 43 Countries: A Test of Cultural Relativity', *International Journal of Psychology* 30(3): 377–400.

Snodgrass, J.G. and J. Corwin 1988. 'Pragmatics in Measuring Recognition Memory: Applications to Dementia and Amnesia', *Journal of Experimental Psychology: General* 117: 34–50.

Souter, G. 1963. *New Guinea. The Last Unknown*. New York: Taplinger Publishing Co.

Spector, P.E., C.L. Cooper, J.I. Sanchez, M. O'Driscoll, K. Sparks, P. Bernin, A. Büssing, P. Dewe, P. Hart, L. Lu, K. Miller, L.R. De Morales, G.M. Ostrognay, M. Pagon, H.D. Pitariu, S.A. Poelmans, P. Radhakrishnan, V. Russinova, V. Salamatov, J.F. Salgado, S. Shima, O.-L. Siu, J.B. Stora, M. Teichmann, T. Theorell, P. Vlerick, M. Westman, M. Widerszal-Bazyl, P.T.P. Wong and S. Yu 2002. 'Locus of Control and Well-Being at Work: How Generalizable Are Western Findings?', *Academy of Management Journal* 45(2): 453–66.

Spickard, P.R., J. Rondilla and D.H. Wright (eds). 2002. *Pacific Diaspora. Island Peoples in the United States and across the Pacific*. Honolulu: University of Hawai'i Press.

Spivak, G. 1988. *In Other Words: Essays in Cultural Politics*. New York, London: Routledge.

Squire, L. 1992a. 'Declarative and Nondeclarative Memory', *Journal of Cognitive Neuroscience* 4(3): 232–43.

———. 1992b. 'The Neuropsychology of Human Memory', *Annual Review of Neuroscience* 5: 241–73.

Squire, L. and E.R. Kandel 2000. *Memory. From Mind to Molecules*. New York: Scientific American Library.

Stadler, M.A. and P.A. Frensch (eds). 1998. *Handbook of Implicit Learning*. Thousand Oaks, CA: Sage Publications.

Stange, E.P. 1986. '"Legitimate" Mysticism in Indonesia', *Review of Indonesian and Malaysian Affairs* 20(2): 76–117.

Stasch, R. 2002. 'Joking Avoidance. A Korowai Pragmatics of Being Two', *American Anthropologist* 29(2): 335–65.

Steiner, J. 1974. 'Innate Discriminative Human Facial Expressions to Taste and Smell Stimulation', in *Odors: Evaluation, Utilization and Control*, ed. W. Cain. New York: New York Academy of Sciences, pp. 229–33.

Steinmetz, G. (ed.). 2005. *The Politics of Method in the Human Sciences: Positivism and its Epistemological Others*. Durham, NC and London: Duke University Press.

Stewart, P.J. and A. Strathern, 1999. '"Feasting on My Enemy". Images of Violence and Change in the New Guinea Highlands', *Ethnohistory* 46(4): 645–69.

_____. 2000. 'Naming Places: Duna Evocations of Landscape in Papua New Guinea', *People and Culture in Oceania* 16: 87–107.

_____. 2002a. 'Transformations of Monetary Symbols in the Highlands of Papua New Guinea', for a special issue of the journal *L'Homme* on money (Questions de Monnaie) 162 April/June: 137–56.

_____. 2002b. *Gender, Song, and Sensibility: Folktales and Folksongs in the Highlands of New Guinea*. Westport, CT and London: Praeger Publishers (Greenwood Publishing).

_____. 2004. *Witchcraft, Sorcery, Rumors, and Gossip*, for New Departures in Anthropology Series. Cambridge: Cambridge University Press.

_____. 2005. 'Body and Mind on the Move: Emplacement, Displacement, and Trans-Placement in Highlands Papua New Guinea', *Asia-Pacific Forum* (Academia Sinica) 27: 205–17.

_____. (eds). 2000. *Identity Work. Constructing Pacific Lives*. Pittsburgh: University of Pittsburgh Press.

_____. (eds). 2003. *Landscape, Memory, and History: Anthropological Perspectives*, for Anthropology, Culture, and Society Series. London and New York: Pluto Press.

Stewart, S. M., M.H. Bond, O. Deeds and S.F. Chung 1999. 'Inter-generational Patterns of Values and Autonomy Expectations in Cultures of Relatedness and Separateness', *Journal of Cross-Cultural Psychology* 30(5): 575–93.

Stichweh, R. 2000. 'Raum, Region und Stadt in der Systemtheorie', in his *Die Weltgesellschaft*. Frankfurt am Main: Suhrkamp Verlag, pp. 184–206.

Strassoldo, R. 1980. 'Centre-Periphery and System Boundary: Culturological Perspectives', in *Centre and Periphery. Spatial Variation in Politics*, ed. J. Gottmann. Beverly Hills and London: Sage, pp. 27–61.

Strathern, A. 1972. *One Father, One Blood: Descent and Group Structure among the Melpa People*. Canberra: Australian National University Press.

_____. 1979. *Ongka. A Self-Account by a New Guinea Big-Man*. Transcribed and translated by Andrew Strathern. London: Duckworth.

_____. 1996. *Body Thoughts*. Ann Arbor: University of Michigan Press.

Strathern, A. and P.J. Stewart (eds). 1998. *Kuk Heritage: Issues and Debates in Papua New Guinea*. The National Museum of PNG and the James Cook University, Centre for Pacific Studies and the Okari Research Group, Department of Anthropology, University of Pittsburgh.

_____. 1999. 'Objects, Relationships, and Meanings: Historical Switches in Currencies in Mount Hagen, Papua New Guinea', in *Money and Modernity: State and Local Currencies in Melanesia*, ed. D. Akin and J. Robbins. ASAO

(Association for Social Anthropology in Oceania) Monograph Series No. 17. Pittsburgh: University of Pittsburgh Press, pp. 164–91.

———. 2000a. *Collaborations and Conflicts. A Leader Through Time*. Fort Worth Texas: Harcourt Brace College Publishers.

———. 2000b. *Stories, Strength and Self-Narration. Western Highlands, Papua New Guinea*. Adelaide, Australia: Crawford House Publishing.

———. 2000c. *Arrow Talk: Transaction, Transition, and Contradiction in New Guinea Highlands History*. Kent, Ohio and London: Kent State University Press.

———. 200d. *Identity Work. Constructing Pacific Lives. Pittsburgh: University of Pittsburgh* Press.

———. 2001. *Minorities and Memories: Survivals and Extinctions in Scotland and Western Europe*. Durham, NC: Carolina Academic Press.

———. 2004. 'Transnationalism, Globalization, and Diaspora: How Can We Deal With These Concepts?', Public lecture IV, presented in the Contemporary Anthropology Lecture Series held at Peking University, China, in October 2004.

———. (n.d.). 'Hagen Settlement Histories: Dispersals and Consolidations', in *Nine Thousand Years of Gardening: Kuk and the Archaeology of Agriculture in Papua New Guinea*, ed. J. Golson and P. Swadling, forthcoming.

Strathern, M. 1988. *The Gender of the Gift. Problems with Women and Problems with Society in Melanesia*. Berkeley: University of California Press.

———. 2004. *Partial Connections. Updated Edition*. Walnut Creek: AltaMira Press.

Strauss, C. 1992. 'What Makes Tony Run? Schemas as Motives Reconsidered', in *Human Motives and Cultural Models*, ed. R.D. D'Andrade and C. Strauss. Cambridge: Cambridge University Press, pp. 191–224.

———. 2005. 'Analyzing Discourse for Cultural Complexity', in *Finding Culture in Talk: A Collection of Methods*, ed. N. Quinn. New York: Palgrave Macmillan, pp. 203–42.

Strauss, C. and N. Quinn 1997. *A Cognitive Theory of Cultural Meaning*. Cambridge: Cambridge University Press.

Stürzenhofecker, G. 1998. *Times Enmeshed. Gender, Space and History among the Duna of Papua New Guinea*. Stanford: Stanford University Press.

Super, C.M. and S. Harkness, 1993. 'The Developmental Niche: A Conceptualization at the Interface of Child and Culture', in *Life Span Development: A Diversity Reader*, ed. R.A. Pierce and M.A. Black. Dubuque: Kendall/Hunt Publishing Co., pp. 61–77.

———. 1997. 'The Cultural Structuring of Child Development', in *Handbook of Cross-Cultural Psychology*, 2nd edition, Vol. 2: *Basic Processes and Human Development*, ed. J.W. Berry, P.R. Dasen and T.S. Saraswathi. Boston: Allyn & Bacon, pp. 1–39.

Sutton, D.E. 2001. *Remembrance of Repasts. An Anthropology of Food and Memory*. Oxford, New York: Berg.

Swets, J. A., R.M. Dawes and J. Monahan 2000. 'Psychological Science Can Improve Diagnostic Decision', *Psychological Science in the Public Interest* 1: 1–26.

Synnott, A. 1991. 'A Sociology of Smell', *The Canadian Review of Sociology and Anthropology* 28: 437–59.

_____. 1993. *The Body Social. Symbolism, Self and Society*. London, New York: Routledge.

Sztompka, P. 1999. *Trust: A Sociological Theory*. Cambridge: Cambridge University Press.

_____. 2001. 'Trust: Cultural Concerns', *International Encyclopedia of the Social & Behavioral Sciences*, ed. N. Smelser and P.B. Baltes. Amsterdam et al: Elsevier, pp. 15913–15917.

Talmy, L. 1978. 'Figure and Ground in Complex Sentences', in *Universals of Human Language*, Vol. IV: *Syntax*, ed. J.H. Greenberg. Stanford: Stanford University Press, pp. 625–49.

Telban, B. 1999. *Dancing Through Time. A Sepik Cosmology*. Oxford: Oxford University Press.

Thomas, N. 1992. 'The Inversion of Tradition', *American Ethnologist* 19: 213–32.

Tiger, L. and R. Fox. 1974. *The Imperial Animal*. London: Paladin.

Tomasello, M., M. Carpenter, J. Call, T. Behne and H. Moll 2005. 'Understanding and Sharing Intentions: The Origins of Cultural Cognition', *Behavioral and Brain Sciences* 28(5): 675–735.

Tonkinson, R. 1982. 'Kastom in Melanesia: Introduction', *Mankind* 13: 302–5.

_____. 1993. 'Understanding "Tradition" – Ten Years On', *Anthropological Forum* 6: 597–606.

_____. 1997. 'Anthropology and Aboriginal Tradition: The Hindmarsh Island Bridge Affair and the Politics of Interpretation', *Oceania* 68: 1–26.

_____. 2000. '"Tradition" in Oceania and Its Relevance in a Fourth World Context (Australia)', FOLK 42: 169–95.

Townley, B. 1993. 'Foucault, Power/Knowledge, and Its Relevance for Human Resource Management', *Academy of Management Review* 18: 518–45.

Trask, H.-K. 1991. 'Natives and Anthropologists: The Colonial Struggle', *The Contemporary Pacific* 3: 159–67.

Triandis, H.C. 1995. *Individualism and Collectivism*. Boulder: Westview Press.

Triandis, H.C. and E.M. Suh 2002. 'Cultural Influences on Personality', *Annual Review of Psychology* 53(1): 133–60.

Troadec, B., C. Martinot and P. Cottereau-Reiss 2002. 'A Cross-Cultural Study of Diversity in Cognitive Development: Categorization and Space', in *New Directions of Cross-Cultural Psychology*, ed. P. Boski, F. Van de Vijver and M. Chodynicka. Warsaw: Polish Academy of Sciences, pp. 243–58.

Trommsdorff, G. 1995. 'Parent-Adolescent Relations in Changing Societies: A Cross-Cultural Study', in *Psychological Responses to Social Change: Human Development in Changing Environments*, ed. P. Noack, M. Hofer and J. Youniss, Berlin: De Gruyter, pp. 189–218.

_____. 2006. 'Parent-Child Relations Over the Life Span. A Cross-Cultural Perspective', in *Parenting Beliefs, Behaviours, and Parent-Child Relations. A Cross-Cultural Perspective*, ed. K.H. Rubin and O.B. Chung. New York: Psychology Press, pp. 143–83.

———. 2007. 'Entwicklung im kulturellen Kontext [Development in cultural context]', in *Enzyklopädie der Psychologie: Themenbereich C. Theorie und Forschung: Serie VII. Kulturvergleichende Psychologie: Band 2. Kulturelle Determinanten des Erlebens und Verhaltens* [Encyclopaedia of Psychology: Subject Area C. Theory and Research: Serial VII. Cross-Cultural Psychology: Vol. 2. Cultural Determinants of Experience and Behavior], ed. G. Trommsdorff and H.-J. Kornadt, Göttingen: Hogrefe, pp. 435–519.

———. in press. 'Intergenerational Relations and Cultural Transmission', in *Perspectives on Cultural Transmission*, ed. U. Schönpflug. Oxford: Oxford University Press.

Trommsdorff, G. and P.R. Dasen 2001. 'Cross-Cultural Study of Education', in *International Encyclopedia of the Social and Behavioral Sciences*, ed. N.J. Smelser and P.B. Baltes. Oxford: Elsevier, pp. 3003–7.

Trommsdorff, G. and C.A. Essau 1998. 'Japanese and German Adolescents' Control Orientation: A Cross-Cultural Study', in *Japan in Transition: Social and Psychological Aspects*, ed. G. Trommsdorff, W. Friedlmeier and H.-J. Kornadt. Lengerich: Pabst, pp. 198–211.

Trommsdorff, G. and W. Friedlmeier 1999. 'Motivational Conflict and Prosocial Behavior of Kindergarten Children', *International Journal of Behavioral Development* 23(2): 413–29.

———. 2004. 'Zum Verhältnis von Kultur und Individuum aus der Perspektive der kulturvergleichenden Psychologie [The Relation Between Culture and the Individual in Cross-Cultural Perspective]', in *Positionen der Kulturanthropologie* [Positions of Cultural Anthropology], ed. A. Assmann, U. Gaier and G. Trommsdorff. Frankfurt am Main: Suhrkamp, pp. 358–86.

Trommsdorff, G. and H.-J. Kornadt 2003. 'Parent-Child Relations in Cross-Cultural Perspective', in *Handbook of Dynamics in Parent-Child Relations*, ed. L. Kuczynski. London: Sage Publications, pp. 271–306.

Trommsdorff, G. and B. Mayer 2005. 'Kulturvergleichende Ansätze [Cross-Cultural Approaches]', in *Handbuch der Persönlichkeitspsychologie und Differentiellen Psychologie. Handbuch der Psychologie* [Psychology of Personality and Individual Differences. Handbook of Psychology], ed. H. Weber and T. Rammsayer. Göttingen: Hogrefe, pp. 220–28.

Trommsdorff, G. and B. Nauck (eds). 2005. *The Value of Children in Cross-Cultural Perspective*. Case studies from Eight Societies. Lengerich: Pabst Science Publishers.

Trommsdorff, G. and F. Rothbaum in press. 'Development of Emotion Regulation in Cultural Context', in *Regulating Emotions: Social Necessity and Biological Inheritance*, ed. S. Ismer, S. Jung, S. Kronast, C. v. Scheve and M. Vandekerckhove. London/New York: Blackwell.

Trommsdorff, G., W. Friedlmeier and B. Mayer in press. 'Sympathy, Distress, and Prosocial Behavior of Preschool Children in Four Cultures', *International Journal of Behavioural Development*.

Trommsdorff, G., B. Mayer and I. Albert 2004. 'Dimensions of Culture in Intra-Cultural Comparisons: Individualism/Collectivism and Family-

Related Values in Three Generations', in *Comparing Cultures: Dimensions of Culture in a Comparative Perspective*, ed. H. Vinken, J. Soeters, and P. Ester. Leiden: Brill Academic Press, pp. 157–84.

Tsing, A.L. 2005. *Friction: An Ethnography of Global Connection*. Princeton, Oxford: Princeton University Press.

Tuhiwai Smith, L. 1999. *Decolonizing Methodologies. Research and Indigenous Peoples*. London, New York: ZED Books.

Tulving, E. 1972. 'Episodic and Semantic Memory', in *Organisation of Memory*, ed. E. Tulving and W. Donaldson. New York: Academic Press, pp. 381–403.

______. 1985. 'How Many Memory Systems Are There?', *American Psychologist* 40: 385–98.

Tulving, E. and J. Psotka 1971. 'Retroactive Inhibition in Free Recall: Inaccessibility of Information Available in the Memory Store', *Journal of Experimental Psychology* 87: 1–8.

Tulving, E. and D.L. Schacter. 1990. 'Priming and Human Memory Systems', *Science* 247: 301–306.

Tulving, E. and D.M. Thomson 1973. 'Encoding Specificity and Retrieval Processes in Episodic Memory', *Psychological Review* 80: 352–73.

Tulving, E., D.L. Schacter and H.A. Stark 1982. 'Priming Effects in Word-Fragment Completion Are Independent of Recognition Memory', *Journal of Experimental Psychology: Learning, Memory, and Cognition* 8: 336–42.

Turner, H.W. 1979. *From Temple to Meeting House. The Phenomenology and Theology of Places of Worship*. The Hague, Paris: Mouton.

Tuzin, D.F. 1972. 'Yam Symbolism in the Sepik: An Interpretive Account', *Southwestern Journal of Anthropology* 28(3): 230–54.

Vail, L. (ed.). 1989. *The Creation of Tribalism in Southern Africa*. London: James Currey.

van de Vijver, F. and K. Leung 1997. *Methods and Data Analysis for Cross-Cultural Research*. Thousand Oaks, CA: Sage.

Van Dijk, T. 1995. 'Discourse Analysis as Ideology Analysis', in *Language and Peace*, ed. C. Schaeffner and A. Wenden. Dartmouth: Aldershot, pp. 17–33.

Van IJzendoorn, M.H. and A. Sagi 1999. 'Cross-Cultural Patterns of Attachment: Universal and Contextual Dimensions', in *Handbook of Attachment: Theory, Research, and Clinical Applications*, ed. J. Cassidy and P.R. Shaver. New York: Guilford Press, pp. 713–34.

Van Maanen, J. 1995 *Representation in Ethnography*. Thousand Oaks: Sage.

van Staden, M. G. Senft, N. Enfield, J. Bohnemeyer and A. Dukers 2001. 'Staged Events', in *Manual for the Field Season 2001. Language and Cognition Group*, ed. S.C. Levinson and N. Enfield. Nijmegen: Max Planck Institute for Psycholinguistics, pp. 115–25.

Vaterrodt-Plünnecke, B., T. Krüger and J. Bredenkamp. 'Process-Dissociation Procedure: A Testable Model for Considering Assumptions about the Stochastic Relation between Consciously Controlled and Automatic Processes', *Experimental Psychology* 49: 3–26.

Virilio, P. 1996. *Fluchtgeschwindigkeit*. Munich and Vienna: Hanser Verlag.

von Stutterheim, C. and M. Carroll 1993. 'Raumkonzepte in Produktionsprozessen', *Kognitionswissenschaft* 3: 1–13.

Vroon, P., A. van Amerongen and H. de Vries 1996. *Psychologie der Düfte: Wie Gerüche uns beeinflussen und verführen*. Zürich: Kreuz Verlag.

Warrington, E. K. and L. Weiskrantz 1970. 'Amnesic Syndrome: Consolidation or Retrieval?', *Nature* 228: 628–30.

Wassmann, J. 1990. 'The Nyaura Concepts of Space and Time', in *Sepik Heritage: Tradition and Change in Papua New Guinea*, ed. Nancy Lutkehaus et al. Durham, North Carolina: Carolina Academic Press. pp. 23–35.

———. 1993. 'When Actions Speak Louder than Words. The Classification of Food among the Yupno of Papua New Guinea', *Quarterly Newsletter of the Laboratory of Comparative Human Cognition* 15(1): 30–40.

———. 1994. 'The Yupno as Post-Newtonian Scientists: The Question of What Is "Natural" in Spatial Description', *Man* 29: 546–66.

———. 1995. 'The Final Requiem for the Omniscient Informant? An Interdisciplinary Approach to Everyday Cognition', *Culture and Psychology* 1: 167–201.

———. 1997. 'Finding the Right Path: The Route Knowledge of the Yupno of Papua New Guinea', in *Referring to Space*, ed. Gunter Senft. Oxford: Clarendon. pp. 143–74.

———. 2001. 'The Politics of Religious Secrecy', in *Emplaced Myth. Space, Narrative, and Knowledge in Aboriginal Australia and Papua New Guinea*, ed. A. Rumsey and J. Weiner. Honolulu: University of Hawai'i Press, pp. 43–72.

———. (ed.). 1998. *Pacific Answers to Western Hegemony: Cultural Practices of Identity Construction*. Oxford, New York: Berg.

———. 1994. '"Hot" and "Cold": Classification and Sorting among the Yupno of Papua New Guinea', *International Journal of Psychology* 29: 19–38.

———. 1998. 'Balinese Spatial Orientation: Some Empirical Evidence for Moderate Linguistic Relativity', *The Journal of the Royal Anthropological Institute incorporating Man* (N.S.) 4: 689–711.

———. 2006. 'How to Orient Yourself in Balinese Space. Combining Ethnographic and Psychological Methods for the Study of Cognitive Processes', in *Pursuit of Meaning. Advances in Cultural and Cross-Cultural Psychology*, ed. J. Straub et al. Bielefeld: Transcript Publishers, pp. 351–76.

Wassmann, J. and P.R. Dasen (eds). 1993. *Alltagswissen. Les Savoirs Quotidiens. Everyday Cognition*. Fribourg: Universitätsverlag.

Weiner, B. 1992. *Human Motivation: Metaphors, Theories and Research*. Newbury Park: Sage Publications.

Weiner, J.F. 1991. *The Empty Place. Poetry, Space, and Being among the Foi of Papua New Guinea*. Bloomington: Indiana University Press.

———. 1995. 'Anthropologists, Historians and the Secret of Social Knowledge', *Anthropology Today* 11(5): 3–7.

———. 1997. '"Bad Aboriginal" Anthropology: A Reply to Ron Brunton', *Anthropology Today* 13(4): 5–8.

Weis, G. and R. Wodak (eds). 2003. *Critical Discourse Analysis: Theory and Interdisciplinarity*. Basingstoke: Palgrave Macmillan.

Weiss, J. 1977. 'Folk Psychology of the Javanese of Ponogoro', Dissertation, Yale University.

Weissenborn, J. 1986. 'Learning How to Become an Interlocutor – The Verbal Negotiation of Common Frames of Reference and Actions in Dyads of 4–14 Year Old Children', in *Children's Worlds and Children's Language*, ed. J. Cook-Gumperz, W.A. Corsaro and J. Streeck. Berlin: Mouton de Gruyter, pp. 377–404.

Weisz, J.R. 1990. 'Development of Control-Related Beliefs, Goals, and Styles in Childhood and Adolescence: A Clinical Perspective', in *Self-Directedness: Cause and Effects Throughout the Life Course*, ed. J. Rodin, C. Schooler and K.W. Schaie. Hillsdale: Erlbaum, pp. 103–45.

Weisz, J.R., F.M. Rothbaum and T.C. Blackburn 1984. 'Standing out and Standing in: The Psychology of Control in America and Japan', *American Psychologist* 39(9): 955–69.

Wellman, H.M. and A.T. Phillips 2001. 'Developing Intentional Understandings', in *Intentions and Intentionality: Foundations of Social Cognition*, ed. B.F. Malle, L.J. Moses and D.A. Baldwin. Cambridge, MA: MIT Press, pp. 125–48.

Werbner, P. and T. Modood (eds). 1997. *Debating Cultural Hybridity. Multi-Cultural Identities and the Politics of Anti-Racism*. London: Zed Books.

Werlen, B. 2000. *Sozialgeographie. Eine Einführung*. Bern, Stuttgart, Vienna: Paul Haupt Verlag.

West, P. 2005. 'Translation, Value, and Space: Theorizing an Ethnographic and Engaged Environmental Anthropology', *American Anthropologist* 107(4): 632–42.

White, G. and J. Kirkpatrick (eds). 1985. *Person, Self, and Experience: Exploring Pacific Ethnopsychologies*. Berkeley: University of California Press.

White, G.M. and L. Lindstrom (eds). 1993. 'Custom Today. Special Issue', *Anthropological Forum* 6 (4): 465–606.

Whitehouse, H. (ed.). 2001. *The Debated Mind. Evolutionary Psychology versus Ethnography*. Oxford: Oxford University Press.

Widlok, T. 1997. 'Orientation in the Wild: The Shared Cognition of Hai//om Bushpeople', *Journal of the Royal Anthropological Institute* 3: 317–32.

———. 1998. 'Cognitive Foundations of Khoisan Common Sense', in *Language, Identity, and Conceptualization among the Khoisan*, ed. M. Schladt. Köln: Köppe.

———. 1999. *Living on Mangetti. 'Bushman' Autonomy and Namibian Independence*. Oxford: Oxford University Press.

Wiedenhofer, S. 1990. 'Tradition, Traditionalismus', in *Geschichtliche Grundbegriffe: Historisches Lexikon Zur Politisch-Sozialen Sprache in Deutschland*, ed. O. Brunner, W. Conze, and R. Koselleck. Stuttgart: Klett-Cotta, pp. 607–50.

Wikan, U. 1995. 'The Self in a World of Urgency and Necessity', *Ethos* 23(3): 259–85.

Wilkins, D. and D. Hill 1995. 'When "GO" means "COME": Questioning the Basicness of Basic Motion Verbs', *Cognitive Linguistics* 6: 209–59.

Williams, R. 1983. *Keywords: A Vocabulary of Culture and Society*, revised edition. New York: Oxford University Press.

Willke, H. 2001. *Atopia. Studien zur atopischen Gesellschaft*. Frankfurt am Main: Suhrkamp Verlag.

Winter, W. 1976. 'Size Constancy, Relative Size Estimation and Background: A Cross-Cultural Study', *Psychologia Africana* 12: 42–58.

Wodak, R. and M. Meyer (eds). 2001. *Methods in Critical Discourse Analysis*. London: Sage.

Wohlt, P. 1978. 'Ecology, Agriculture and Social Organization: The Dynamics of Group Composition in the Highlands of New Guinea', Ph.D. dissertation, University of Minnesota.

Wolf, E. 1982. *Europe and the People without History*. Los Angeles: University of California Press.

Wysocki, C.J., J.D. Jr. Pierce and A.N. Gilbert 1991. 'Geographic, Cross-Cultural, and Individual Variation in Human Olfaction', in *Smell and Taste in Health and Disease*, ed. T.V. Getchell et al. New York: Raven Press, pp. 287–314.

Yan, W. and E.L. Gaier 1994. 'Causal Attributions for College Success and Failure: An American Asian Comparison', *Journal of Cross-Cultural Psychology* 25: 146–58.

Yonelinas, A.P. 1994. 'Receiver-Operating Characteristics in Recognition Memory: Evidence for a Dual-Process Model', *Journal of Experimental Psychology: Learning, Memory, and Cognition* 20: 1341–54.

Young, M. 1988. *The Metronomic Society: Natural Rhythms and Human Timetables*. Cambridge, MA: Harvard University Press.

Young, M.W. 1983a. *Magicians of Manumanua. Living Myth in Kalauna*. Berkeley: University of California Press.

———. 1983b. '"Our Name Is Women; We Are Bought with Limesticks and Limepots": An Analysis of the Autobiographical Narrative of a Kalauna Woman', *Man* 18: 478–501.

Index